MODELING THE MECHANICAL RESPONSE OF STRUCTURAL MATERIALS

MODELING THE MECHANICAL RESPONSE OF STRUCTURAL MATERIALS

Proceedings of a symposium sponsored by the Structural Materials Committee, held at the 127th Annual Meeting of The Minerals, Metals & Materials Society in San Antonio, Texas, February 15-19, 1998.

Edited by:

Eric M. Taleff
University of Texas
Austin, Texas

Rao K. Mahidhara
Cypress Semiconductor
San Jose, California

A Publication of
TMS
Minerals • Metals • Materials

A Publication of The Minerals, Metals & Materials Society
420 Commonwealth Drive
Warrendale, Pennsylvania 15086
(412) 776-9000

Printed in the United States of America
Library of Congress Catalog Card Number 97-75880
ISBN Number 0-87339-392-9

TMS
Minerals • Metals • Materials

TABLE OF CONTENTS

Superplasticity and Microstructural Evolution

Fracture Processes

DEDICATION

The editors dedicate this volume to Prof. Oleg D. Sherby and Prof. Amiya Mukherjee in thanks for introducing us to the area of mechanical properties.

PREFACE

This volume includes papers presented at the Topical Symposium entitled "Modeling the Mechanical Response of Structural Materials," held at the 127th Annual Meeting of The Minerals, Metals & Materials Society in San Antonio, Texas, February 15-19, 1998. The purpose of these proceedings is to pull together a diverse set of modeling work performed in many areas concerned with the mechanical response of structural materials. Modeling efforts, whether numerical analysis using finite element techniques, constitutive modeling of materials, or the modeling of microstructural evolution, share common themes which are of interest across the range of studies into mechanical properties. Communication of these themes among researchers concerned with different problems can provide fertilization for new creative ideas and future advances.

This volume is divided into five sections loosely based on several themes in the work presented. The first section, "Numerical Simulations," considers the application of numerical analysis to model mechanical response. The second section, "Constitutive Models and Experimental Evaluation," contains a diverse set of papers addressing various aspects of mechanical response in ambient conditions. The third section, "Creep Deformation," considers the behavior of materials at elevated temperatures. The fourth section, "Superplasticity and Microstructural Evolution," covers modeling the behavior and microstructural features of superplastic materials. The fifth and final section, "Fracture Processes," presents the modeling of fracture in materials.

The editors wish to thank the authors for making this volume successful through their hard work in preparing the manuscripts contained herein. The editors dedicate this volume to their mentors, Prof. Oleg D. Sherby and Prof. Amiya Mukherjee, who opened their eyes to the exciting world of mechanical properties.

Eric M. Taleff
University of Texas
Austin, Texas

Rao K. Mahidhara
Cypress Semiconductor
San Jose, California

October 31, 1997

Numerical Simulations

NUMERICAL SIMULATION OF SUPERPLASTIC FORMING

R.D. Wood and J. Bonet

Institute for Numerical Methods in Engineering
Department of Civil Engineering
University of Wales, Swansea SA2 8PP, UK

Abstract

The numerical simulation of superplastic formed diffusion bonded components using the finite element method has been available to industry for a number of years. However it remains the case that practitioners in the field, that is, those who make the products, generally remain sceptical of the possibilities afforded by the ability to simulate the SPF process. This paper will explain what SPF simulation can provide and illustrate this with a number of applications.

Differences between the tradition flow formulation and an improvement known as the incremental flow formulation will be discussed together with a brief mention of constitutive equation considerations.

It is reasonably well known that the finite element simulation of superplastic forming (FE/SPF) can provide a history of the forming process including the developing and final distribution of material thickness. However what is less understood is that the method can yield pressure time cycles that attempt to optimise the forming process by controlling some function of the strain rate so that superplastic behaviour is maintained.

Finally problems and opportunities relating to the use of FE/SPF simulation are discussed.

Modeling the Mechanical Response of
Structural Materials
Edited by E.M. Taleff and R.K. Mahidhara
The Minerals, Metals & Materials Society, 1997

1 Introduction

Superplastic forming (SPF) is invariably associated with the manufacture of blow formed thin and, more recently, thick sheet components made from either titanuim or aluminium alloy. Typically, thin sheet components with initial thicknesses of up to 2 mm have been formed using this procedure. More recently, however, components with thicknesses of up to 7 mm are also being manufactured in this fashion rather than using other traditional techniques such as casting.

The numerical finite element (FE) simulation of SPF allows the engineer to predict the complete deformation of the forming material during the whole forming cycle, [1]. This enables the final thickness distribution to be ascertained thereby determining those areas where excessive thinning has taken place. For SPF diffusion bonded celluar components simulations predict the development of contact between bonding parts and the motion of the cells. In some situations it may be necessary to chemi-etch the initial blank to provide thicker areas that will eventually be positioned at locations where reinforcement or fixing of the component is required. The determination of the chem-etch pattern so that the final desired thickness distribution is obtained is particularly difficult and experimentally costly. Such problems can easily be addressed at an early stage of the design cycle using SPF simulation thereby reducing the need for extensive experimentation. In addition to these geometrical predictions is the ability of the method to provide the pressure cycle required to ensure that the strain rate distribution in the forming component is such that the material remains within the optimum superplastic range, [2]. Compared to automotive sheet forming SPF is a slow process, however the introduction of high strain rate SP materials together with the prediction of optimum pressure cycles could enable forming times to be minimised.

Although SPF is a continuous, the numerical simulation evaluates the process at a discrete series of consecutive values in time. Two possible ways in which the process can be formulated have been proposed, [1]. Either the continuum governing equations are discretized in time using standard time stepping schemes; or the continuum kinematic quantities can be approximated over a finite time increment to yield approximate incremental governing equations. The former is known as the *flow formulation* whilst the latter is known as the *incremental flow formulation.* Traditionally the flow formulation has been popular for SPF simulation [3] and, indeed, metal forming simulation in general. However, as discussed in the kinematics section below, it suffers from a number of disadvantages.

The finite element formulation of the SPF problem is complicated and in the present setting it is not particularly helpful to even attempt a mathematical summary. However it is important to attempt an explanation of the essential difference between the types of flow formulation and emphasise why the incremental formulation is preferable. For full details the reader may consult a number of papers written by the authors which discuss all aspects of the formulation and which, in addition, make full refernce to other work on the topic. Reference 1 is a review paper containing 63 references to finite element and non finite element numerical simulation of SPF together with a summary of the flow and incremental flow formulations. Reference 2 concentrates on the forming pressure prediction aspects of FE/SPF simulation.

2 Flow and incremental flow kinematics

Initially the finite element simulation of SPF was formulated using the flow formulation, [1,3] in which the nodal velocity is the primary kinematic variable and the motion of the deforming material, given by the geometrical positions of the nodes, is obtained by time integration of the velocity. However the flow formulation is not ideal when geometric constraints abound. Firstly, boundary conditions relate to position and not velocity. Secondly superplastic material flow is incompressible which is a geometrical constraint not well represented, in the flow formulation, by constraining the volumetric strain rate to zero. However, the material is strain rate dependent and velocities are required for the constitutive behaviour. This conflict between the constitutive and boundary/volume constraint demands is resolved using the incremental flow formulation developed by Bonet, [4,5], whereby the geometry is the main kinematic variable and the strain rate is approximated in terms of the incremental changes in geometry. All governing equations now become functions of the geometry, including the volume constraint. Using a viscoplastic potential expression of the constitutive equations, a variational statement of the problem can be established which rigorously includes incompressibility and contact boundary constraints. Such a methodology sets the incremental flow formulation rigorously in a standard finite deformation context.

The rate of deformation and the incompressibility constraint are the key kinematic items in the constitutive equations discussed in §3. Consequently the difference between the flow formlation and the incremental flow formulation is exemplified by examining the expressions for these items.

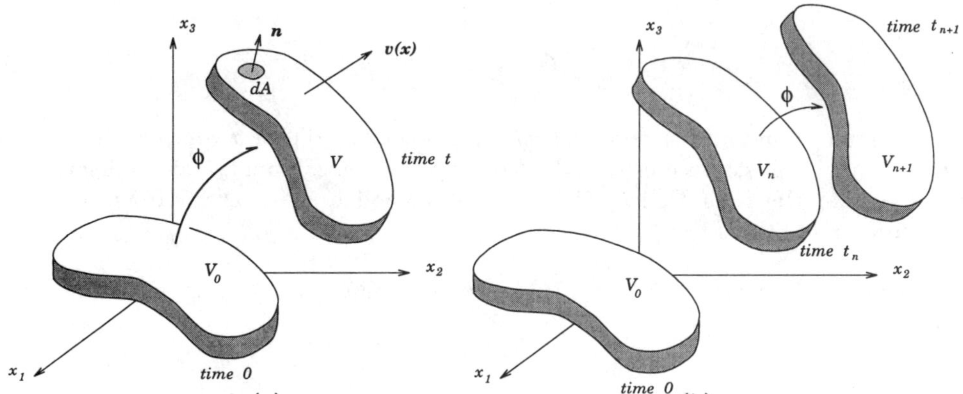

Figure 1 Initial (o), previous (n) and current (n+1) configurations

Consider the deformation of a given volume of superplastic material as shown in Figure 1(a). The position of the sheet at any given time t is defined by a time dependant mapping ϕ which gives the current coordinates $\boldsymbol{x}$ of a particle initially at position $\boldsymbol{X}$ as,

$$\boldsymbol{x} = \phi(\boldsymbol{X}, t) \tag{1}$$

The time derivative of this mapping defines the particle velocity vectors $\boldsymbol{v}$ and the symmetric part of the gradient of the velocity over the current position of the body defines the rate of deformation tensor as,

$$\boldsymbol{d} = \frac{1}{2}(\boldsymbol{\nabla v} + (\boldsymbol{\nabla v})^T) \tag{2}$$

where $\boldsymbol{\nabla} = \partial/\partial \boldsymbol{x}$ With velocity the primary kinematic variable the incompressibility condition implies the the trace of $\boldsymbol{d}$ must vanish, that is,

$$\operatorname{tr} \boldsymbol{d} = 0 \tag{3}$$

The von Mises equivalent strain rate required by the generalized constitutive equations is given as,

$$\dot{\bar{\varepsilon}}^2 = \tfrac{2}{3}\boldsymbol{d}' : \boldsymbol{d}' \ ; \quad \boldsymbol{d}' = \boldsymbol{d} - \tfrac{1}{3}I_d \boldsymbol{I} \ ; \quad I_d = \operatorname{tr} \boldsymbol{d} \tag{4a, b, c}$$

Alternatively, in the incremental flow formulation, two consecutive positions $\boldsymbol{x}_n$ and $\boldsymbol{x}_{n+1}$ of the forming sheet are considered at forming times t_n and $t_{n+1} = t_n + \Delta t$ respectively, as illustrated in Figure 2(b). The finite deformation over this timestep and the corresponding volume change can be described by the incremental deformation gradient tensor $\boldsymbol{F}$,

$$\boldsymbol{F} = \frac{\partial \boldsymbol{x}_{n+1}}{\partial \boldsymbol{x}_n} = \boldsymbol{\nabla}_n \boldsymbol{x}_{n+1} \tag{5}$$

where $\boldsymbol{\nabla}_n = \partial/\partial \boldsymbol{x}_n$ denotes the gradient at configuration n. Here the position at step n, rather than the initial state at time $t = 0$, is taken as the reference configuration. In the incremental scheme the rate of deformation tensor $\boldsymbol{d}$ is approximated over a finite time increment, Δt, in purely geometrical (i.e. non rate) terms as,

$$\tilde{\boldsymbol{d}} = \frac{1}{2\Delta t}(\boldsymbol{C} - \boldsymbol{I}) \ ; \qquad \boldsymbol{C} = \boldsymbol{F}^T \boldsymbol{F} \tag{6a, b}$$

This equation defines the *incremental rate of deformation* tensor $\tilde{\boldsymbol{d}}$ which in the limit, as the time increment Δt becomes smaller, converges to the continuum rate of deformation $\boldsymbol{d}$. Note that the right Cauchy-Green tensor $\boldsymbol{C}$ and hence $\tilde{\boldsymbol{d}}$ are tensors defined at configuration (n).

In contrast to the rate equation (3), incompressibility is now expressed directly in geometric terms as,

$$J = 1 \ ; \qquad J = \det \boldsymbol{F} = dV_{n+1}/dV_n \tag{7a, b}$$

The von Mises equivalent strain rate is similarly approximated in terms of the right Cauchy-Green tensor $\boldsymbol{C}$ to give,

$$\dot{\bar{\varepsilon}}^2 = \frac{1}{3\Delta t^2}(III_C^{-1/3} I_C - 3) \ ; \quad I_C = \operatorname{tr} \boldsymbol{C} \ ; \quad III_C = \det \boldsymbol{C} = J^2 \tag{8a, b, c}$$

The difference between the flow and incremental flow formulations is now revealed by the two central expressions for the rate of deformation tensor given by equations (2) and (6), $\boldsymbol{d}$ is a function of velocity which requires integration in order to yield the motion of the forming material, whilst $\tilde{\boldsymbol{d}}$ is directly a function of the deformation, with division by Δt yielding the approximation to the rate.

3 Constitutive equations

Superplastic constitutive equations are obtained from one dimensional tensile or compressive tests and take the generic form,

$$\dot{\varepsilon} = \Phi(\sigma, g, T) \; ; \qquad \sigma = \Phi^{-1}(\dot{\varepsilon}, g, T) \tag{9a, b}$$

where σ is the Cauchy flow stress, g the grain size and T the temperature. Observe that the grain size evolves during forming and comprises both static (temperature dependent) and dynamic (strain induced) components, consequently a rate of grain growth evolution rule is necessary.

It is shown in reference [5] that the one dimensional constitutive equations can be generalised to the three dimensional case by introducing a viscoelastic potential, $\psi(\dot{\bar{\varepsilon}}, g, T)$, from which the Cauchy stress tensor, $\boldsymbol{\sigma}$, is found from the deviatoric, $\boldsymbol{\sigma}'$ and hydrostatic pressure components, σ_m, respectively as,

$$\boldsymbol{\sigma} = \boldsymbol{\sigma}' + \sigma_m \boldsymbol{I} \; ; \quad \boldsymbol{\sigma}' = \frac{\partial \psi(\dot{\bar{\varepsilon}}, g, T)}{\partial \boldsymbol{d}} \tag{10a, b}$$

where σ_m will emerge from the enforcement of the incompressibility condition. In the flow formulation the above equation yields the Cauchy stress whereas in the incremental flow formulation $\boldsymbol{\sigma}$ is related (somewhat indirectly [4,5]) to the incremental rate of deformation $\tilde{\boldsymbol{d}}$ by,

$$\boldsymbol{\sigma}' = J^{-1} \boldsymbol{F} \boldsymbol{S}' \boldsymbol{F}^T \; ; \quad \boldsymbol{S}' = \frac{\partial \psi(\dot{\bar{\varepsilon}}, g, T)}{\partial \tilde{\boldsymbol{d}}} \tag{11, a, b}$$

Insofar as $\tilde{\boldsymbol{d}}$ enshrines the essence of the incremental algorithm, the relation between $\boldsymbol{\sigma}$ and $\tilde{\boldsymbol{d}}$, given by equations (10a) and (11), is known as an algorithmic constitutive equation.

4 Pressure cycle prediction

In order to derive the benefit of superplasticity it is necessary to control the variation of the forming pressure so that maximum ductility is maintained. This usually implies adjusting the pressure so that the strain rate sensitivity index m is a maximum throughout the forming operation. Algorithms for achieving this, [2,5], are based, preferably, on calculating the pressure, $p(t)$, that maintains a rate of energy dissipation $\dot{E}_0(t)$ in the forming material where,

$$\dot{E}_0(t) = \int_{V_n} \bar{\sigma}(\dot{\bar{\varepsilon}}_0) \dot{\bar{\varepsilon}}_0 \, dV_n \tag{11}$$

where $\bar{\sigma}$ is the equivalent Cauchy stress. It is usual to have some idea of a target equivalent strain rate $\dot{\bar{\varepsilon}}_0$ that will provide a maximum strain rate sensitivity, at least on average over the forming material, thus enabling $\dot{E}_0(t)$ to be evaluated. (Note: in this case, evolving grain size will change $\bar{\sigma}(\dot{\bar{\varepsilon}}_0)$ and thus $\dot{E}_0(t)$). However there is evidence, [6] to suggest that maximum ductility may be achieved by specifying a constant $\bar{\sigma}$ rather than $\dot{\bar{\varepsilon}}_0$. Furthermore it is conjectured that simply maintaining a constant rate of energy dissipation, $\dot{E}_0$, may provide the optimum pressure cycle. Whatever the case, the algorithm given in [2] is sufficiently versatile to explore these opportunities.

5 Finite element considerations

In any realistic SPF simulation many thousands of elements will be required to adequately approximate the complex shape of the forming material. In addition the forming process is highly geometrically and materially nonlinear both requiring substantial computation effort, typically many hundreds of linear solutions to many thousands of equations. These factors mean that the computations required per finite element must be minimised which, in turn, means that integration over the volume of the element must be as simple as possible. For FE/SPF simulation of thin sheet simple three node membrane elements suffice, [7], but for thick sheet either shell or three dimensional elements must be employed. The authors have not attempted the intermediate shell formulation as the range of applicability is judged to be to limited for practical use. Instead they have opted for an eight node solid element with mean quadrature and hour glass control, [5].

The physical problem associated with chemi-etched blanks is analogous to choosing an initial finite element mesh distribution that has the correct location and density of elements to enable accurate modelling of those regions that eventually attain high strain and curvature. Obviously such regions are not known *a piori*. For thin sheet forming it is possible to devise a mesh enrichment procedure that monitors both strain and curvature and increases the density of the mesh where necessary, [7]. Such a scheme is computationally very efficient since during the early stages of forming a fine mesh is generally not required, moreover, if a very fine mesh were established it is likely that the elements would be so small as to cause numerical difficulties with respect to time step size. Thin sheet forming simulation with some form of mesh adaptivity is essential for accurate predictions of thickness distribution.

6 Examples

Double cell diffusion bonding

This artificial example concerns the forming of a Ti-6Al-4V two cell diffusion bonded component, see Figure 2. The die size is 305×305×38 mm. with the single web initiated by a weld line 76 mm. from one side. The CPU time was 45 mins. on a SPARC-ELC workstation. Such computational effiency is due to the mesh enrichment which resulted in an initial mesh of 800 elements culminating in a final mesh, concentrated at the corners, of 5000 elements.

Forming of a double box

This application has been included in order to illustrate the ability of the techniques described above to simulate the forming of realistic components. A 5 mm thick blank of dimensions 120×120 mm^2 is formed into a double square box of compartment heights of 30 and 60 mm. The material is Ti-6Al-4V and the initial grain size is 6μm. Symmetry of the problem enables only a quarter of the problem to be modelled. The computational blank mesh consists of 4218 tri-linear hexahedral elements and 6582 mesh nodes. Two layers of elements have been taken through thickness of the sheet. The pressure cycle is computed during the numerical simulation of the forming process by imposing a target strain rate of 1.5×10^{-4} s^{-1}. The resulting pressure cycle is shown in Figure 3. The deformed finite element meshes at various times are shown in Figure 4. The complete run took 156 time steps involving 280 Newton-Raphson iterations. The CPU time on a SUN-SPARC-1000 was 31 hours and 17 minutes and the final computed forming time was 2 hours 6 minutes.

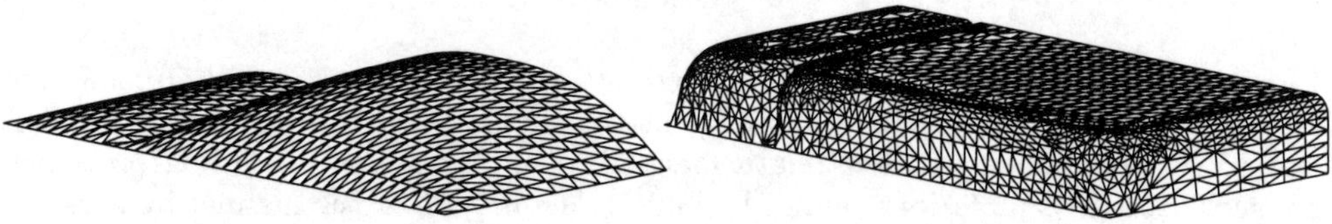

Figure 2 Diffusion bonded double box shapes at various times

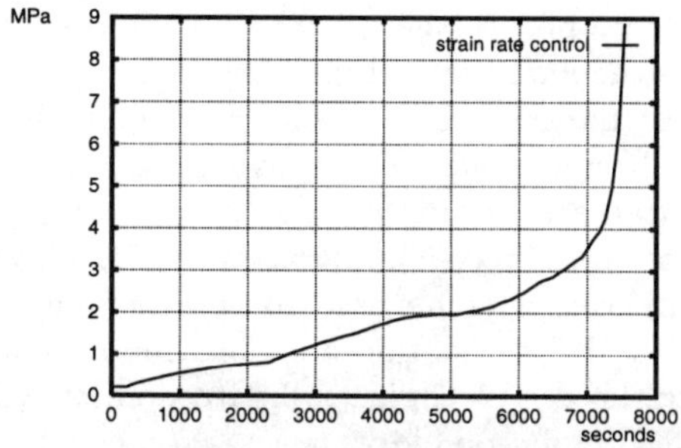

Figure 3 Double box pressure cycle

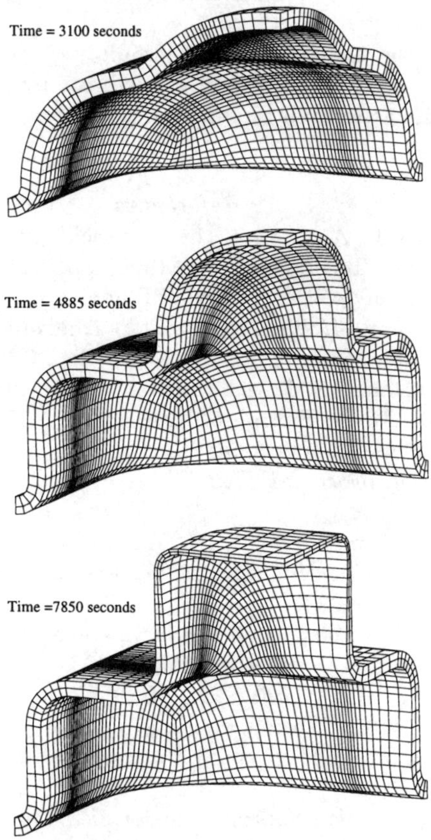

Figure 4 Double box shapes at various times

7 Discussion

Clearly the finite element method has much to offer the SPF practioner however a number of factors mitigate against the adoption of this design tool by industry. The finite element formulation for SPF is mathematically complex and the solution procedures are such that the simulations cannot be undertaken in a *black box* manner by personnel who do not have, at least, a rudementary knowledge of the finite element method and the way in which the solution algorithm handles the highly nonlinear nature of the problem. This issue is developed further in [8]. Thirty years ago when structural FE analysis was in its infancy the analysts had the benefit of exact solutions to validate their simulations. The nonlinear nature of SPF does not admit such luxury and experimental checks are essential. Furthermore the constitutive equations are complex involving the development of state variable evolution equations. These experimental activities are very costly particularly tests on actual components which invariable compete for time on machinery that would otherwise be used for production. However if these issues can be addresses then FE/SPF simulation does provide a great opportunity for cheap experimentation enabling the design engineer the freedom to explore creative ideas that do not rely on the slow accumulation of expertise.

8 Acknowledgement

The work reported by the authors on SPF simulation was largely funded by British Aerospace Commercial Aircraft Ltd. and carried out by the Institute for Numerical Methods in Engineering in the Civil Engineering Department at the University of Wales, Swansea, U.K.

9 References

1. Wood, R.D. and Bonet J., *A review of the numerical analysis of superplastic forming*, J. Materials Processing Technology, **60**, 1996, pp. 45–53. This review contains refernces to work by many others on FE/SPF anaylsis.
2. Bonet, J., Wood, R.D. and Collins, R., *Pressure control algorithms for the numerical simulation of superplastic forming*, Int. J. Mech. Sci., **36**(4), 1994, pp.297–309.
3. Argyris, J.H. and St. Doltsinis, J., *A primer on superplasticity in natural formulation*, Compt. Meths. Appl. Mech. Engrg., **30**, 1984, pp. 83–132.
4. Bonet, J. *The incremental flow formulation for the numerical analysis of plane stress and thin sheet forming processes*, Comp. Meth. Appl. Mech. Engrg., **114**, 1994, pp.103–122.
5. Bonet, J., Bhargava, P. and Wood, R.D., *Finite element analysis of the superplastic forming of thick sheet using the incremental flow formulation* Int. J. Num. Meth. Engrg. **40**, 1997, pp. 3205–3228.
6. Collins, R. *Finite element analysis of a superplastic tensile test*, M.Phil. Thesis, Civil Eng. Dept., Univ. of Wales, Swansea, C/M/280/93.
7. Bonet, J., Bhargava, P. and Wood, R.D., *Finite element simulation of 3-dimensional superplastic forming with diffusion bonding*, in Proc. Int. Conf. *NUMIFORM92 Numerical Methods in Industrial Forming Processes*, 1992, pp. 843–849, Ed. by Chenot et al., Balkema, 1992.
8. Wood, R.D. and Bonet, J., *Superplastic forming simulation - A case study for virtual manufacturing*, in Proc. Virtual Manufacturing Conf., The Royal Aeronautical Society, London, 1996, pp. 7.1–7.7, ISBN 1 85768 013.

Finite Element Modelling of Superplastic Forming of Titanium Helmet Articles

Sami M. El-Soudani*, Michel Jamjian, James Mackiewicz***, and Kenneth Newell**[†]

* **Dr. El-Soudani,** Principal Investigator, is a Senior Engineering Specialist, Boeing Defense and Space Group, North American Aircraft Division, Seal Beach, California

** **Mr. Jamjian,** is an Engineer, Engineering Systems International, ESI North America, San Diego, California

*** **Mr. Mackiewicz,** is Technical POC, U.S. Army Research Laboratory, Natick, Massachusetts

[†] **Dr. Newell,** Consultant, Formerly with Rockwell Science Center, Rockwell International Corporation, Thousand Oakes, California

ABSTRACT

Superplastic forming (SPF) of single-sheet, and diffusion-bonded multi-sheet structures has been in use for some time in certain military and commercial, light-weight structural applications. Ample opportunities for Industrial usage of this process so far have been limited to the aerospace and automotive fields. Ballistic-resistant helmets for the army and Law-enforcement personnel impose even more stringent demands of light-weight and high-performance structural integrity compared to other military and aerostructures applications. A serious challenge to the application of the SPF process remains to be the final product thickness uniformity, particularly in situations involving high depth-to-width ratios of the final superplastically formed (or deep drawn) products. Helmet shell fabrication by this method challenges the limits of superplastic forming of single-sheet shell structures, as it often requires surface-to-projected-area ratio in the range of 2.5 to 3, thus exceeding that of a hemisphere. Prior attempts to superplastically form hemispherical pressure vessel halves revealed a natural trend of final formed part thickness tapering down to a minimum in the dome area by nearly a factor of 2 to 3. Trial-and-error methods for reducing the global thickness taper, and possible thinning due to localized necking from using non-optimized parameters may prove costly and time-consuming. Using a modified explicit finite element code originally developed by the ESI Group in Paris, for commercial metal-stamping simulation known as PAM-STAMP, an attempt is made in this paper to simulate the helmet SPF process, whereby the interplay of various material forming characteristics, initial preform design, and processing parameters (temperatures, forming gas pressure-time schedules, etc.) can be subject to experimentation and process design with a computer. Preliminary trial runs have been completed, whereby the results of a measured degree of success, and the needs for further refinement of SPF process modelling are discussed. Successful implementation of this technology could benefit areas other than ballistic-resistant helmet fabrication, such as a multitude of compound curved aerodynamic titanium and aluminum shell structures, domes, nose cones, fuel tanks, pressure vessels, etc.

Modeling the Mechanical Response of
Structural Materials
Edited by E.M. Taleff and R.K. Mahidhara
The Minerals, Metals & Materials Society, 1997

Introduction

Process modelling of titanium helmet superplastic forming (SPF), as developed in this investigation, is comprised of the following mutually complementing three component development activities: (a) Superplastic material modelling, (b) Finite element model development and numerical simulation methodology "or code" definition, and (c) refined model application for process control aimed at meeting certain preselected design goal requirements for the subject component. The purpose of this paper is to demonstrate the viability of this general approach to advanced process design, albeit some further refinements may still be needed in dealing with more complex physical process modelling aspects, such as friction loading conditions, microstructural effects (grain size), product form banding , etc. In the following sections various components of this investigation will be examined separately, however before doing so our technical approach will be outlined in some detail, giving a seven-step sequence for the execution of these three components of the investigation.

Technical Approach to Helmet Shell Superplastic Forming Modelling

Helmet shell forming from an initially flat plate or circular membrane subject to gaseous pressure involves biaxial and triaxial states of stress, strains and strain rate distributions. On the other hand, key superplastic material properties (to be discussed in the next section) are usually measured in an isothermal uniaxial tensile test environment. Validation of the application of such parameters to the stress and strain tensor components of the complex forming process is essential to the viability of finite element and finite difference modelling of helmet shell forming. The objective of initial process simulations, therefore, is to force the analytical code to agree with certain observed behavior of actual trial runs within acceptable percent errors. Once such a goal has been achieved the fine tuned model may be used to predict thickness profiles for other processing conditions. Still better, a viable model should be capable of working backwards to provide the fabricator with the initial plate thickness required and/or its thickness profile requirements for yielding a final formed part thickness distribution under the action of an optimum pressure-time diagram given only the basic SPF material properties and desired processing temperature and cycle duration. At this stage of development the finite element model and associated forming code would be driving the fabrication process. With the above goals in mind the following sequence of seven steps summarizes the technical approach pursued in this program for helmet shell SPF process modelling :

(a) **Step 1** - Develop material data base for the given alloy (Ti 6242S) using uniaxial stepped strain rate tests and regress such data base to quantify the material strain rate sensitivity of the flow stress over a selected broad range of temperatures.

(b) **Step 2** - Develop a digitized file of the forming tool . In the present study such a file was developed from the combination of the PASGT helmet outer surface CAD file (ideally conforming to the female tool inner surface) combined with additional corner and other transitional radii measurements directly read off the cast female die surface and /or a fully formed helmet surface (with its outer rim area).

(c) **Step 3** - Conduct initial forming trials recording initial and final plate thickness distributions in 3-D using empirically selected initial pressure-time cycles at a given reasonably optimum temperature. The latter may be deduced from the material data base in step 1.

(d) **Step 4** - Using PAM-STAMP™ code conduct several trial runs using the data base in steps 1 and 2 for predicting the observed forming behavior in step 3, thus verifying or validating the applicability of the chosen material parameters in step 1.

(e) **Step 5** - Using FEM forming simulation run data base establish optimum pressure-time profiles aiming at appropriate strain rate control and providing for realistic forming cycle time duration while avoiding localized thinning (necking instability).

(f) **Step 6** - Conduct further forming trials to verify optimized finite element model predictions focussing in particular on final minimum wall thickness (both value and location over the helmet dome surface).

(g) **Step 7** - Examine finite element / finite difference model predictive viability for forming trials under processing conditions other than those proposed by the optimized FEM code. Should there be certain deficiencies detected, define as applicable any further physical model development/upgrade needs (e.g. friction effects, microstructural refinement, thermomechanical processing, etc.)

Superplastic Material Modelling

This component of advanced process design focusses on defining and measuring the applicable material SPF properties, among which a most prominent characteristic is the strain rate sensitivity of the flow stress at the optimum processing temperature (often quantified by two material coefficients so-called m-value and K parameter). This aspect of material modelling is also concerned with the relevance of m and K values normally measured in a uniaxial tensile test to the actual multiaxial stress and strain states, as they develop and progressively change over time in the formed part (in the case at hand, an antiballistic titanium helmet shell). Superplasticity is the ability of a metal to elongate extensively under low applied stresses and elevated temperatures without localized thinning. Under appropriate conditions of temperature and strain rates, certain metallic alloys deform to very large plastic strains of up to 1000% without fracturing. This elongation under superplastic flow conditions is at least one order of magnitude greater than that available in most conventional forming operations. It was this particular feature that made the superplastic forming process especially attractive for application to the forming of antiballistic titanium helmet shells from flat plates, in view of the required deep draw in this operation (final to initial surface area ratio of 2-3 which is substantially greater than that of a hemisphere).

Various alloys are known to exhibit superplastic behavior. The industrially important alloys include titanium, aluminum and iron as principal constituents, with titanium being the alloy of choice for helmet shells. Necessary conditions for superplasticity are ideally:

- a very fine grain size material (on the order of 10 µm),
- a high temperature (greater than half the melting temperature),
- a carefully controlled, low strain rate ($\dot{\varepsilon} = 0.0001$ to $0.01\ s^{-1}$).

A grain size of 10 µm or less may not be readily achievable in titanium plates of a thickness substantially greater than 0.125 inches without resorting to special measures of grain refinement. The Ti 6242S plate used in this investigation was 0.4 inches thick, and its metallographic section analysis revealed a reasonably equiaxed grain structure consisting of alpha-stabilized beta phase with a mean grain size of approximately 25 µm. While it may be desirable to use a finer grain size, such a microstructure is anticipated to respond isotropically under appropriate conditions of SPF sheet metal forming.

Modelling of Superplastic Uniaxial Tensile Material Behavior

Superplastic flow is a creep phenomenon which for uni-axial loading is approximately described by the equation:

$$\sigma = K\dot{\varepsilon}^{m}\varepsilon^{n}, \qquad (1)$$

where σ is the stress, ε the strain, and $\dot{\varepsilon}$ the strain rate. K, m and n are material parameters. n is the strain hardening exponent, taken equal to 0 in the following examples. This is a good approximation since n is small in reality. The main influencing factor is the exponent m which measures the strain rate sensitivity of the material. The extent of the thinning in the tensile specimen is strongly related to m. A large value of m is desirable to obtain more uniform thickness distribution after forming. In practice, the parameter m varies with the strain rate, typically exhibiting a maximum value. For best forming conditions, it is important that the maximum strain rate in the sheet be regulated to stay in the optimum range where the m value is high (typically 0.5 or higher approaching unity). With a negligible value of n in Equation (1) this relation assumes a form familier to metallurgists,

$$\sigma(\bar{d}, \varepsilon, \dot{\varepsilon}, T) = K\,\dot{\varepsilon}^{m(\bar{d}, \varepsilon, \dot{\varepsilon}, T)}, \tag{2}$$

where $\bar{d}$ = average grain size, ε = true plastic strain, T = test temperature, and K is a material constant. The stepped strain rate uniaxial tensile test data obtained at a given temperature must be reduced into true stress-strain rate format ($\sigma - \dot{\varepsilon}$). The strain rate sensitivity factor (m) plot as a function of strain rate ($\dot{\varepsilon}$) is then computed by differentiating the log σ versus log $\dot{\varepsilon}$ curve at small intervals of $\dot{\varepsilon}$. The material constant K is derived from the intercept value corresponding to the linear interval or tangent to log σ versus log $\dot{\varepsilon}$ curve obtained at the temperature in question. This approach to material data base development is not often the most convenient or adaptable to mathematical modelling of material behavior.

A crucial aspect of SPF testing is the representation of the σ vs. $\dot{\varepsilon}$ behavior in a constitutive form, thus, making material response to variation in temperature and applied loads amenable to finite element analysis without the use of look-up tables or others forms of interpolation. The recommended constitutive equation for this purpose is based on a simplified form of a glide-controlled creep rate equation for the analysis of post power-law response to high applied stresses at elevated temperatures given by:

$$\dot{\gamma} = A \exp\left(\frac{\beta\sigma}{RT} - \frac{Q}{RT}\right) \tag{3}$$

where β is a stress factor and Q is the activation energy. In order to apply this formulation to a wider range of stress levels and temperatures for the purpose of representing the stepped strain rate data obtained for the Ti-6242S coupons, it is necessary to expand the rate equation and add two product terms associated with stress and temperature. The final form of the constitutive equation is obtained from a phenomenological relation for the transient creep response as:

$$|\dot{\varepsilon}| = \alpha + \beta_1 \ln|\sigma| + \beta_2 \ln(T) + \frac{\beta_3|\sigma| - \beta_4}{RT} \tag{4}$$

where the coefficients α and β_1 thru β_4 are extracted from a multi-linear least squares fit to the flow stress-strain rate data, taken from tensile coupons

Development of the viscoplastic constitutive material model involves the linear regression of the transformed strain rate $y = \ln(\dot{\varepsilon})$ upon multiple values of x as:

$$y = \alpha + \beta_1 x_1 + \beta_2 x_2 + \ldots + \beta_n x_n, \tag{5}$$

where $x_1, x_2, \ldots, x_n$ are n-different variables. These variables are transformations of the measured flow stress and corresponding temperature. As a constitutive viscoplastic material model, it is assumed that the measured flow stress and temperature are known precisely, but that the strain rate is subject to random error, being normally distributed about the regression line with constant variance σ^2. In the actual stepped strain rate test, the opposite is the case since the flow stress is the unknown measured quantity. Nonetheless, we wish to find estimates a, b_i of the parameters α, β_i, together with their variances and covariance's, and also to find an estimate s^2 of σ^2. In this paper details of the least squares regression analysis method will be deleted for brevity, and the results will be given below without proof. Using this phenomenological approach for fitting data to a physical relation, the coefficients α, β_i are extracted from the entire set of flow stress versus strain rate data over the range of temperatures associated with the SPF forming process. As shown by the color triangles in Figure 1, testing was performed at 825°C (1517°F), 850°C (1562°F), 875°C (1607°F), 900°C (1652°F) and 925°C (1697°F) and numerically fit to the constitutive relationship (2) shown by the solid lines corresponding to each temperature. The least squares fit shown in the Figure 2 is given by the coefficients α, β_i as:

$$\left.\begin{aligned} \alpha &= 1349.934 \\ \beta_1 &= 0.89015 \\ \beta_2 &= -165.52 \\ \beta_3 &= 0.1193 \\ \beta_4 &= 456382.0 \end{aligned}\right\} \qquad (6)$$

From the data in Figure 1 the material property values of m and K shown in Table I were extracted from measured slopes and intercept values, respectively. These material property values determined at the selected operating temperatures, namely 1650°F and 1700°F were also used in checking out preliminary simulations with finite element analyses to be discussed in the following sections. At this point it should be noted that strain rates below 10^{-4} [1/sec] were ruled out for helmet forming primarily from the standpoint of excessive processing cycle duration and low productivity. A desirable forming cycle duration should be in the range of 1 to 3 hours. The data in Table I derived from Figure 1 indicate that for the more practical range of strain rates between 10^{-4} and 10^{-3} [1/sec] the corresponding m-values vary from 0.5 to 0.8. These values are not unusual for Ti 6242S plate with an average grain size of 25 μm.

Table I - A Sample of Material Superplastic Forming Parameters Derived from Regression Analyses in Figures 1 and 2 and Used in Finite Element Modelling of Ti 6242S Helmet Shell Forming Behavior

Test Temperature		ε	K	m	σ
[°F]	[°C]	[1/sec]	[psi]		[psi]
1650	900	10^{-4}	85,295,400	1.203	1315
		5×10^{-4}	1,192,600	0.693	6150
		2×10^{-3}	254,700	0.500	11390
1700	925	10^{-4}	268,793,600	1.375	850
		5×10^{-4}	2,545,700	0.820	5000
		2×10^{-3}	204,900	0.465	11390

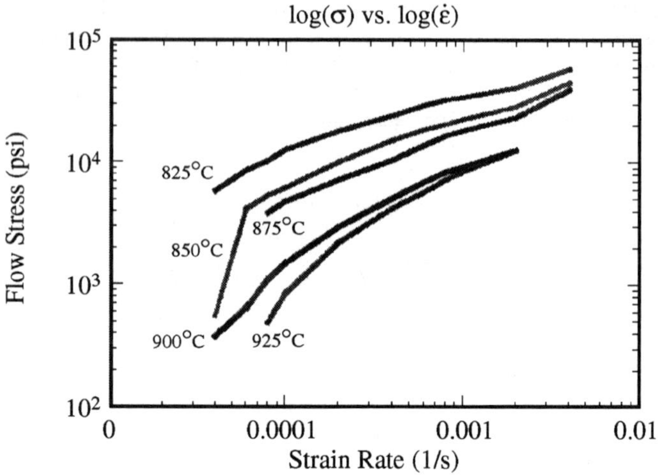

Figure 1 - Results of stepped strain rate testing performed at 825°C (1517°F), 850°C (1562°F), 875°C (1607°F), 900°C (1652°F) and 925°C (1697°F)

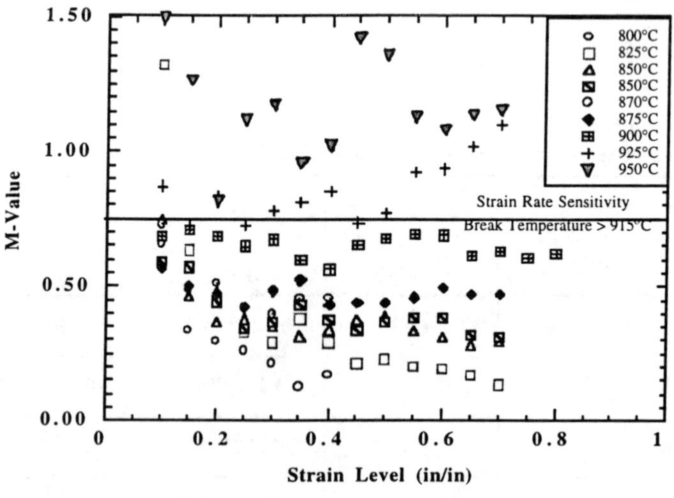

Figure 2 - Initial-strain-level dependence of a set of m-values for Ti 6242S alloy plate 0.4 inches thick (least square fit per Equations 4 and 6 verifying strain dependence per Equation 2)

Modelling of Superplastic Multiaxial Tensile Material Behavior

The superplastic sheet is modeled as a non-linear visco-elastic Maxwell material with the following constitutive equations:

$$s_{ij} = 2\mu \dot{e}^{p}_{ij}$$

$$s_{ij} = \sigma_{ij} - \frac{1}{3}\sigma_{kk}\delta_{ij}$$

$$e_{ij} = \varepsilon_{ij} - \frac{1}{3}\varepsilon_{kk}\delta_{ij}$$

$$\sigma_{ij} = \lambda\varepsilon^{e}_{kk}\delta_{ij} + 2G\varepsilon^{e}_{ij}$$

$$\varepsilon_{ij} = \varepsilon^{e}_{ij} + \varepsilon^{p}_{ij}$$

$$\varepsilon^{p}_{kk} = 0$$

$$\mu = \frac{1}{3}K\dot{\bar{e}}^{p\,m-1}$$

where σ_{ij} is the Cauchy stress, ε_{ij} the total strain. The total strain can be divided into an elastic part ε^{e}_{ij} and a plastic part ε^{p}_{ij}. For superplastic forming, the elastic strain is negligible when compared to the plastic strain and can be neglected if no springback evaluation is needed. K and m are the superplastic material constants already described in the previous section. G is the shear modulus and λ the lame constant. $\bar{e}^{p}$ is the equivalent plastic strain. The above equations lead to the previous flow relation:

$$\bar{\sigma} = K\dot{\bar{e}}^{p^{m}} \tag{7}$$

where $\bar{\sigma}$ is the equivalent Von-Mises stress.

Helmet SPF Process Finite Element Modelling with PAM-STAMP™ Code

In this second component of the investigation finite element model development focusses on finite element representation of the formed part geometry, equilibrium, constitutive laws, and validation of the applicability of material SPF properties defined in the previous Section to the process of helmet expansion. The analysis uses appropriate iterative schemes for incremental membrane element deformation assessment and correlation with observed actual trial runs for fine tuning of model assumptions as appropriate.

Model Definition

The superplastic forming of a titanium Ti-6242S Helmet is simulated using PAM-STAMP™ which is an explicit non-linear finite element code, developed by Engineering Systems International. Two algorithms are implemented in PAM-STAMP™ to allow the superplastic forming of aluminum and titanium alloys. The first algorithm prescribes the maximum allowable strain rate and regulates the pressure applied to the blank to achieve this strain rate. This yields a minimum forming time and the corresponding pressure history profile. Also, the blank thinning is obtained as a result of the numerical simulation. An alternative algorithm sets the gas flow rate history used to apply the pressure on the material resulting in part thickness and plastic strain rate distributions. The finite element shell material model includes membrane, bending and out-of-plane shear effects leading to a better characterization of the material behavior under forming conditions. Both algorithms are used to simulate the forming of a helmet. As noted earlier in the material modelling Section, the goal is to find a forming pressure time history that will result in a minimal thinning effect and a minimal forming time while predicting the complete final thickness distribution of the formed part.

The pressure-time histories from three different numerical simulations that resulted in the best thickness profiles were supplied to Boeing North American Inc. to validate the material parameters originally provided to ESI by Boeing per Figures 1 and 2 and Table I. Experimental tests were performed by Boeing North American at Murdock Inc. manufacturing development facilities (currently acquired by GKN Westland Aerospace) using two pressure-time histories. While the first experimental test resulted in a failure of the blank due to high strain rates, the second test showed good agreement with the predicted numerical results in Ti 6242S alloy. The third trial using Ti 6Al-4V alloy 0.5-inch thick plate also suffered rupture, apparently due to poor superplastic forming microstructure of that particular plate.

Solution Scheme

In order to minimize the thinning and the forming time, a specific value of the strain rate to be achieved in the blank is desired. A constant pressure is seen to develop a variable strain rate, thus a variable forming pressure is necessary to maintain a constant strain rate. The pressure history is dependent on the strain history, consequently an automatic control process was incorporated in PAM-STAMP™ to regulate the pressure required to achieve a specified strain rate once it is reached in a part of the blank. Once the designer specifies a strain rate, both the pressure history and the total forming time are obtained as a result of the numerical simulation. This loading option is referred to as automatic pressure regulation to obtain maximum strain rate at prescribed value.

Another loading option, which can be used to retroactively check a pressure history, involves the prescription of the pressure-time history or the prescribed flow rate history of the gas in the fluid cell. The output of the simulation is a check on the formability of the sheet to the desired final shape.

In both loading options, stress, strain, strain rate and thickness contours are obtained from the numerical prototyping. Also, the tearing of the sheet can be detected if a corresponding criterion is specified.

CAD to PAM-STAMP™ Mesh Modelling

In order to run a PAM-STAMP™ superplastic simulation, the following data are needed: a CAD file of the die or a mesh of the die, material properties for the blank (if not available, then experimental tests to qualify the material model), the process physical conditions from boundary conditions to pressure histories and maximum pressure. In our case, a CAD file of the helmet geometry was made available (Figure 3).

The CAD data was transformed to a finite element mesh using an automatic mesher. This creates a mesh of shell elements that represent the material of the helmet. In the present case, no CAD data of the die was available, instead a formed helmet was used to measure the missing geometry of the die, i.e. the radius of entry, the connection of the radius to the helmet, and the position of the gas seal. But since the die CAD geometry was not available and manual measurements were done, several inconsistencies were detected between the real die used in the testing and the meshed die used in the simulations (Figure 4). Some of the differences involved the radius of entry, and the geometry of the helmet. Note that once the CAD file is available, the meshing is done automatically using one of the meshers CAD-CAR™ or Delta-Mesh™ supplied by ESI. The size of the elements used for the die should be chosen such that they accurately represent the geometry. The die is modelled as a rigid undeformable body (see Figure 4).

The blank is modelled as a circular plate with variable thickness (patent pending Reference [1]). The size of the elements are chosen to accurately represent the thinning and the different curvatures of the die. The variable plate thickness is accounted for in the numerical simulation. Adaptive meshing can be used to refine the mesh during the simulation.

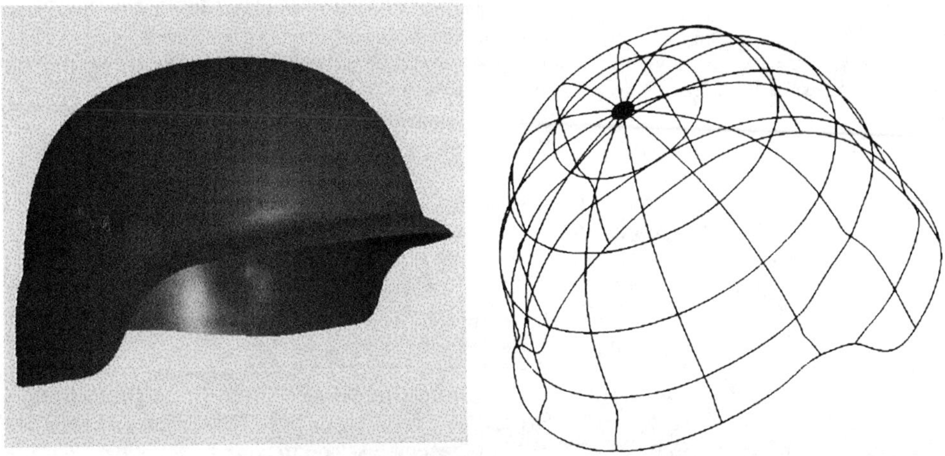

Figure 3 - CAD picture of the PASGT helmet

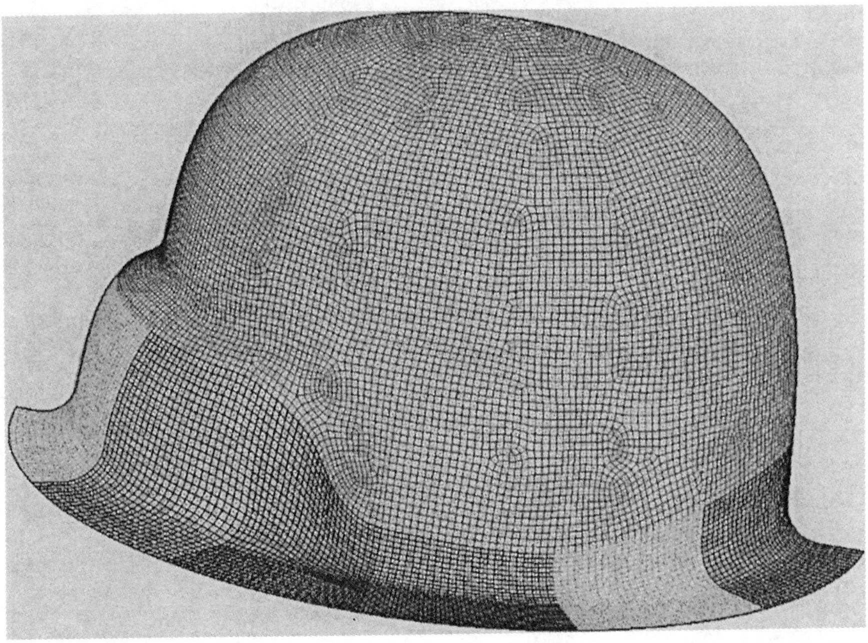

Figure 4 - Die mesh from helmet CAD and hand measurements

Once the mesh of the die and the blank are available, the pre-processor of PAM-STAMP™, GENERIS™, can be used to define material parameters, boundary conditions representing the gas seal, fluid cell modelling the gas cell of the press, process conditions such as maximum strain rate or pressure-time history input and output of variables for the results. These results can be viewed with PAM-VIEW™, the post-processor to the solver. Progressive animation plots showing the progress in time of the forming process as well as contours and time-history curves can be generated. Some of the variables that can be saved for viewing include the thinning, the maximum strain rate, the pressure-time history, the strain state in the blank, and the time of forming.

Use of Material Data Base in Combination with Preliminary Trial Runs in SPF Process Modelling

The superplastic material model in PAM-STAMP™ is defined by two variables K and m. These variables are obtained from test results and are temperature and grain size dependent (see Equation 2). They also depend on the strain rate of forming. As a first step Boeing North American supplied material data generated under a subcontract with the Rockwell International Science Center to characterize the Ti 6242S material (see Figures 1 ad 2 and Table I). These tests are usually uni-axial tests and do not take into account the complexities of the real three dimensional loading situation. To validate the K and m measurements generated at the Rockwell Science Center, two test simulations performed at Murdock Inc. were used. In the first test, the part was not completely formed and the height attained was measured. Also the pressure-time history used to form the part was supplied. The height at the end of the forming cycle was 4.5 inches (see Figure 5). In the second test, a different pressure-time history was applied resulting in complete forming of the helmet. The pressure cycle is shown in Figure 6.

A thickness profile of the formed part was also measured in order to compare with the numerical simulation results. Several numerical simulations were run with different values of K and m to find the combination that gave the best agreement with the test results. While such correlations are deleted here for brevity good agreement was obtained verifying the validity of the material data base usage.

Numerical Simulation Results with Optimized Processing Parameters

The third component of the helmet SPF process modelling dealt with ranking its predictive viability as a process control tool. Should this be achieved the model may be used to guide further SPF processing applications aimed at meeting certain part design goals (such as minimum thickness, maximum weight, or both, etc.). In order to maximize the minimal thickness in the helmet obtained from the numerical simulation, several factors were considered and varied in different simulations to study the effect on the thickness. These factors include the maximum strain rate, the thickness profile of the blank and the pressure-time history corresponding to the gas rate in the press. Three runs that projected the best process conditions and thickness profile were chosen, and a new pressure-time history was determined for the Boeing-North-American-preselected SPF temperature and thickness profile of the blank. The first and second runs were for Ti 6242S 0.4-inch thick plate, while the third run was for a 0.45-inch thick Ti 6Al-4V plate. Two of the three FEM-developed pressure-time diagrams were tested experimentally to try and validate the numerical simulation and the material parameters chosen, i.e. K and m. The first run using Ti 6242S plate was not successful as it involved very high initial strain rates and led to failure of the blank before complete forming of the helmet. The second run also on Ti 6242S was completed successfully and with good agreement with the numerically predicted thickness values. The third run using Ti 6Al-4V plate was unsuccessful due to what appeared to be a poor SPF microstructure of the 0.5-inch thick plate of this alloy. In the interest of brevity only results of the second Ti 6242S successful run will be shown here.

Model Verification - Run II Results : Run II was performed using the same thickness profile blank as Run I but with a maximum strain rate calculated by PAM-STAMP™. A gas rate was specified with a maximum pressure that can be obtained on this press, namely 600 psi. The maximum strain rate predicted by the numerical simulation is shown in Figure 7, and the associated pressure profile resulting from this run is presented in Figure 8. The predicted final formed part thickness profile for this run is shown in Figure 9

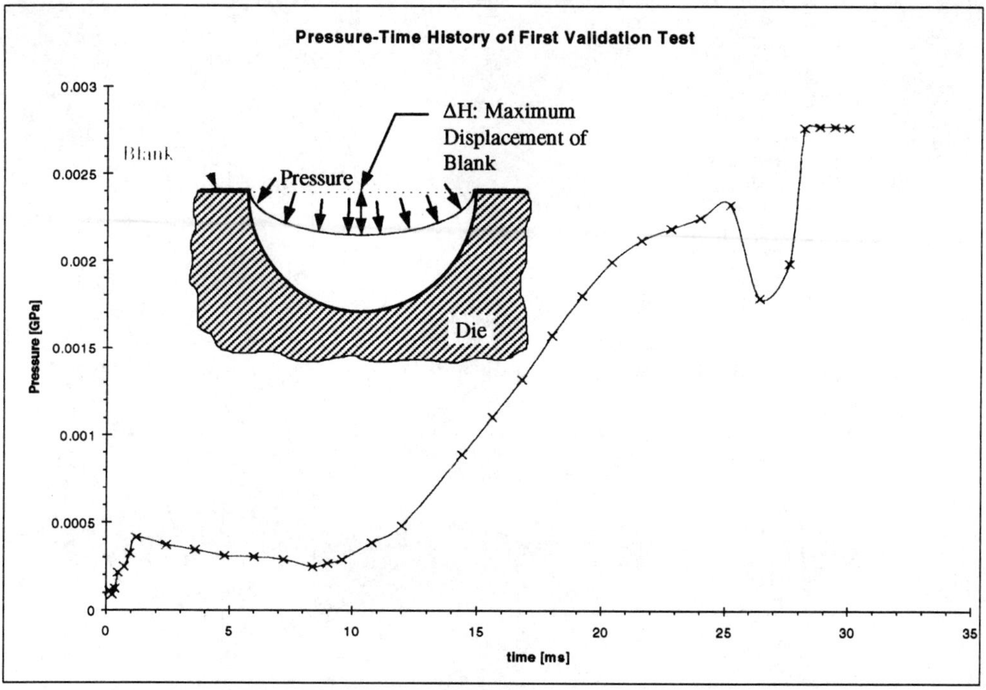

Figure 5 - Pressure-time history of first validation test with incomplete formed helmet

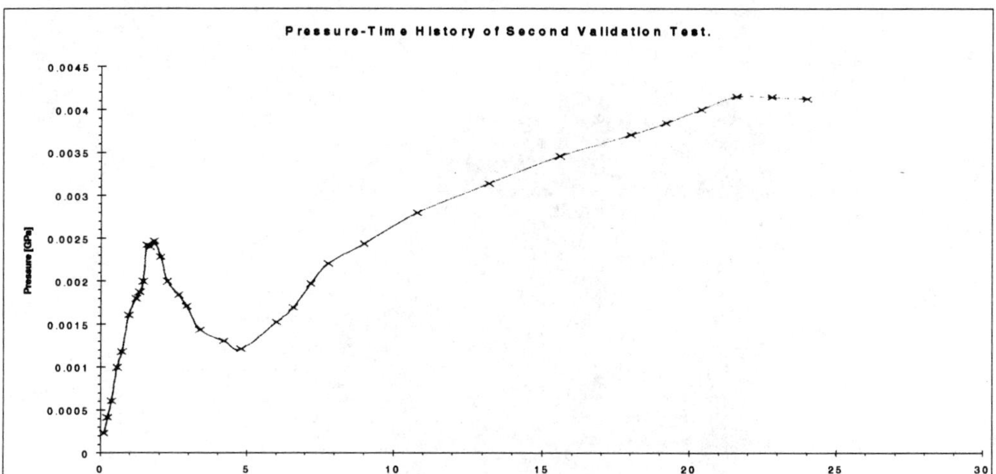

Figure 6 - Pressure-time diagram of second validation test

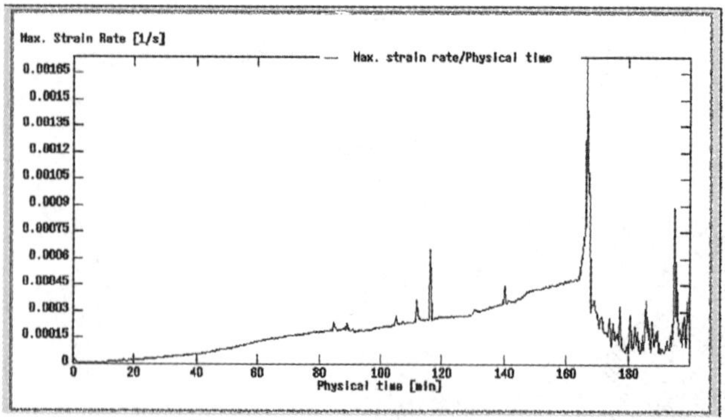

Figure 7 - Predicted maximum strain rate for superplastic forming helmet run II in Ti 6242S

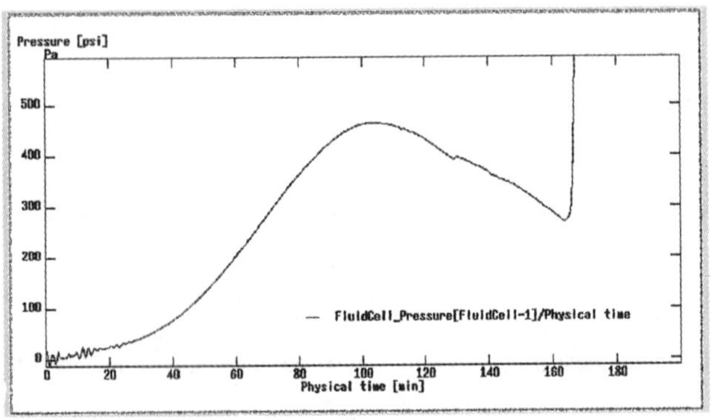

Figure 8 - Pressure-time history for superplastic forming run II in Ti 6242S

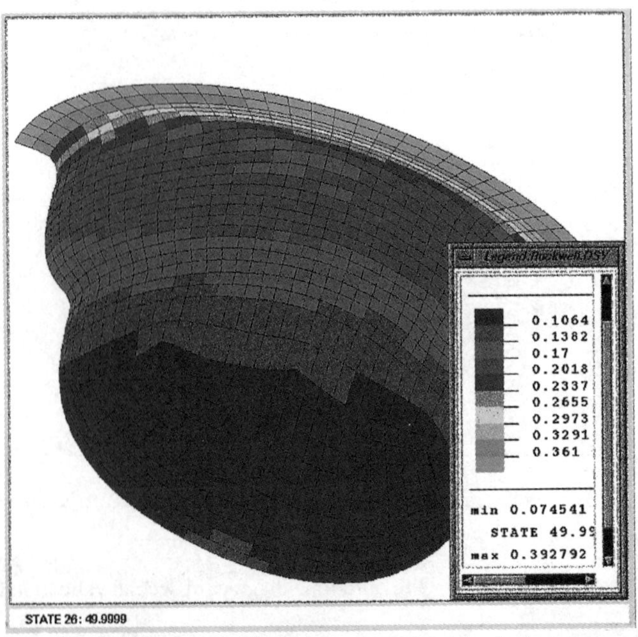

Figure 9 - Final thickness profile at the end of superplastic forming run II in Ti 6242S

The predicted pressure cycle shown in Figure 8 was then used for correlation with a superplastic forming trial. The helmet formed completely with a thickness profile that agrees well with the numerically predicted one. The minimum Thickness predicted was 0.074 inches compared to the experimental value of 0.08 inches. Note that the strain rates achieved in this run are 10 times less than the values in Run I (see Figure 9). Figure 10 shows a photograph of one of the formed Ti 6242S helmet shells.

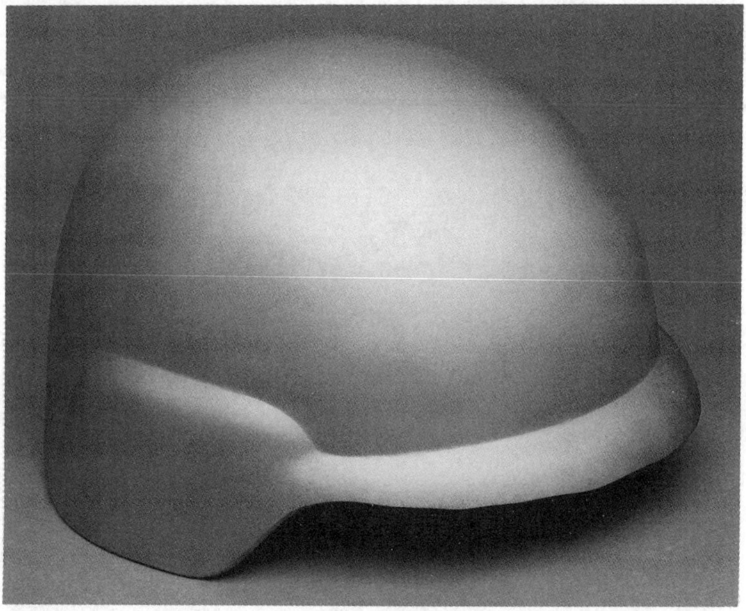

Figure 10 - A photograph of a superplastic fully formed Ti 6242S PASGT helmet

Discussion and Conclusions

The model developed herein for helmet superplastic forming showed encouraging preliminary results achieved thru the use of PAM-STAMP™ together with a reasonably valid material data base development methodology. Despite the measured success in predicting minimum thickness there are several physical modelling issues that still require further follow on work. the location of rupture failures for the most part were near curvature radii transitions, which may be pointing to local friction effects on part thinning . Also the consistent failures of all parts formed in Ti 6Al-4V at both low and high strain rates may be indicative of a non-superplastic microstructure. Finally, as a result of these optimization trials the natural thickness taper associated with the helmet deep drawing appears to have been alleviated relative to prior contract trials, but not completely eliminated. Further refinement of the titanium plate microstructure may bring about some further improvement in natural thickness taper alleviation.

References

1. Sami M. El-Soudani, "Ballistic Resistant Helmet," U.S. Patent Office Disclosure Docket No 95L080, Boeing North American

2. Sami M. El-Soudani, "Superplastic Forming of Antiballistic Titanium Helmet Shells," Technical Report NA-97-1305, Boeing North American Inc. , Seal Beach, CA , 1997

THE EFFECT OF FORMING

IN THE DESIGN OF DEEP-DRAWN STRUCTURAL PARTS

Guy C. Marron and Pascal F. Patou

Sollac, Usinor Group
CRPC, 13776 Fos-sur-Mer
France

Abstract

During the forming of sheet steel, numerous secondary physical phenomena occur and accompany the change of shape. The most obvious effect is the thickness variation which induces some variations in in-service stresses. However other effects also take place, such as residual stress, strain hardening and damage, Young's modulus variation etc. All these effects have an influence on admissible stress and rigidity. Traditionally, the designer does not take into account these effects, which are considered as being negligible. Not even the thickness variations which can reach approximately 20% are considered. The aim of this paper is to classify these secondary effects according to their influence on the behaviour of the part in service, on its fatigue life and therefore on weight saving. For this purpose, the results of numerical stamping simulation are introduced in an analytical model coupled to an FEM analysis code. The fatigue criterion used in this demonstration is that of Dang Van. The risk of fatigue failure can be estimated with accuracy using this approach, which takes into consideration the actual mechanical properties of the material, at any point in the part. Examples of applications in the automotive industry are detailed. They show how profitable this approach can be towards weight reduction. This approach can also be useful in other domains, such as crash resistance. It is necessary to take into account the secondary effects mentioned which occur during metal forming, in order to optimise the use of the steel. The result is safety improvement, increased fatigue life, improved crash resistance, and also weight and cost reduction.

Modeling the Mechanical Response of
Structural Materials
Edited by E.M. Taleff and R.K. Mahidhara
The Minerals, Metals & Materials Society, 1997

Introduction

During the forming of sheet steel, numerous secondary physical phenomena occur and accompany the change of shape. The most obvious effect is the thickness variation which induces some variations in in-service stresses. However other effects also take place, such as residual stress, strain hardening and damage, Young's modulus variation etc. All these effects have an influence on admissible stress and rigidity. Traditionally, the designer does not take into account these effects, which are considered as being negligible. However recent progress in numerical simulation allows these phenomena to be introduced. The aim of this paper is to classify these secondary effects according to their influence on the behaviour of the part in service, on its fatigue life and therefore on weight saving.

Effect of thickness variations

Origin

One ground feature of the deep-drawing process is to provide undevelopable shapes, because of « membrane » strain. The result is a significant variation of the sheet thickness, particularly in the domain of stretching. The origin of this variation is the incompressibility of the metal. Of course the extreme values of thickness strain depend on material ductility, however the typical range is [-0.25, +0.15] in terms of normal strain ε_3.

Why these variations should be taken into consideration ?

Designers making FEM analysis of structural parts often neglect the thickness gradient. However, in-service stress is above all a function of the quantity of material present. Generally speaking, the higher the metal volume, the lower the stress level. In the case of membrane loading, the stress is inversely proportional to the thickness. Therefore not taking into account a 20% thickness reduction will induce a 25 % overestimation of the stress. In the case of pure bending loading, stress is inversely proportional to the square of the thickness. The same approximation will then induce an error of 56 %.

Example showing how important the effect of thickness variations can be

Let us consider the example of a wheel disc [ref. 1-2-3]. One of the most critical zones is the upper part of the hat - see figure 1 - which is loaded in both bending and membrane modes. The stamping sequence introduces a thickness variation in the critical area. The design of the stamping sequence has considerable influence value of the thickness, see figure 2. Good process design can increase the final thickness, whist a poor process design can decrease it. Figure 3 highlights the influence of thickness reduction upon the in-service Von Mises stress. Indeed, if the thickness is considered to be equal to that of the initial blank, the calculated stress will be 300 MPa, whilst reaching 360 MPa when the actual thickness is taken into account.

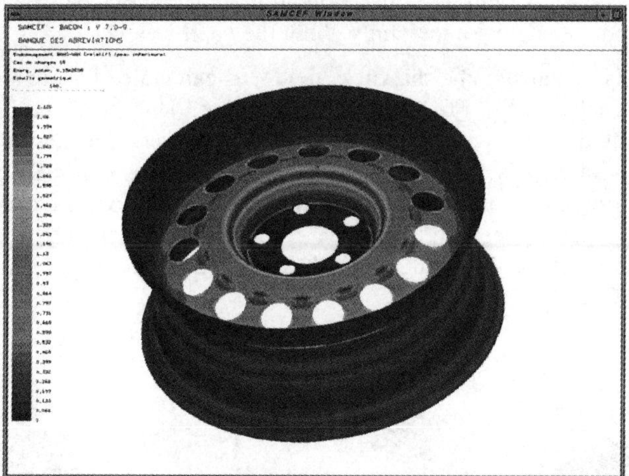

Figure 1 - Critical fatigue area of a wheel disc - Fatigue analysis (Dang Van's criterion).

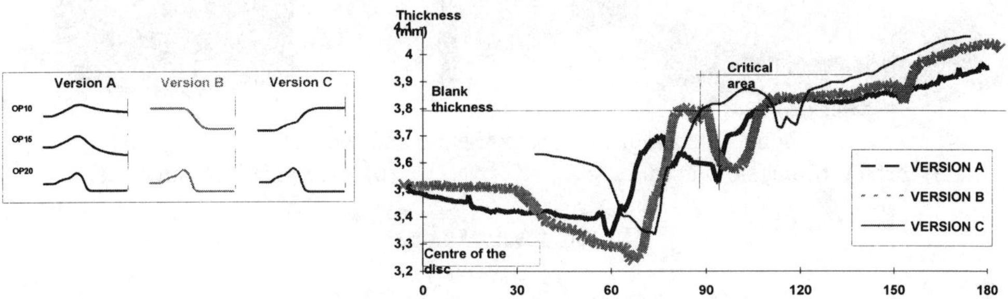

Figure 2 - Influence of the stamping sequence on thickness distribution.

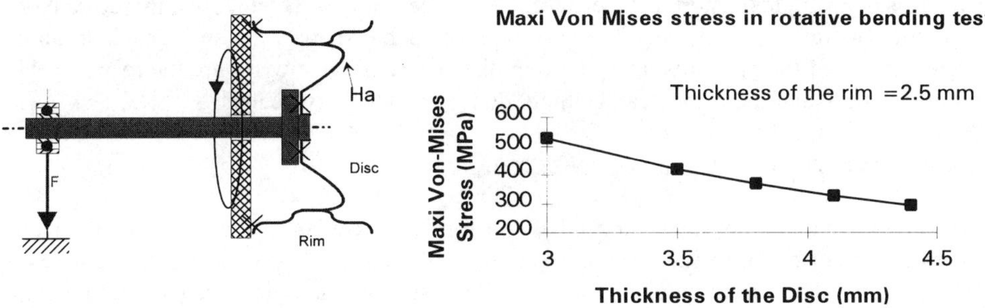

Figure 3 - Von Mises' stress from FEM analysis function of the thickness.

<u>How this is achieved ?</u>

It could be thought that a simple correction be the best solution to this problem : the correcting factor would be something like $(t_0/t)^k$, where t_0 is the thickness of the initial blank, t the thickness as estimated from deep-drawing data, and k an exponent between 1 and 2, depending on the relative level of membrane and bending forces. Unfortunately, this approach is imprecise, as it does not take into consideration all the interactions between zones of different rigidity. Moreover, there is great uncertainty about the *t* and *k* values.

A better approach is to transfer the thickness field - as calculated by numerical simulation of metal forming - directly to the FEM analysis modelling. This principle appears to be quite simple to apply but this is easier said than done... However, the meshing from numerical simulation of deep-drawing is very dense, while it is much less dense for the FEM analysis. Consequently, an operation of « rezoning » must be carried out in order to transfer the thickness field, from dense meshing to a wider spaced meshing. Note the fact that during this operation, the mass must not be changed. Such a method has been used in the case of water heater end cap, see figure 4.

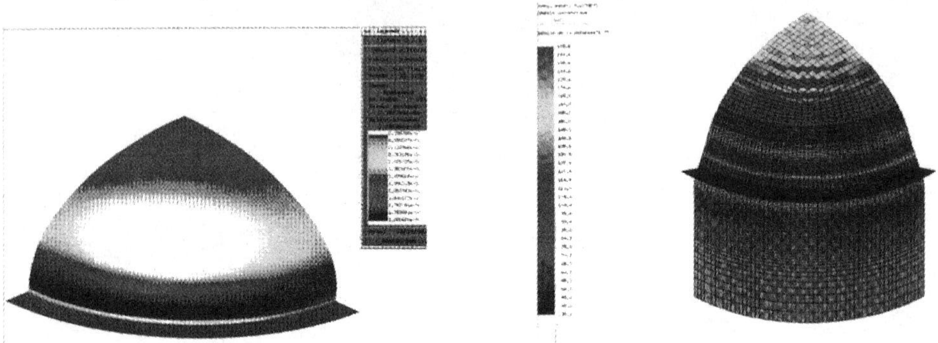

Figure 4 - Example of a water heater end cap [ref. 4] :
(a) metal forming simulation, (b) in-service FEM analysis.

Effect of residual stress

Origin

Residual stress results from the elastoplastic behaviour of metals. Springback does not induce residual stress in a simple tension test, because the sample shape is developable, therefore there is no bending. However in the general case of undevelopable shapes, the springback strain of the different zones of the part must merge. Long distance residual stresses are the result of this « accommodation », and their level can be high, up to ± 200 MPa typically.

Why these stresses should be taken into consideration ?

Residual stresses should be superimposed on in-service stresses by the designer to avoid serious errors. The effect of residual stresses can be beneficial if they reduce the resulting level of in-service stress, or detrimental in the opposite case. Technologies such as laser shock hardening can be used with success [ref. 5].

Example showing how important the effect of residual stress can be

Residual stresses were measured in the critical zone of a wheel disc, using X-ray diffraction and Barkhausen noise, see figure 5. A typical value was 130 MPa, to be compared to the in-service stress due to the rotating bending moment applied to the disc hub, which was 200 MPa in this particular case. Note that the residual stress resulting from metal forming was also superimposed on that resulting from the rim, through an effect of radial compressive fretting.

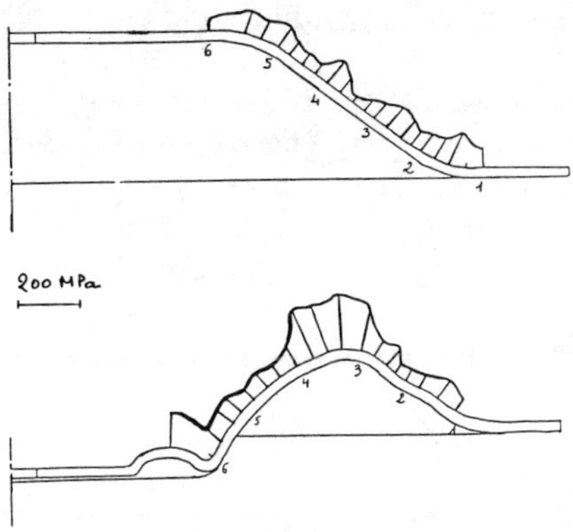

Figure 5 - Von Mises residual stresses, as measured by the Barkhausen noise method, on the two first steps of the stamping of a wheel disc.

Another classical example is the case of very deep gas cylinders such as acetylene cylinders. The level of residual stresses is very high, the proof being that brittle fracture is sometimes observed after the last step of deep-drawing. The case is the same for some compressor housings, where the fracture occurring at brazing is the symptom of a significant level of residual stress.

How this is achieved ?

The simplest way to take these effects into consideration would be to concentrate on a single zone, known as critical. A localised measurement of residual stresses would then be enough in a first approach. However, critical zones of in-service and residual stresses are generally not located at the same place. This is the reason why a full analysis of residual stresses should be done. As in the previous paragraph (on thickness variations), the numerical simulation of metal forming can be of use and commercial software can now provide a rather good estimation of residual stresses. However, the same problems are met and an operation of « rezoning » must be carried out in order to transfer the thickness field, from dense meshing to a wider spaced meshing. Note that some residual stresses - known as type III - can be eliminated during the fatigue life of the part [ref. 6]. Another difficulty is that, even if the thickness distribution can be described by a simple scalar field, the residual stress distribution must be described by a full 3D tensor. The final modelling must be mechanically balanced after rezoning.

Effect of strain-hardening and damage

Origin

There are two particular phenomena attached to metal forming : strain hardening and damage, whose influence on fatigue is in opposition. The strain hardening created by the plastic deformation has a consolidating effect, therefore it is particularly evident that a beneficial effect on fatigue life is induced. Damage can be associate to the increase of the volume of micro-cavities, thus damage has the opposite effect, it decreases the fatigue resistance.

Why these phenomena should be taken into consideration ?

Each area of the formed part is subject to specific deformation. Thus the characteristics of the material are different at each point. In order to have optimal design, the designer has to know the admissible stress at each point of the part. This is to say, the fatigue data which is usually obtained from basic material has to be modified in order to give the designer authentic data taking into account the fact that the material has been formed.

Example showing how important the effect of strain-hardening and damage can be

This effect can be illustrated by a simple trial. A cup with a flat top has been formed (Marciniak trial, see figure 6), in DD14 forming steel, and is deformed in an expansion mode. The strain obtained is almost 0.4. Sample was taken by water jet cutting in order not to modify microstructure characteristics due to temperature increase. Some fatigue trials were made in an alternative bending mode. The results give a fatigue resistance of 214 MPa (scatter = 10 MPa) after forming. The results obtained before forming were only 173 MPa (scatter = 3 MPa). Thus the deformation leads to an increase of 24 % of fatigue resistance. These experimental trials confirm that this phenomenon should be taken into account in order to improve the design.

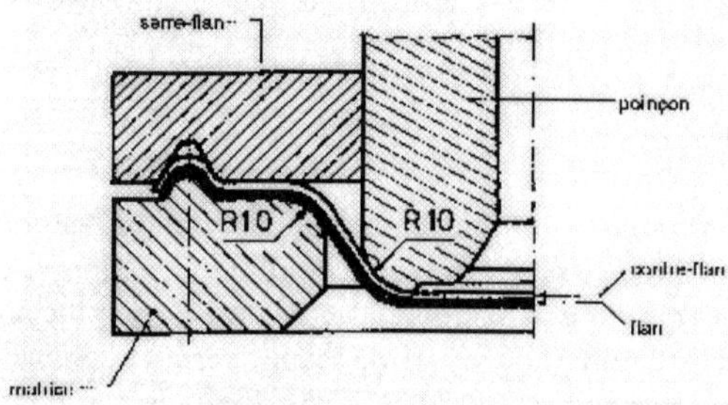

Figure 6 - Marciniak trial.

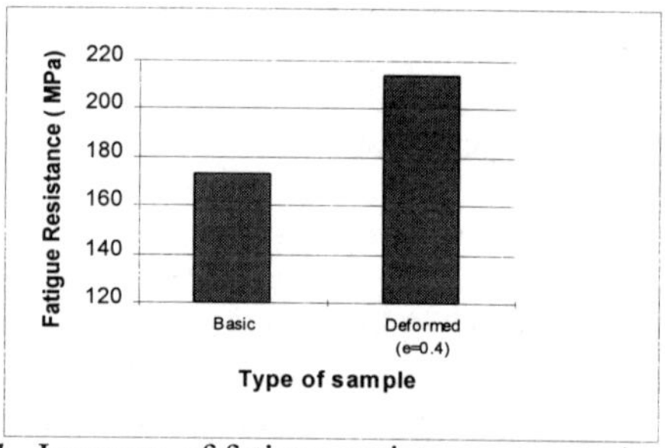

Figure 7 - Increase of fatigue resistance vs. strain.

How this is achieved ?

The procedure is to couple forming process numerical simulation, linear finite element software with a fatigue post-processor. Numerical analysis permits the evaluation of the consequence of metal forming on fatigue life by the use of the Dang Van criterion. Some words about the Dang Van criterion [ref. 7] : the multiaxial Dang Van fatigue criterion is based on a local approach to fatigue damage : fatigue cracks will initiate in slip bands of unfavourably oriented grains. Considering a given part submitted to fatigue loading, the position and plane at which the mean shear stress is maximum can be determined. In this plane, the local instantaneous shear stress at a given point M can be expressed as :

$$\tau(t) = T(t) - Tm$$

where T(t) is the shear stress at instant t and Tm is the re-centered mean shear stress. At « t », the hydrostatic stress Ph is :

$$Ph(t) = \text{trace}\,[\sigma(t)]/3$$

In the plane (Ph, τ), the endurance domain, at a given number of cycle, is defined by two straight lines, symmetrical about the hydrostatic pressure axis. These straight lines are defined by : $|\tau(t)| = b - a.Ph(t)$ where a and b are characteristics of the metal, determined by uniaxial fatigue testing, using three different loading ratios. The condition for no crack initiation is :

$$|\tau(t) + a\,Ph(t)| < b$$

Some Dang Van straight lines are given in figure 8 (number of cycles : 2.10^6). It can be seen that the relative positions of the steel grades depend on the hydrostatic stress : under compression (Ph<0) or low tension, the fatigue resistance increases with the steel grade. This is not true at higher tension levels, where Soldur 355 becomes more resistant than Solform 550.

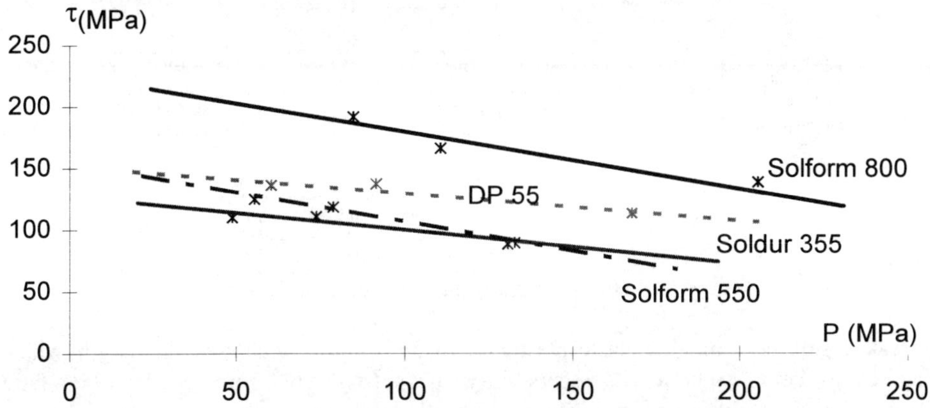

Figure 8 - Dang Van straight lines determined by uniaxial fatigue tests.

The hardening and damage will modify the Dang Van curves by moving them away or closer to the stress path, therefore the risk will be higher or lower, see figure 9.

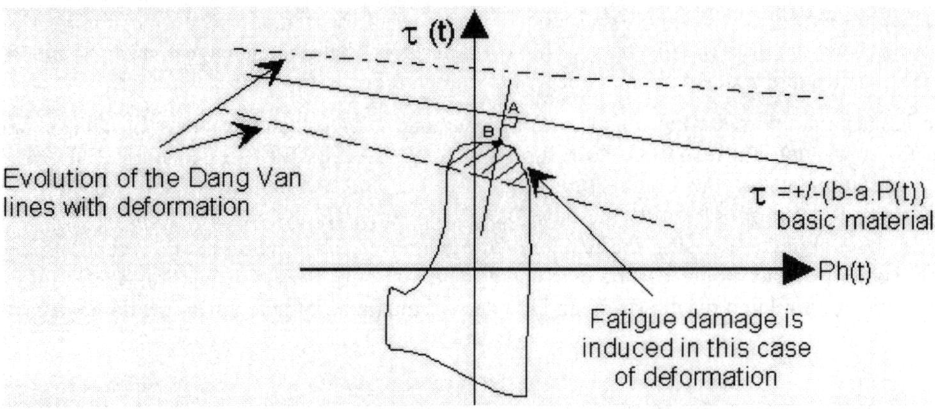

Figure 9 - Evolution of the Dang Van lines - Effect on the application of the Dang Van criterion - The risk (r) is equal to the distance AB.

Each point of the formed part has its own plastic deformation, the Dang Van lines have to be calculated for each element of the numerical model of the formed part. The overall architecture of such a tool is represented in figure 10. The proposed calculation of the fatigue resistance after forming is based on the mechanical analysis of J. Lemaitre [ref. 8]. By this method, it is possible to determine the fatigue resistance by taking into account plastic flow, damage and the type of fatigue load. The basic ideas and hypothesis are explained hereafter.

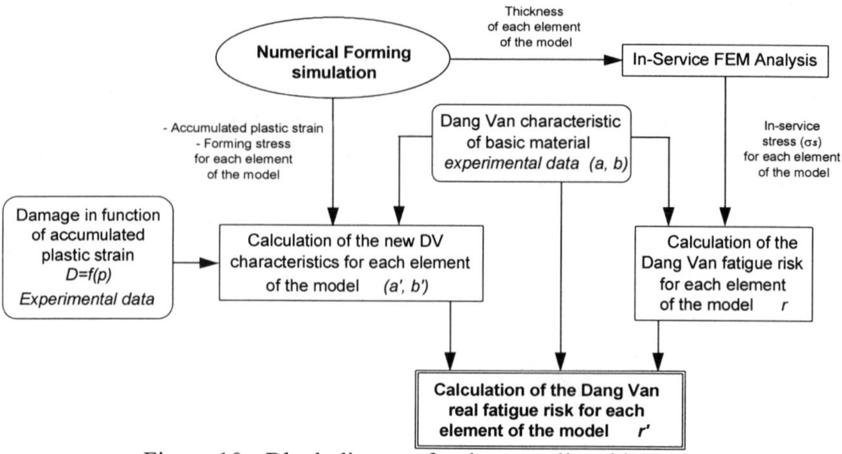

Figure 10 - Block diagram for the overall architecture.

The weakness of the damaging elasto-plastic inclusion is in the metal's yield strength, which is considered as being equal to its reversibility stress (σr). This corresponds to its fatigue resistance for an infinite number of cycles. The others characteristics of the inclusion are as those of the matrix surrounding the inclusion. In a simple model, the hypotheses are :

- *isotropy* : the evolution of the damage is considered as being isotropic. The final damage is a spherical cavity.
- *independence* from the strain path mode : the damage is evaluated using the plastic deformation, the forming stress and the characteristics of the damage curve being in a tension mode. The strain path can modify the value of the damage because of the triaxiality factor and the equivalent stress. The hypothesis is that the damage curve, obtained in tension, is supposed as being independent whatever the strain mode.

The input parameters are the metal forming characteristics (plastic deformation and damage) and the initial fatigue resistance corresponding to a specified mode of loading. Calculations give an estimation of the fatigue limit after forming and a repetition of the procedures for different fatigue modes permits the new Dang Van line calculation after forming. Figure 11 shows the architecture. Using the results of the numerical forming simulation, the damage in each element of the model can be calculated. The damage (Do) is a function of the forming stress, the accumulated plastic strain and the data of the material taken from the damage curve.

Do = function [σF(p), Po, D=f(p), c, E, ν]

Do	:	damage at the end of the forming operation
σF(p)	:	forming stress tensor
p (t, M)	:	accumulated plastic strain function of time and position
D = f(p)	:	damage function of accumulated plastic strain (experimental data)
Po	:	cumulated strain at the end of the forming operation
ν	:	Poisson's coefficient
E	:	Young's modulus

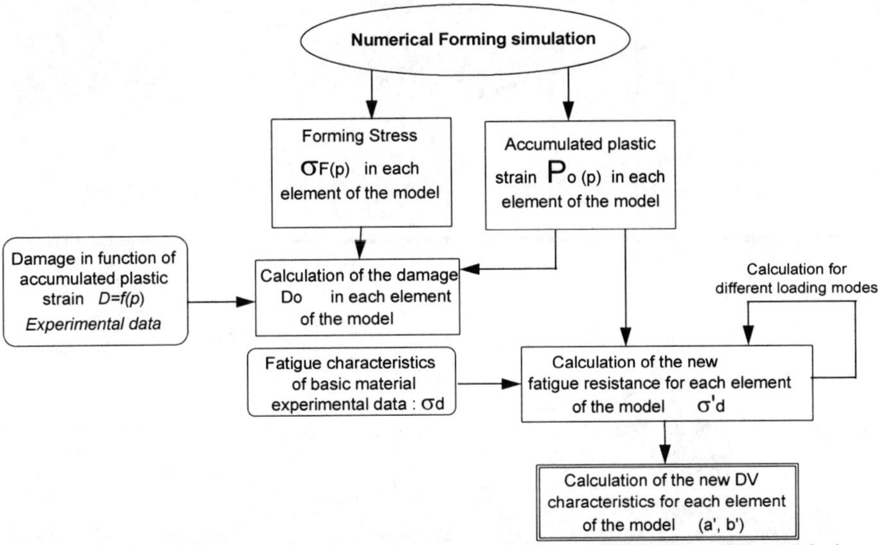

Figure 11 - Block diagram of the calculation of the new Dang Van characteristics.

Knowing the accumulated plastic strain (Po) and the damage value (Do) and the initial fatigue characteristics, it is possible to calculate the real fatigue characteristics in each point of the formed part by performing a 3D calculation.

$\sigma'd$ = function (σd, Do, Po, Dc, D=f(p), c, E, ν)

$\sigma'd$	:	real fatigue resistance after the forming operation
σd	:	fatigue resistance before forming operation
Dc	:	critical damage, above this value, the fracture is inevitable
c	:	linear cinematic hardening coefficient

This calculation has to be done three time in order to model three modes of fatigue loading (loading ratios = -1 : repetitive tension; 0.1 : waving tension; -2 : compressive tension). The shape of the different equations performing the real fatigue resistance after the forming operation ($\sigma'd$) is always the same. The coefficients of the Dang Van line are identified by using a linear regression by the calculated values of the fatigue resistance. Knowing the Dang Van risk calculated with the Dang Van line corresponding to the basic material, the estimation of the real fatigue risk can be calculated by taking into consideration the properties of the metal after forming. This method avoids large calculations as each element of the model will have specified Dang Van characteristics. A simple geometrical post processor recalculates the real Dang Van fatigue risk (r' = A'B) knowing the risk (r = AB) without taking into account the effect of the deformation, and the Dang Van characteristics (a, b) and (a', b') before and after deformation.

The introduction of the hardening and damage effects in the numerical analysis of fatigue permits a better forecast. The model presented is not a perfect approach, needed improved, however it does provide some ideas towards a better design.

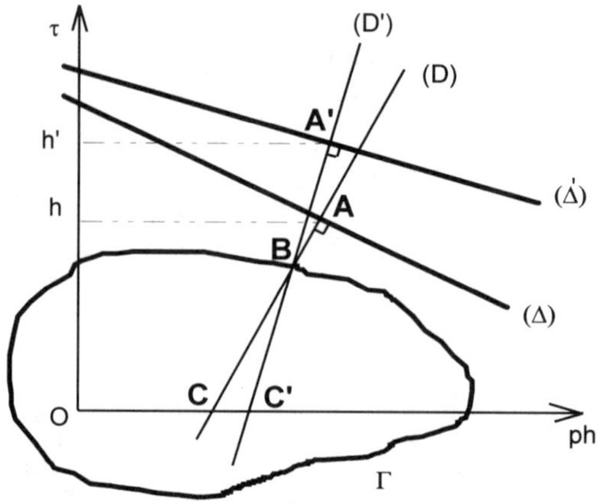

Figure 12 - Geometric passage between the Dang Van lines before and after forming.

Effect of Young's modulus variation

Origin

Among the different factors which influence Young's modulus, the strain hardening, the damage, the crystallographic texture are the more obvious. Strain hardening has an effect on Young's modulus because of the influence of the insterstitials, dislocations, strain velocity and temperature. Damage is a factor which has an influence on Young's modulus. It has been demonstrated that light porosity has little influence on Young's modulus, but heavy porosity can have great influence (case of sintered materials). All these analyses show that the initial value of Young's modulus determination requires great precaution. As each measurement method is more or less sensitive to the different metallurgical factors, there is no absolute method. The measurement has to be adapted to the problem. For different problems such as fatigue (light elastic deformation at high frequency), oligocyclic fatigue (high elastic and plastic deformation), rigidity (buckling), damage characterisation, rupture mechanics... the value of Young's modulus to be taken into account has to result from an experimental method appropriate to each case and specific to the application of the Young's modulus value.

Young's modulus decreases with plastic deformation. The decrease can often reach 20 to 25 %. The figure 13 shows the evolution of Young's modulus function of plastic deformation. This has been measured by using a strain gauge on a DD14 steel grade in 2 mm thickness.

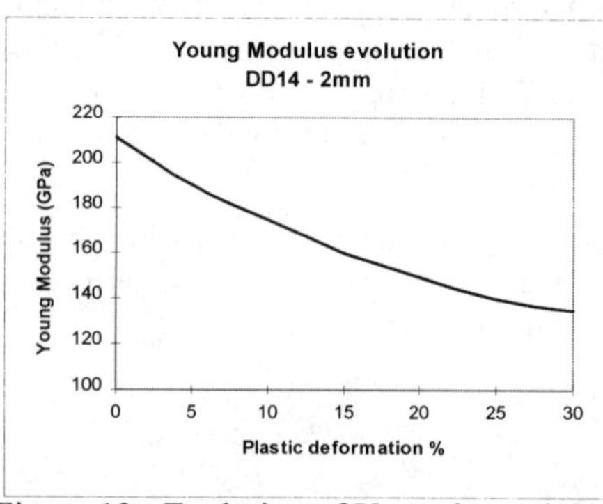

Figure 13 - Evolution of Young's modulus.

Why Young's modulus should be taken into consideration ?

The final rigidity of the formed part depends on its shape, its thickness distribution and its Young's modulus distribution. When a part is designed with rigidity constraints (i.e. the domain of small elastic deformation), it is important to take into account Young's modulus variations and also thickness variations. Of course a forming operation can bring-about a decrease in thickness as well as that of Young's modulus (E); the design of the part can be reviewed if both of these effects have not been taken into account in the design calculations. For a beam in flexion with one end fixed (cantilever), the deflection is given by the following :

$D = -F.L^3 / (3.E.I)$, where :

D : deflection
F : force
L : length of the beam
I : moment of inertia

Therefore a decrease of 20% in Young's modulus has the effect of increasing the deflection by 25%. Springback is a phenomenon where the effect of Young's modulus variations has considerable effect. The lower Young's modulus, the higher the springback.

Example showing how important the effect of Young's modulus variations can be

The example of tube bending illustrates how important the effect of Young's modulus variation is on the springback. However, the bending operation introduces a strain state on the outer surfaces of the tube. It is not rare to find an elongation of 20%. The consequence is a Young's modulus variation effecting springback. Figure 14 shows the springback effect as a function of Young's modulus in the case of the bending of a tube.

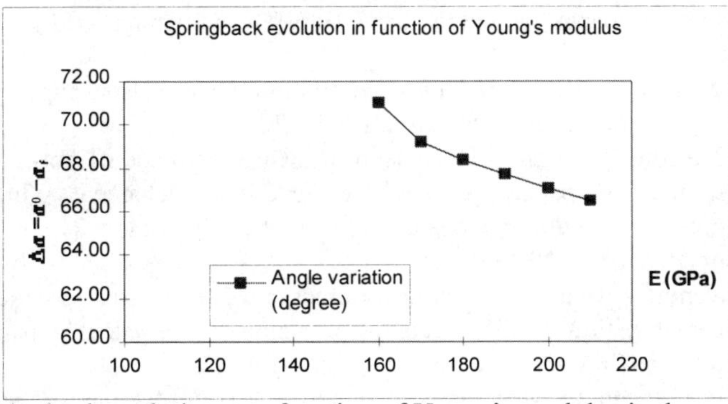

Figure 14 - Springback evolution as a function of Young's modulus in the case of 180° tube bending (diameter = 32 mm, thickness = 1.2 mm, bending radius = 1.2 m).

The effect of Young's modulus variation on springback can reach 5 degrees. Of course other factors like thickness evolution, etc., have to be introduced to obtain an accurate forecast of the springback value.

How this is achieved ?

At present, it is very difficult to take into account the Young's modulus variations. No software exists and the difficulty is to have good measurements of this phenomenon permitting the decorrelation of the different factors. Nowadays the designer has to integrate the effect of the variation of this phenomenon manually.

Conclusion

This paper shows how important the influence of what are known as secondary factors can be, such as thickness variations, residual stress, strain-hardening, damage, and Young's modulus variations. The combination of all these effects may be balanced statistically, whilst inducing significant error... which can prove to be very expensive. Today it is more and more accepted that the designer should control all these secondary effects. The industrial challenge is of course weight and cost reduction, but also significant time saving in the design stage. For example, by using such technologies as software coupling, the automotive industry could save a full wave of prototypes, with a cost expressed in millions of dollars. The near future will probably see the intensive development of such techniques, under the increasing demands of designers.

Acknowledgements

We are grateful to Professor Jean Lemaitre of the *Ecole Normale Supérieure de Cachan*, for his precious help, and to Daniel Gnaedig for his efficient contribution to this study. Our thanks are also due to the people of the Research Centre of Hot-Rolled Steel Products of the SOLLAC company.

References

1. G. Marron, P. Teracher, « The application of high strength hot rolled steels in auto wheels », Journal Of Metals, July 1996, 16-20.
2. G. Hibon, G. Marron, P. Patou, « Light wheels made from high strength steels for the automotive industry », paper presented at the 19th IDDRG Congress, Eger, Hungary, 10th June 1996.
3. G. Rigaut, G. Marron, « Weight reduction in automotive wheels », paper presented at the Symposium Auto, Cannes, France, 5th June 1997, 172-191.
4. G. Hibon, G. Marron, P. Patou, « How deep -drawn parts made of hot-rolled sheet steels should be designed ? », paper presented at the 37th Mechanical Working and Steel Processing Conference, Hamilton, Canada, 24th October 1995, 115-121.
5. European Patent (PCT n° WO97/28914).
6. J. Lu, JF. Flavenot, « Relaxation des contraintes résiduelles sous chargement cyclique, influence des caractéristiques mécaniques et prévision par le calcul », paper presented at Journées de Printemps SFM-ATT, Paris, France, May 1987.
7. K. Dang Van, B. Griveau, O. Message, « On a new multiaxial fatigue limit criterion : theory and application » Biaxial and Multiaxial Fatigue, London, Mechanical Engineering Publications, 479-496, 1989.
8. Jean Lemaitre, A course on Damage Mechanics, Springer Verlag 1992.

Finite Element Modeling of the Formation of Dislocation Cell Structures

H. Feng and M.N. Bassim

Metallurgical Sciences Laboratories, Department of Mechanical and Industrial Engineering, The University of Manitoba, Winnipeg, MB, Canada R3T 5V6

Abstract

Theories of work hardening have been based on the occurrence of dislocation cell structures whose formation has significant influence on material properties in metals. In recent decades, considerable interest has been shown in understanding the evolution of these dislocation cell structures.

The dislocation cell structure was studied by simulating the distribution of the elastic strains after unloading in each grain. The changes of the elastic strains and their magnitude are usually related to dislocation density introduced by non-uniform plastic deformation. Therefore, the distribution of dislocations is indirectly reflected by the changes of the elastic strains after unloading.

In the present work, the occurrence of the structures during tensile deformation was studied using finite element (FE) modeling. The deformation within several grains was considered in this FE-model. In each grain, different anisotropic elastic moduli, yield strengths and work hardening rates were taken into account. A uniaxial tensile deformation was approached. The distributions of the elastic strains after unloading were calculated. Such distribution was found to be similar to the mosaic blocks. These calculated results indicated that anisotropic materials property in metals is one of the reasons influencing the formation of dislocation cell structures.

Modeling the Mechanical Response of
Structural Materials
Edited by E.M. Taleff and R.K. Mahidhara
The Minerals, Metals & Materials Society, 1997

Introduction

Considerable interest has been shown in understanding the evolution of dislocation structures in deformed metals and the role of these structures in defining the mechanical properties of the materials such as the fatigue behaviors, fracture, wear, etc. This interest has been exhibited in four conference publications [1-4] as well as two notable review papers by Kuhlmann-Wilsdorf [5-6] relating dislocation structures to the mesh length theory of work hardening.

Low energy dislocation structures (LEDS) from predominantly in plastically deformed metals since they can accommodate a large dislocation density and at the same time, screen out their internal stresses creating regions of stress free materials. Calculations of the stresses and energy of such structures were performed by Bassim [7] and reported in a series of publications by Bassim and Kuhlmann-Wilsdorf [8,9].

More recently, dislocation structures were studied by different approaches including dissipative structure and synergetics [10], molecular dynamics [11], finite element (FE) modeling [12,13], geometric methods [14], mesh-length theory [15] and Taylor lattice [16]. It was further found in copper that there are significant differences in the dislocation structures observed in metals either during torsion (shear) loading or drawing (and rolling) [17-18], and, even for the same metal, these differences affect the work hardening behavior of the material.

The main objective of this present study is to extend these experimental earlier studies by investigating the formation of cell structures using modeling of the elastic strain distribution after tensile plastic deformation. It was earlier found that the dislocation density is related to the change of elastic strains [19-20]. If the distribution of the elastic strains appears as a mosaic block, its corresponding dislocation structures will show as cell structures. Thus, it is possible to investigate indirectly the occurrence of dislocation structures by modeling the distribution of the plastic strains after unloading.

Finite Element Modeling

Dislocation cell structures observed by experiments

Typical dislocation structures observed during plastic deformation of copper specimens (99.94% Cu) in torsion and in tension are shown in Figs. 1a and 1b, respectively. The TEM samples were cut in the longitudinal direction, similar to the loading direction. For the torsion specimen, at strain $\gamma = 400\%$, the structure is characterized by the formation of dislocation walls (Fig. 1a). In the tension specimen, at strain of 217% (Fig. 1b), we observe the presence of three sets of dislocation walls.

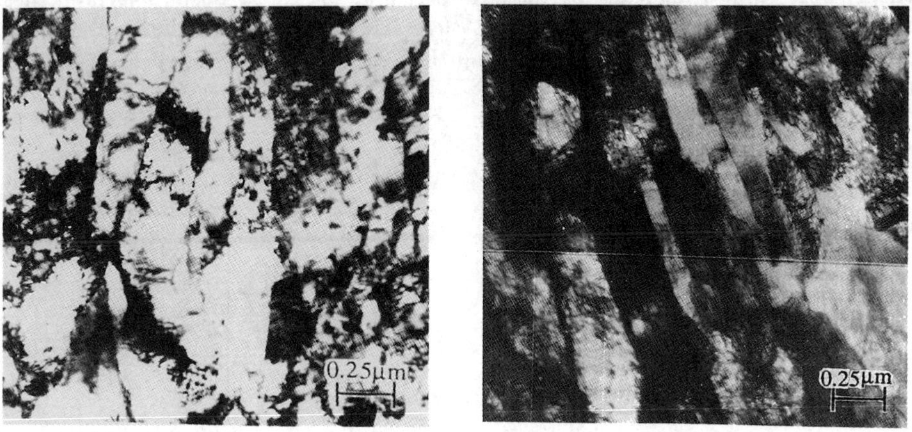

(a) (b)

Fig. 1 Dislocation cell structures observed in copper after torsion at stain of 440% (a) and after tension at strain of 217% (b), respectively.

FE-model

As is well known for metals with only a single phase, if there are dislocation structures, there are certainly some elastic strains after unloading. If the distribution and density of dislocations in metals are known, the corresponding distributions of the elastic strains after unloading can be calculated approximately by FE-analysis, see for example, Ref. [20]. Inversely, the distribution of dislocations and their densities can also be estimated approximately by the elastic strains obtained after plastic deformation and unloading [19]. From continuous mechanics, the dislocation density is essentially determined by the deviation of elastic strain [19]. Therefore, it is qualitatively correct to conclude that the dislocation density is either higher or lower when the elastic strain changes sharply or smoothly, respectively. This concept is applied in the following analysis.

Fig. 2a shows the idealized grains found in metals. For reasons of geometric symmetry, the FE-model was selected as the area surrounded by the dashed square (Fig. 2a). The FE-network was constructed by dividing this area into 4x8 plane elements (Fig. 2b). A stress-free state was defined as the initial condition. The bottom and left boundaries were fixed in the x- and y-directions, respectively. The same displacement of the nodes at the right boundary along the x-direction was assumed during deformation. This assumption held also for the displacements of the nodes at the top boundary along the y-direction. A tensile deformation was simulated by loading a tensile stress of 300 MPa at the top boundary of the FE-network.

The properties of the material for each grain such as elastic moduli (E_x and E_y), yield strengths (τ_x and τ_y) and work hardening rates (K_x and K_y) are found to be anisotropic in different crystal directions as shown schematically by the arrows in Fig. 2a. Those anisotropic properties are defined in each grain by a local coordinate system. The angle between the local x'-axis and system x-axis are indicated by θ_i as shown in Fig. 2a, where i =1, 7, since 7 grains were included in the FE-model.

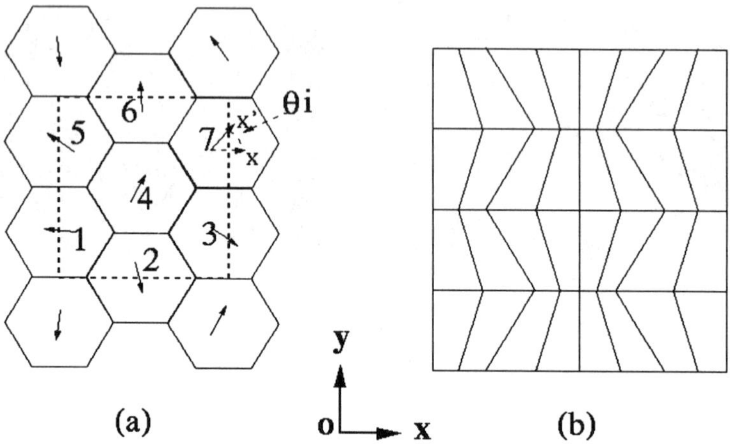

Fig. 2 (a) Idealized representation of the grain structure in a metal. The dotted square represents the area examined by FE-approach. The arrows in each grain represent the local x' coordinates. θ_i represents the angle between the local x'-axis and the system x-axis, where i = 1, 7, indicating the number of grains. (b) the 4x8 finite element network.

Input data

The main input data used in the present work are the following: the elastic moduli (E_x and E_y), the work hardening rates (K_x and K_y) and the yield strengths (τ_x and τ_y). All of those input data are referred to the value in their local coordinates. Their values were taken as E_x = 100 GPa, E_y = 250 GPa, K_x = 20 GPa, K_y = 30 GPa, τ_x = 200 MPa, τ_y = 300 MPa. Another part of input data were the angles (θ_i), where $\theta_i = \theta_1, \theta_2, ...\theta_7$, indicating the angles of the corresponding grains (1 to 7). Those angles were randomly selected as follows. $\theta_1 = 45^o$, $\theta_2 = 5^o$, $\theta_3 = 55^o$, $\theta_4 = 135^o$, $\theta_5 = 90^o$, $\theta_6 = 30^o$, $\theta_7 = 75^o$. All of the present FE-modeling was performed by the FE-code ANSYS 5.3.

Results and Discussion

Distributions of elastic strains after unloading

Fig. 3 shows the distributions of the equivalent elastic strain (Fig. 3a), the components of the elastic strains in the x- and y-directions (Figs. 3b and 3c), when the sample is strained under a tensile stress of 300 MPa and after unloading. In Fig. 3a, the lightest and the darkest areas represent these with highest (0.058%) and lowest (0.0040%) equivalent elastic strains, respectively. Between the lightest and the darkest, the distribution was shaded into 10 levels.

It can be seen from Fig. 3a that, after unloading, the equivalent elastic strain does not return to zero, in spite of the fact that the externally applied stress is removed. Different equivalent elastic strains can be found between the grain boundaries and within each grain. Namely, several mosaic blocks can be found in the distribution of the equivalent elastic strains after tensile plastic deformation and after unloading. Similar mosaic blocks can also be found in the distributions of the components of elastic strains in the x- and y-directions, respectively (Fig. 3b and 3c)

(a) (b) (c)

Fig. 3 The distributions of equivalent elastic strain after unloading (Fig. 3a), the components of the elastic strains in the x- and y-directions (Fig. 3b and 3c), when a sample is strained under a tensile stress of 300 MPa and after unloading.

Influence of anisotropic material properties on dislocation cell structures

As mentioned in the section of the FE-model, the dislocation cell structures can be related qualitatively by the changes of the elastic strains. The location of the highest dislocation density should be there with the sharpest change of the elastic strains. Based on this concept, it is shown from Fig. 3 that dislocation cell structures are formed after the tensile plastic deformation, due to the mosaic blocks of the distributions of the elastic strains (Fig. 3). For those elastic strain distributions, the dislocation density is relatively higher between the boundaries of mosaic blocks and relatively lower within the blocks. Therefore, the mosaic blocks of the elastic strains represent the dislocation cell structures approximately.

Generally speaking, there are no macroscopic elastic strains in the sample after the uniaxial tensile deformation. However, there are elastic strains on a microstructure scale according to the present calculation. The reason for this is due to the anisotropic material properties, which cause a non-uniform plastic deformation within each grain under loading. After unloading, elastic strains can not return to zero because of the non-uniform plastic strains, which are not reversible during unloading. It should be pointed out that a similar mosaic blocks in the distributions of the elastic strains can be found after plastic deformation and unloading, if other anisotropic material properties (E_x, E_y, K_x K_y τ_x and τ_y) are considered as input data.

Other possible influences on dislocation cell structures

Considering the anisotropic material properties in each grain as mentioned above, one can find that the dislocation cell structures (i.e., mosaic blocks in the distributions of elastic strains) can be formed after tensile plastic deformation. Besides this factor, there can be some other reasons being responsible for the formation of the dislocation cell structures such as the changes of yield strengths and the change of the active slip system in every grain. The possible reasons can be stated as the followings.

First, let's discuss the possible influence of the changes of yield strengths in each grain (i.e. in every elements) on dislocation cell structures. At the beginning of the tensile deformation, there are no dislocations within each grain, because there is no plastic deformation and the initial stress state is zero. Once the plastic deformation occurs, non-uniform plastic deformation can be induced within each grain due to the anisotropic material properties, leading to the mosaic block of elastic strains (dislocation cell structures) as shown in Fig. 3. For further plastic deformation to occur, the micro-yielding strengths in each grain are changed, because of the different dislocation density within each grain and of the dependence of yield strengths (τ_{yield}) on local dislocation density (ρ) ($\tau_{yield} = \tau - \tau^* = \alpha G b \sqrt{\rho}$), where τ^* is the friction stress, τ applied stress, α a constant, G shear modulus, b Burgers vector. Therefore, the yield strength as input date within each grain (in every element) should be changed with the evolution of the dislocation cell structures during tensile deformation. Consequently, the increased inhomogeneity of the yield strengths may lead to a finer mosaic block of elastic strains (dislocation cell structures), compared to the case with only anisotropic material properties as input date (Fig. 3). Those two processes happen simultaneously and repeatedly, resulting in more and more fine mosaic blocks in the distribution of elastic strains (dislocation cell structures).

Another reason such the change of the active slip systems may also affect the formation of dislocation cell structures. Normally, dislocation movement on multiple slip systems dominates plastic deformation in metals. It is reasonable to assume the change of the active slip system because of the following reasons. Firstly, there are multiple slip systems in metals (<110> {111} for FCC, <111> {110} for BCC). These properties provide the basic possibility for the variation of the active slip system during plastic deformation. Secondly, the active slip planes can be changed due to dislocation pile-ups and interactions during the evolution of deformation. Obviously, the change of the active slip system leads to increase the inhomogeneity of the plastic deformation, resulting in more possibility of forming the mosaic blocks in the distribution of the elastic strains after plastic deformation and unloading.

Conclusions

The formation of the dislocation cell structures can be simulated approximately by FE-modeling. The dislocation cell structures is indirectly related to the non-uniform distribution of the elastic strains after plastic deformation and unloading. Anisotropic material property is one of the reasons for forming dislocation cell structures in metals.

Acknowledgments

The authors would like to acknowledge the Natural Science and Engineering Research Council for support of this investigation.

References

1. M.N. Bassim, W. Jessu, D. Kuhlmann-Wilsdorf and and H. Wilsdorf, ed., Low Energy Dislocation Structures I, (Elsevier, 1986).
2. M.N. Bassim, W. Jessu, D. Kuhlmann-Wilsdorf and G. Shi, ed., Low Energy Dislocation Structures II, (Elsevier, 1986).
3. G. Kostorz, H.A. Caldeson and J.L. Martin, ed., Low Energy Dislocation Structures III, (1993).
4. M.N. Bassim, W. Jessu and D. Kuhlmann-Wilsdorf, Eds., Low Energy Dislocation Structures IV, (Physica Status Solidi I, 1995).
5. N. Hansen and D. Kuhlmann-Wilsdorf, "Low Energy Dislocation Structure due to Unidirectional Defect at Low Temperatures", Mater. Sci. Eng., 81 (1986), 141-162.
6. D. Kuhlmann-Wilsdorf, "Theory of Plastic Deformation: - Properties of Low Energy Dislocation Structures", Mater. Sci. Eng., 113 (1989), 1-41.
7. M.N. Bassim, "Dislocation Subboundaries, A Theoretical Study", Ph. D. Thesis, University of Virginia, 1970.
8. M.N. Bassim and D. Kuhlmann-Wilsdorf, "Stresses of hexagonal Screw Dislocation Arrays", Cryst. Lattice Defects, 4 (1973), 9
9. M.N. Bassim and D. Kuhlmann-Wilsdorf, "Computer Studies of dislocation networks and Cells", Physica Status Solidi A, 16 (1973), 241.
10. Zhimin Zou et al., "Evolution Model of Dislocation Patterning for Creep and Its Computer Simulation", Scripta Metall., 31 (1994), 515-520.
11. T. Kaplan et al., "Simulation of the Dislocation Array at Ge/Si Interface", Defect-Interface Interaction Materials Research Society Symp. Proceedings, Published by Materials Research Society, Pittsburgh, PA, USA, 319 (1994), 153-157.
12. N. Hansen et al., "Thermomechanical Behavior and Finite Element Modeling of Metal Matrix Composites", Journal De Physique, 3 (7) (1993), 1705-1710.
13. H. Balke and Y. Estrin, "Micromechanic Modeling of Shear Banding in Single Crystals", International Journal of Plasticity, 10 (2) (1994), 133-147.
14. X.F. Fang and W. Dahl, "Investigation of the Formation of Dislocation Cell Structures and the Strain Hardening of Metals by Computer Simulations", Mater. Sci. and Eng., A164, (1993), 300-305.
15. M.N. Bassim, "Mathematical Prediction of Dislocation Cell Size with Strain Using the Mesh-length Theory of Work Hardening", Mater. Sci. Eng., A113 (1989), 367-371.
16. D. Kuhlmann-Wilsdorf and C. Laird "Dislocation Behavior in Fatigue II. Friction Stress and Back Stress as Inferred from an Analysis of Hysteresis Loops", Mater. Sci. Eng., 37 (1979), 111.
17. C.D. Liu and M.N. Bassim, "Dislocation Substructure Evolution in Torsion of Pure Copper", Metall. Trans., 24A (1993), 361-367
18. M.N. Bassim and C.D. Liu, "Dislocation Cell Structures in Copper in Torsion and Tension", Mater. Sci. Eng. A164 (1993), 170-174.
19. E. Kroener, Kontinuumstheorie der Versetzungen und Eigenspannungen, Springer-Verlag, Berlin-Goettingen-Heidelberg, 1958.
20. H. Feng, H. Bierman and H. Mughrabi, "Microstructural - Based 3D Finite Element Modeling of Lattice Misfit and Long - Range Internal Stresses in Creep - Deformed Nickel-Base Superalloy Single Crystals. Mater. Sci. Eng., A214 (1996), 1-16.

MODELING THE MECHANICAL REPONSE

OF A Nb-1Zr/218W COMPOSITE SYSTEM

Louis J. Ghosn and S.V. Raj

NASA Lewis Research Center
21000 Brookpark Rd.
Cleveland, Ohio 44135

Abstract

Finite element analyses were performed to model the mechanical behavior of various fiber volume fractions of a Nb-1Zr/218W composite system. Due to the lack of reliable experimental data on the 218-W fibers at the temperatures and loading rates of interest, the constituent mechanical properties were back calculated from the composite response for one fiber volume fraction. For verification, the same calibrated properties were applied to composites with other fiber volume fractions. Good agreement was observed between the predicted and measured composite stress/strain behaviors at various temperatures and loading rates. The present analyses also provided insight into the build-up of residual stresses in the matrix and fibers after cool down to room temperature from the processing temperatures, as well as the subsequent stress state resulting from heat-up to the testing temperatures.

Modeling the Mechanical Response of
Structural Materials
Edited by E.M. Taleff and R.K. Mahidhara
The Minerals, Metals & Materials Society, 1997

Introduction

NASA expects that future space missions involving the International Space Station, long duration space exploration flights and human habitation of the moon and Mars will require large amounts of electrical power far exceeding 100kW [1]. An analysis of the potential energy sources for these applications suggests that nuclear power is likely to be the most feasible for attaining the desired goals [1]. It is currently envisioned that nuclear power will be generated using molten lithium-cooled reactors with fuel pins made of Nb-1(wt.%)Zr/ 218 W composites. As a result, NASA/DOD/DOE have funded a joint demonstrator project to construct a ground-based reactor system to demonstrate the concept under the SP-100 program. The present study was undertaken as part of this investigation to establish and evaluate processing methods for fabricating the composite system [2]. Several composites containing different fiber volume fraction, V_f, varying between 0 and 0.7 were fabricated and evaluated for their mechanical properties between 1400 and 1600 K. However, in order to predict the long term mechanical behavior of these composites for the intended applications, it is necessary to develop modeling techniques using the relatively short term experimental data generated in the laboratory.

Various analytical procedures have been used by previous investigators to study the mechanical response of composite systems with known a priori constituent properties [3-5]. Typical solutions extend from the application of simple elasticity theory for concentric cylinders [3-4] to actual finite element analysis of sub-models of repeated unit cells [5]. The finite element technique used here assumes a concentric cylinder unit cell model of a single 218 W fiber embedded in a cylindrical Nb-1Zr matrix. The present paper reports the results and insights gained from analyzing the mechanical behavior of Nb-1Zr/218W composites with different V_f deformed at 1400 and 1500K under strain rates varying between 1.6×10^{-5} to 1.6×10^{-3} s^{-1}.

Finite Element Procedure

Axisymmetric meshes of concentric cylinders containing a single fiber encased by a cylindrical matrix are generated for various values of V_f ranging from 0.1 to 0.70. The 218 W fiber radius (R_f) equals 98 μm. The outer radius (R_o) of the Nb-1Zr matrix cylinder encasing the fiber is a function of the fiber volume fraction considered given by $R_o^2 = R_f^2/V_f$. The room temperature material properties of the 218 W fiber are [6]: Young's modulus, E_f= 160 GPa, Poisson's ratio, υ_f = 0.3, thermal expansion coefficient, $\alpha_f = 4.5\times10^{-6}$ K^{-1}, ultimate tensile strength, σ_u^f = 1.8 GPa, and yield stress σ_y^f = 750 MPa. The matrix room temperature properties for the Nb-1Zr are taken [7] to be: for Young's Modulus, E_m = 102GPa, Poisson's ratio, υ_m = 0.38, thermal expansion coefficient, $\alpha_m = 7.3\times10^{-6}$ K^{-1}, ultimate tensile strength, σ_u^m = 300 MPa, and yield stress, σ_y^m = 246 MPa. Elastic-plastic rate dependent constituent material properties are assumed in the present analysis making use of the capability of the ABAQUS general purpose finite element program [8]. Material properties at higher temperatures, if not available in [6-7], are calibrated from experimental composite stress/strain behavior at a given temperature for a particular fiber volume fraction. Once calibrated, the same properties are then applied to other fiber volume fractions without any re-adjustments.

Only, eight eight-noded axisymmetric elements are used: four along the fiber and four along the matrix, Fig. 1. The eight noded axisymmetric elements are deemed necessary to capture the sharp gradients in the matrix near the interface given the low number of elements used. The nodes along the fiber/matrix interface are duplicated to prevent oscillation of the stresses due to discontinuities in the axial and tangential directions. Hence, the opposing duplicate nodes are tied to each other to force displacement continuity along the interface. Symmetry boundary conditions are applied on the bottom nodes in the axial direction and along the fiber mid-planes in the radial direction. The top nodes of both the fiber and matrix elements are tied together in the axial direction to simulate a uniform axial strain condition (generalized plane strain condition). Due to the coefficient of thermal expansion mismatch between the fiber and the matrix, the residual stresses are first calculated due to cool down from the processing temperatures prior to loading.

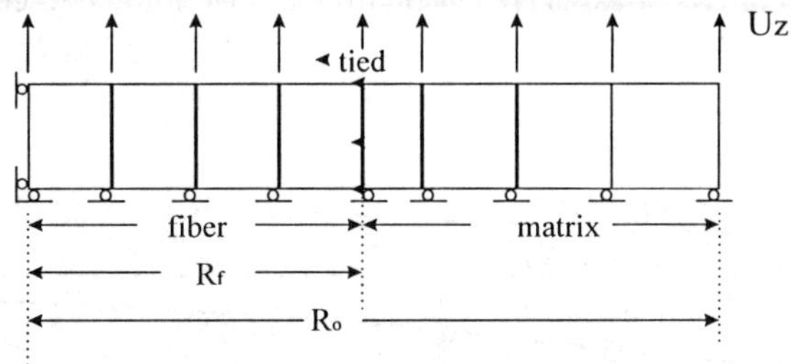

Figure 1 - Typical axisymmetric mesh.

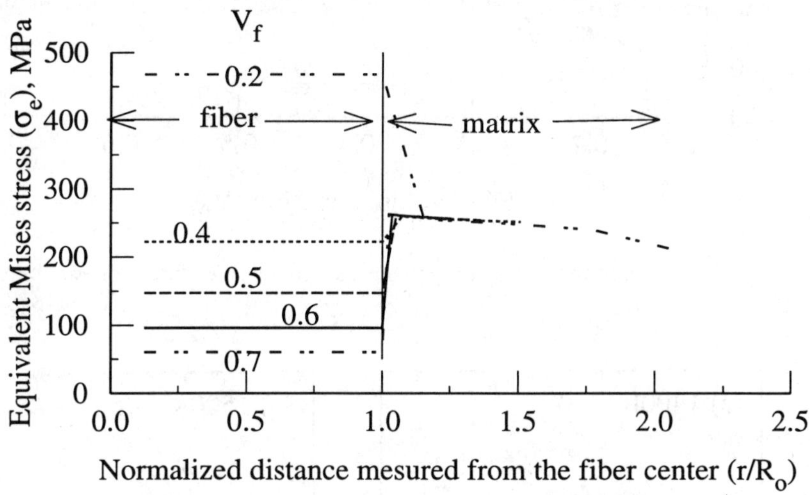

Figure 2- Von-Mises stresses versus radial distance for various fiber volume fractions.

Results and Discussions

Residual Stresses

The variations in the Von-Mises equivalent stresses (σ_e) after cool down from the processing to room temperature as a function of radial distance are shown in Fig. 2 for various fiber volume fractions, given here for completeness as a function of the active stress components:

$$\sigma_e = \frac{1}{\sqrt{2}}\sqrt{(\sigma_{rr}-\sigma_{\theta\theta})^2+(\sigma_{rr}-\sigma_{zz})^2+(\sigma_{\theta\theta}-\sigma_{zz})^2} \qquad (1)$$

The Von-Mises stresses are uniform in the fiber due to the cylindrical model assumed. While the stresses in the matrix are equal for the fiber volume fractions examined, ranging from 0.2 to 0.7, the Von-Mises stresses in the fiber (σ_e^f) decrease rapidly (from 470 to 60 MPa) with increasing fiber volume fractions. The maximum Von-Mises stresses in the matrix (σ_e^m) seems to be at the fiber/matrix interface equal to 300 MPa, well above the matrix yield stress of 246 MPa. On the other-hand, the fiber Von-Mises stresses are well below yield and decrease rapidly with increasing fiber volume fractions. The actual stress components at the fiber/matrix interfaces are plotted in Fig. 3 as a function of fiber volume fraction. All the stress components in the fiber and matrix increase with increasing fiber volume fractions. While all three stress components in the fiber are compressive, only the radial stresses in the matrix are compressive. The axial and hoop matrix stresses are tensile for the fiber volume fraction examined.

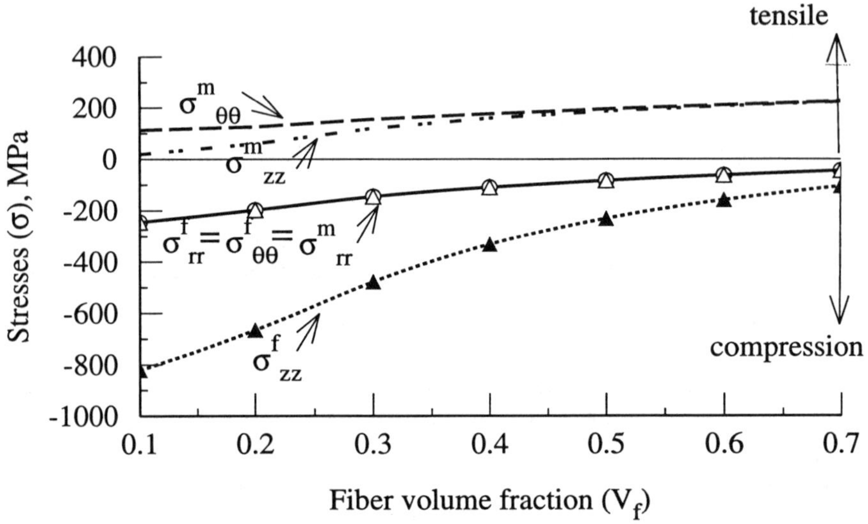

Figure 3 - Variations of the stress components as a function of fiber volume fraction.

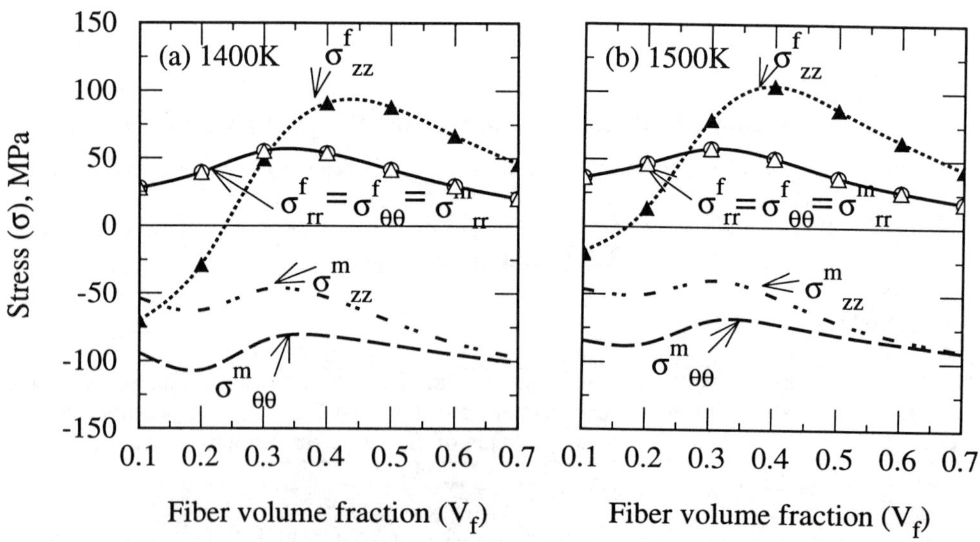

Figure 4- Stress components after heat-up to a) 1400K and b) 1500K.

Heating up Cycle

After finishing the processing cycle, the composite specimens are heated up to the testing temperatures of 1400 and 1500K before loading. During the heat-up cycle, the residual stresses in the constituents relax (Fig. 4). The residual stress components after heat-up to 1400 and 1500K are shown in Figs. 4 (a) & (b), respectively. An important observation is that the radial stresses in the fiber and matrix are now tensile due to the prior yielding of the matrix during the cool down cycle. Although small, these tensile radial stresses may cause debonding of the weakest segments of the fiber/matrix interfaces. It is important to note that an elastic analysis does not capture the calculated reversals in direction of the stress components. Further more, the axial residual stresses in the fibers (σ_{zz}^{f}) have a pronounced maximum around 0.4 fiber volume fraction with values of 90 and 105 MPa at 1400 and 1500K, respectively. The magnitudes of σ_e^{f} with V_f, shown in Figs. 5 for both test temperatures, also show a maximum at $V_f = 0.4$. This stems from the formulation of the Von-Mises equivalent stress that is always a positive number proportional to the square root of the square to the difference of the principal normal stresses given in Eq. 1. For the fiber, σ_{rr}^{f} is equal to $\sigma_{\theta\theta}^{f}$ and around a fiber volume fraction of 0.25 these stresses are also equal to the axial stresses (σ_{zz}^{f}), rendering the Von-Mises stresses null. As the fiber volume fraction decreases ($V_f < 0.25$), the fiber axial stresses decrease sharply and even become compressive causing the effective Von-Mises stresses to increase again to higher values. These calculations show that the state of stresses upon heat-up has an important effect on subsequent loading. It is significant to note that a Nb-1Zr/ 218 W with a $V_f \cong 0.3$ has the lowest fiber Mises stress (σ_e^{f}). Further more, it is desirable to use Nb-1Zr/218W in composites with $V_f < 0.3$ where the residual fiber stress components are minimum.

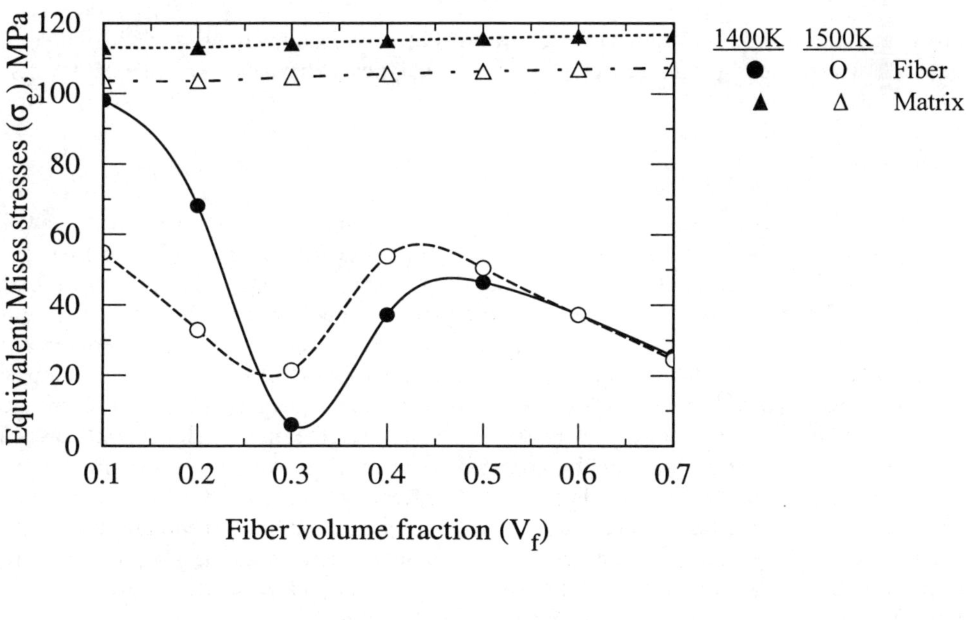

Figure 5- Von-Mises stress after heat up as a function a the fiber volume fraction.

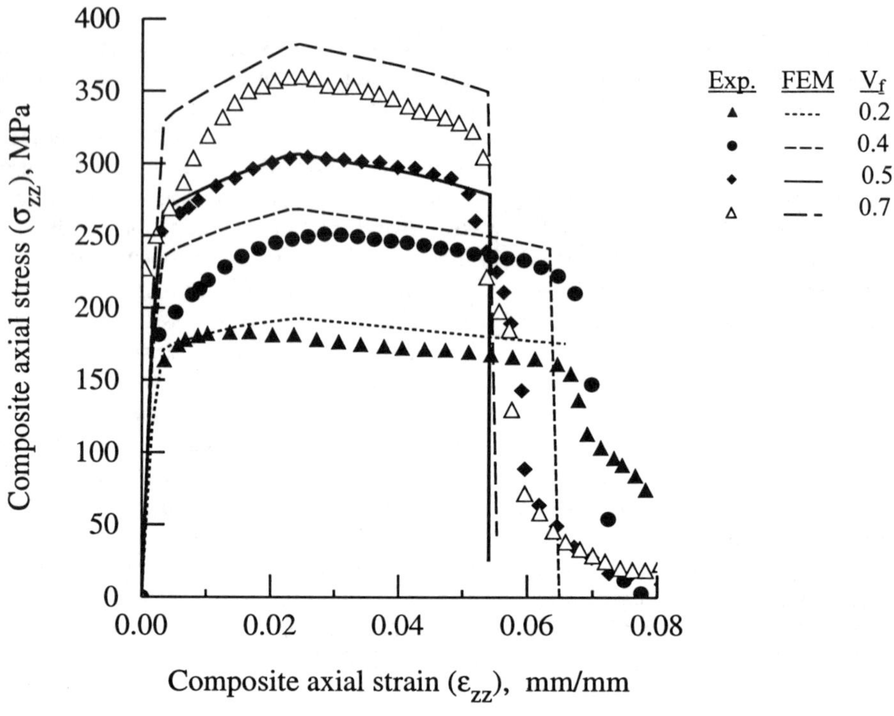

Figure 6-Composite stress-strain curves for various fiber volume fractions at 1400K under an engineering strain rate of 1.6×10^{-5} s^{-1}: analytical (lines) and experimental (symbols).

Fixed Grip Loading

Following the heat-up cycle, several Nb-1Zr/218 W composite specimens were loaded in tension along the fiber direction under various strain rates ranging from 1.6×10^{-3} to 1.6×10^{-5} s^{-1} at 1400 and 1500K. The stress strain curve for $V_f = 0.5$ shown in Fig. 6 was used for calibrating the model. These calibrated values were used to predict the behavior for the other fiber volume fractions (Fig. 6). Overall, a reasonable agreement is observed between the experimental and the predicted composite stress-strain curves. The predicted onset of yielding for the composites is not captured properly for $V_f =$ 0.4 and 0.7. Duplicate experimental tests have also revealed a large scatter in the yield stresses of these composites for the same volume fractions. The calculated ultimate stress (σ_u) for the composites appears to follow the experimental data closely for all the volume fractions (Fig. 7). Good agreement is observed between the predicted and measured ultimate tensile strength values, for a wide range of fiber volume fractions and loading rates. These results indicate that the ultimate strengths of the composites increase linearly with V_f. The observed linear variation is somehow expected given the fact that the fiber strength is much larger than the matrix tensile strength.

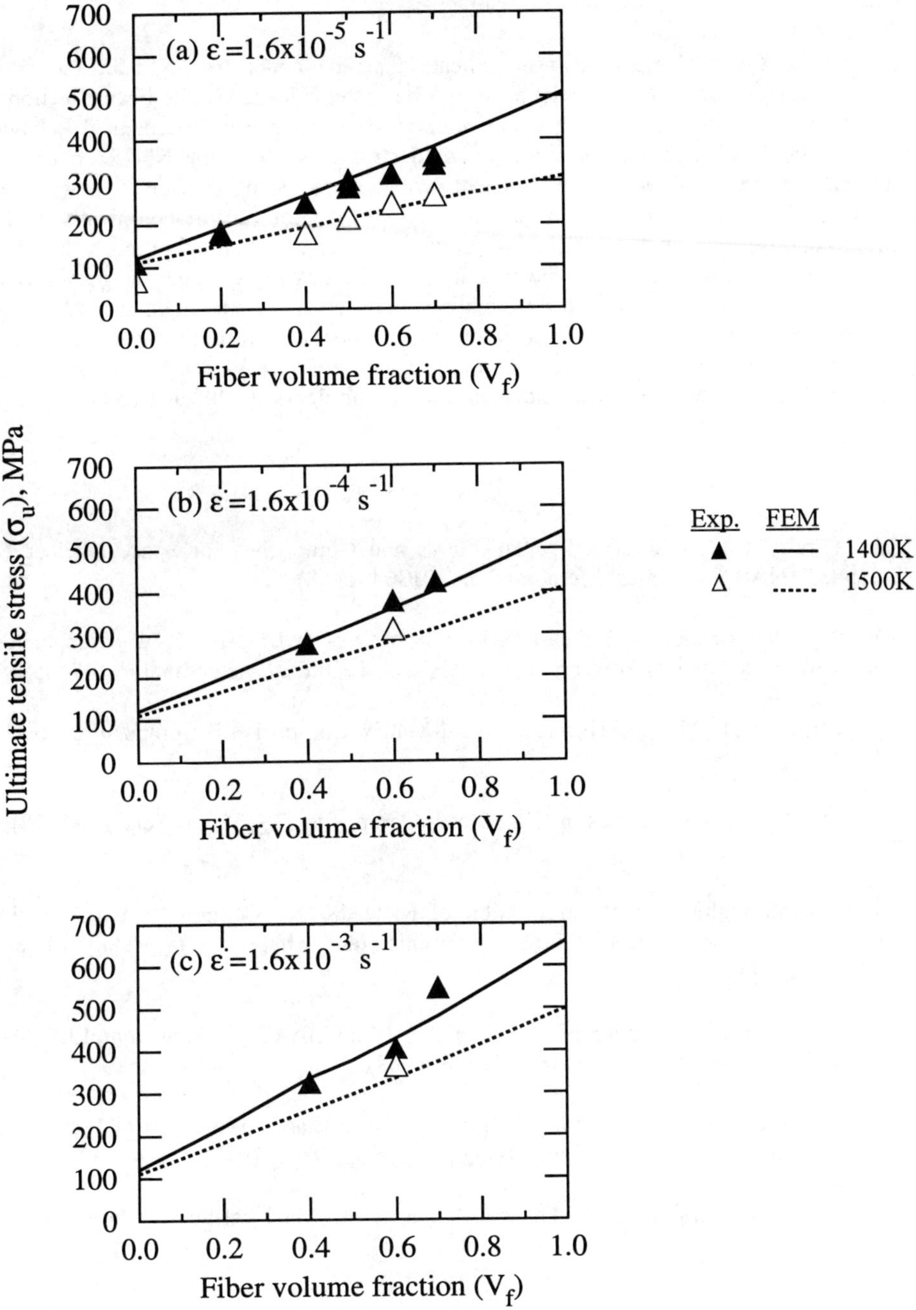

Figure 7-Composite ultimate tensile strength versus the fiber volume fraction under three different engineering strain rates at 1400 and 1500K: analytical (lines) and experimental (symbols).

Conclusions

Results of the finite element analysis indicate that the concentric cylinder model is an appropriate idealization for unidirectional composite systems loaded in the fiber direction. It is important to include the processing cycle prior to determining the mechanical behavior of certain composite systems with low matrix yield stresses such as the Nb-1Zr matrix. The residual stresses in the fibers and matrix can switch signs during heat-up if matrix yielding occurs on cool down. Upon heat-up, the fiber stresses do not vary monotonically with fiber volume fractions, but have a maximum around 0.4 fiber volume fraction. The present methodology used to back calculate the constituent properties at the higher temperatures seems to work well given the excellent correlation between the experimental and analytical stress/strain behavior at various fiber volume fractions and loading rates. The present calculations indicate that Nb-1Zr/218W composites containing a $V_f < 0.3$ are desirable in the intended application due to the low residual stresses in the fibers at 1400 and 1500 K.

References

1. R.H. Titran, et al., "Refractory Metal Alloys and Composites for Space Nuclear Power Systems," (NASA Technical Memorandum 101364, 1988)

2. S.V. Raj, "Fabrication and Characterization of Nb-1(wt.%)Zr/218 W Composites for Advanced Space Nuclear Power Systems," (NASA Technical Memorandum 107465, 1997)

3. T.W. Chou, et al., "Fiber-Reinforced Metal-Matrix Composites," Composites, 16 (1985), 187-206.

4. B. Harris, "Shrinkage Stresses in Glass/Resin Composites," J. Mater. Sci., 13 (1978), 173-177.

5. J.R. Brokenbrough, et al., "Deformation of Metal-Matrix Composites with Continuous Fibers: Geometrical Effects of Fiber Distribution and Shape," Acta Metall. Mater., 39, (1991), 735-752.

6. H.M. Yun, "Tensile Behavior of Tungsten and Tungsten-Alloy Wires from 1300 to 1600 K," (NASA Technical Memorandum 101446, 1988).

7. R.J. Farraro, and R.B. McLellan. "High Temperature Elastic Properties of Polycrystalline Niobium, Tantalum, and Vanadium," Metall. Trans. A, 10A, (1979), 1699-1702.

8. MSC/ABAQUS Version 5.6-1, Theory Manual, (The MacNeal-Schwendler Corporation, 1996).

Atomic and Grain Level Modeling of Polycrystal Deformation [1]

P. Dang N. Chandra

Department of Mechanical Engineering, FAMU-FSU College of Engineering,
Florida A&M University, Florida State University
Tallahassee, FL 32310, U.S.A.

Abstract

Modeling of superpalstic deformation mechanisms is carried out in two different scales of atoms and grains. The length scale, computational methods at each level are introduced. In the grain level modeling, a micromechanical model is developed from the constituent grain level to the level of polycrystalline bulk materials, to study the stress-strain rate relations. Using this approach, the influence of temperature and grain size on the high-temperature deformation behavior of superplastic materials are predicted over a wide range of strain rates. In the atomistic simulation, interatomic potentials using Embedded Atom Method (EAM) are used in conjunction with molecular statics and dynamic calculations. Atomistic simulations are performed on a series of grain boundary structures in aluminum, and the energies associated with each of their equilibrium configurations are computed. The equilibrium grain boundary structures are obtained using energy minimization technique. The temperature and applied stress effects on grain boundary sliding (GBS) and migration are analyzed in details by using molecular dynamics calculations.

[1]Work supported by Army Research Office (Drs. Andrew Crowson and Wilbur Simmons).

Modeling the Mechanical Response of
Structural Materials
Edited by E.M. Taleff and R.K. Mahidhara
The Minerals, Metals & Materials Society, 1997

INTRODUCTION

The behavior of materials can be modeled at three different hierarchical levels with varying spatial scales. A structure or a specimen with a scale greater than $10^{-3}m$ represents the macroscopic scale where the principle of continuum mechanics is generally used. At this level, only materials macrostrcuture and global properties are considered; At the other extreme, atomic scale spans the lengths of a few nanometers ($10^{-9} - 10^{-6}m$), which is compatible with the length scale of crystalline defects (e.g. vacancy, impurity, dislocation, grain boundary and interface). Though at this level the microstructure is difficult to directly related to materials macro-level property, it provides very useful information for the high level (mesoscopic) study. Such information is vital for understanding the materials behavior and guiding materials design; In between the two levels lies the mesoscopic scale where the action of individual grains are ($10^{-6} - 10^{-3}m$) modeled. It is at this level, that materials property is directly related to the structure.

In this work, a detail descriptions are given to grain-level modeling of superplastic flow stress – strain rate behavior and atomic-level simulation of grain boundary structure, sliding and migration.

MICRO-CONTINUUM MODELING OF SUPERPLASTIC BEHAVIOR

The high-temperature deformation behavior of superplastic materials is characterized by a sigmoidal curve (regions I, II and III) spanning about seven to eight decades of strain rate in the log σ - log $\dot{\epsilon}$ plot [1]. A micromechanical polycrystalline model developed at grain-level [2, 3] was extended to describe the material behavior in all the three regions by incorporating a threshold stress term (denoted by σ_* in the slip plane level) [4]. This micromechanical is developed from the grain level to the level of the aggregate in an explicit manner (using self-consistent method). It computes the overall strain rate as the sum of the strain rates contributed by diffusions (lattice and boundary diffusions) and dislocations. In this work, the single-phase micromechanical polycrystalline model is modified to describe the material behavior of dual-phase materials. The developed method considers the individual elastic, inelastic and thermodynamic properties and crystalline structure (and hence the corresponding slip systems) of each of the two phases, and computes the overall response by using self-consistent method. Ti-6Al-4V and Zn-22Al alloys are selected as the model two-phase materials to validate the dual-phase micromechanical theory. Numerical solutions are compared with available experimental data, and the theory is used to predict the effect of grain size and temperature on the superplastic flow behavior in these materials.

During a constant strain rate test, the aggregate is further loaded with $d\bar{\sigma}$ at each incremental step. Then, with the assumed elastic isotropy for each constituent grain (denoted as A), we have,

$$d\sigma_A = d\bar{\sigma} - 2\mu_A [1-\beta] \left[\frac{\mu_A}{(\mu_A - \mu)\beta + \mu} \right] (d\epsilon_A^{sp} - d\bar{\epsilon}^{sp}) \tag{1}$$

Where matrix and inclusion undergo the stress-free incremental strains $d\bar{\epsilon}^{sp}$ and $d\epsilon_A^{sp}$, $d\sigma_A$ is the stress increment of phase A, μ_A is the shear modulus of phase A, μ is the shear modulus of the aggregate, and $\beta = 2(4-5\nu)/15(1-\nu)$, ν being Poisson's ratio of the aggregate. Equation (1) provides the self-consistent variation of internal stress in the grain for each incremental step during a constant strain rate test.

The overall superplastic deformation rate can be considered to be controlled by these accommodation processes. In this work, the overall strain rate, $\dot{\epsilon}_{acc}$, is considered as the GBS accommodations due to diffusion and dislocational creep as shown below:

$$\dot{\epsilon}_{acc} = \dot{\epsilon}_{diff.} + \dot{\epsilon}_{disloc.} \tag{2}$$

Diffusion flow occurs as a result of the diffusion of atoms through the crystal lattice and along grain boundaries. The resolved normal stress acting on a slip plane has been found to direct the diffusion [6]. On the other hand, dislocation movement (e.g. glide) is directly related to the slip system (such as the resolved shear stress acting on a slip plane along the slip direction). Thus, the behavior of slip systems serves as a convenient method to study the constitutive equations at micro level and the micro-macro transitions in modeling superplastic deformation. The accommodation due to diffusional flow for kth slip system in each phase is given by

$$\overset{(k)}{\dot{\epsilon}}_{diff.} = a_l[\frac{1}{Td^2} exp\left(\frac{-Q_L}{RT}\right)(\overset{(k)}{\sigma_n} - \sigma_*) + r\frac{1}{Td^3} exp\left(\frac{-Q_B}{RT}\right)(\overset{(k)}{\sigma_n} - \sigma_*)]. \tag{3}$$

where k varies from 1 to 12 for the FCC {111} <$\bar{1}$10> slip systems, 1 to 12 for the BCC {110} <$\bar{1}$11> slip systems and from 1 to 6 for HCP {0001} <11$\bar{2}$0 > and {$\bar{1}$100} <$\bar{1}\bar{1}$20> slip systems. Q_L and Q_B are the activation energy of lattice diffusion and grain boundary diffusion respectively, T is the absolute temperature, d is the grain diameter, R is the universal gas constant, σ_n is the resolved normal stress acting on the slip plane, σ_* is the threshold stress term at the slip plane level, $r = 3.3\delta D_{0B}/D_{0L}$ is a materials constant [3, 7] and the remaining constant a_l is a fitting parameter to be determined from the experimental data.

The accommodation equation due to dislocation movement can be written as:

$$\overset{(k)}{\dot{\gamma}}_{disloc.} = a_s \frac{1}{T} exp\left(\frac{-Q_L}{RT}\right)(\overset{(k)}{\tau})^n, \tag{4}$$

where τ is the resolved shear stress acting on the slip plane, a_s is a material constant to be determined from experimental data of the specific phase under consideration, and n is the stress exponent.

Both the resolved normal stress σ_n and resolved shear stress τ of the k-th slip plane are related to the local stresses σ_{ij} of the grain being considered:

$$\overset{(k)}{\sigma_n} = b_i b_j \sigma'_{ij}, \tag{5}$$

$$\overset{(k)}{\tau} = \frac{1}{2}(b_i n_j + b_j n_i)\sigma_{ij}, \tag{6}$$

where b_i and n_i are respectively the i-th component of the unit slip direction and slip-plane normal of the -th slip system, and σ'_{ij} is the deviatoric component of the local stress field.

Equations (3) and (4) relate the local stress and strain rate for a given slip system. In these equations, the parameters Q_L, Q_B and r are different for different phases and their values are available in the literature [7]. The remaining fitting parameters a_l, a_s and σ_* are selected by matching the known experimental data. We use the same values of these parameters for both the phases since the superplasticity in dual-phase materials occurs only at a specified combination of the constituent phases while each phase by itself may not exhibit superplasticity, and also the fitted experimental data are obtained from the dual-phase material and not from its single constituent phase.

Once the strain rate due to the k-th slip system is calculated from equations (3) and (4), the superplastic strain-rate components of the considered grain can be obtained by the summation of the contributions of all the slip systems as following:

$$\dot{\epsilon}^g_{ij} = \sum_k (b_i b_j - \frac{1}{3}\delta_{ij}) \overset{(k)}{\dot{\epsilon}}_{diff.} + \sum_k \frac{1}{2}(b_i n_j + b_j n_i) \overset{(k)}{\dot{\gamma}}_{disloc.} \tag{7}$$

where k varies depending on the crystalline structure of the phase being considered. The incremental superplastic strain of a grain is given by

$$d\epsilon^g_{ij} = \dot{\epsilon}^g_{ij} dt. \tag{8}$$

Once $d\epsilon_{ij}^{g}$ is determined for all the grains, the incremental superplastic strain of the polycrystalline aggregate can be calculated from the orientational average which is the numerical average of the incremental superplastic strains of all the grains. Symbolically, this average can be expressed as

$$d\bar{\epsilon}_{ij}^{sp} = \{d\epsilon_{ij}^{g}\}. \tag{9}$$

Equations (1) and (9) are therefore the required relations to describe a dual-phase polycrystalline superplastic behavior.

Modeling flow behavior of Ti-6Al-4V

The model described above is first applied to Ti-6Al-4V, the most widely used superplastic titanium alloy. Nearly all the industrial and the state-of-the-art of titanium SPF is based on this alloy [8]. In general, there are two phases in this alloy: α (H.C.P. structure) and β (B.C.C. structure). Depending on the processing conditions, the relative amount of each phase can vary from 0 to 100 %. In order to model any dual-phase material, the fitting parameters a_l, a_s and σ_* have to be determined. This was done by applying the model to the (α+50%β) Ti-6Al-4V alloys. Hamilton [8] has reported the experimental results for (α+50%β) alloys with various grain sizes (6.4–20 μm) at 1200 K. In our model, the known parameters are Q_L=150 KJ/mol, $Q_B = 97$ KJ/mol and $r = 1.38\text{e-}6$ for α phase and Q_L=153 KJ/mol, $Q_B = 153$ KJ/mol and $r = 9.38\text{e-}10$ for β phase respectively [7]. n is taken as 4.4 for both phases as suggested for titanium alloys under the condition of dislocation climb control [7]. The remaining material constants a_l, a_s and σ_* are matched as a_l=8.0e-7, a_s=5.0e0 and σ_*=0.03 MPa based on experimental data and the model. With these materials constants, the overall agreement between the prediction and experimental counterparts for the two phases is reasonable, as shown in Figure 1 (a). *It should be noted that the simulation is done by merely inputting the required grain size value in equation (3) and using the same materials constants (including σ_*) for all grain sizes.* This demonstrates that the model is able to predict the mechanical behavior of dual-phase superplastic materials with only a limited number of experimentally determined constants.

This model is further applied to study the temperature effect on the $\sigma - \dot{\epsilon}$ relations of ($\alpha + \beta$) Ti-6Al-4V alloys. It is well known that the concentration of the β phase generally increases with temperature, and therefore the phase ratio can change with temperature [8]. The experimental data of $\sigma - \dot{\epsilon}$ relations in Ti-6Al-4V alloys are available at temperatures 1120 K (with 76% α phase), 1145 K (with 70% α phase), 1197 K (with 60% α phase) and 1218 K (with 45% α phase) [8]. In order to model the temperature effect, the phase ratio is taken into account in our model by changing the ratio of numbers of grains for both phases (n_B/n_A). Using the same materials constants (except σ_*) obtained from the simulations of (α+50%β) alloys, the simulation are carried out for the four different temperatures. As shown in Figure 1 (b), the predictions appears to be quite reasonable with the adjusting parameter σ_*. In agreement with the experimental observation of the aggregate threshold stress, the micro-level threshold stress (σ_*) used in the model was found to be dependent on temperature.

Modeling the flow behavior of Zn-22%Al

The model is also applied to Zn-22%Al alloy for the study of temperature effect on the σ-$\dot{\epsilon}$ curves. The thermo-mechanical property of this material at 423K, 473K and 503K has been reported by Mohamed et al [9]. The two phases in Zn-22%Al alloy that result from the eutectoid reaction are: α-Zn phase (H.C.P. structure) and α-Al phase (F.C.C. structure). The known constants in the model are Q_L=91.7 KJ/mol, Q_B =60.5 KJ/mol, r= 3.30e-9 and n=4.5 for α-Zn phase and Q_L=142 KJ/mol, $Q_B = 84$ KJ/mol, $r = 9.71\text{e-}10$ and n =4.4 for α-Al phase respectively [6]. To obtain the fitting parameters (a_l, a_s and σ_*), the numerical simulation is carried out for the temperature at 473K and grain size at 2.5 μm. The material constants (a_l=5.0e-2 and a_s=2.0e4) and σ_*=1.0 Mpa (temperature dependent parameter) are obtained which best fit this set of experimental data. The numerical simulation is subsequently carried out for other temperatures by varying only the threshold stress, σ_* and keeping a_l and a_s as constants. The simulation results compared with the experimental data are shown in Figure 2 (a).

The sigmoidal curve is usually divided into three regions based on the strain-rate sensitivity

m, defined as the slope of the curve, i.e., $d(log\sigma)/d(log\dot{\epsilon})$. As can be seen from the figure, m continuously varies throughout the regions, with a maximum value of m occurring in region II (superplastic region). Region I and region III, at the lowest and highest ranges of $\dot{\epsilon}$ respectively, exhibit low values of m. Figures 2 (b) shows the computed m curves at T= 423 K. As can be seen from the figure, the m curve has the expected bell shape. Most of the phenomenological models [e.g.10] attempt to model only region II, and hence m vs. $\dot{\epsilon}$ will be a horizontal line (corresponding to $1/n$), contrary to experimental observations.

ATOMISTIC SIMULATION OF GRAIN BOUNDARY

In the grain-level modeling described above, GBS is considered to be the dominant strain producing mechanism during superplastic deformation. GBS is treated as the accommodation mechanism responsible for the large strain, but no special attention was given to the exact nature of the sliding motion. GB structures and mobility in superplastic flow have not been as well defined due to the experimental difficulty of measurements. In recent years, atomistic computer simulations are increasingly used to study such problems. We will conduct atomistic simulations to study GB structure, energy, GBS and migration by applying molecular statics (energy minimization) and molecular dynamics. The Embedded Atom Methods (EAM)-type interatomic potential of Al developed by Oh and Johnson [11] are used in the simulation.

A series of symmetric tilt boundaries with different misorientation angles are studied in aluminum. Aluminum is selected as the model crystal since aluminum-based superplastic alloys have been widely studied and used. The computational crystal was generated based on the orientation of a given grain and the symmetry between that and the adjacent grain across the boundary plane modeled by the coincident site lattice (CSL). There are three factors used to describe a CSL: the axis [hkl] about which the rotation occurs; the rotation (misorientation) angle θ, and the reciprocal of the density of coincident sites in the (hkl) net, denoted as Σ. Since grain boundaries are extended defects in two dimensions, but inhomogeneous in the direction normal to the grain boundary plane, periodic boundary conditions are applied in 2-D plane of the interface. In the grain boundary normal direction , free-surface boundary conditions are imposed. Consequently, the crystals are designed to be large enough to remove the free surface effects on the grain boundary structure.

Figure 3 (a) shows the grain boundary energy (E_{gb}) for all the tilt grain boundaries studied in this work, plotted as a function of the misorientation angle θ. As can be seen from the plot, three energy cusps for "special" angles are observed in this work, which correspond to three twin boundaries: $\Sigma3(1\bar{1}1)$, $\Sigma3(1\bar{1}2)$ and $\Sigma11(1\bar{1}3)$. Our simulation result is reasonably consistent with the experimental result conducted by Otsuki and Mizuno [12], as shown in Figure 3 (b). The evidence for the existence of all the three energy cusps is shown though the $\Sigma3(1\bar{1}2)$ is not so apparent. Comparing these two sets of results, besides the excellent agreement of the shape of the energy vs. misorientation plots, the absolute value are also in a close range.

Molecular Dynamics simulation is next performed to study grain boundary mobility under applied stress. To promote the grain boundary mobility, a force (0.02 eV/ $\mathring{A}$) is applied in the x direction (to the right) on the atoms in upper-half of the bicrystals (body force) under the periodic boundary conditions applied in x- and z- directions. In order to perform Molecular Dynamics simulation at a chosen temperature, the relaxed structures were first expanded uniformly in all directions to achieve an appropriate lattice parameter obtained using Quasi-harmonic Approximation. During the simulations, the atoms near the free surfaces (4 outmost layers at each side along y-direction) were constrained to move only in the plane parallel to the GB interface. This assures the Molecular Dynamics simulation were performed at a constant volume condition. The temperature of the simulation is set by assigning to each atom a random velocity according to the Boltzman distribution. During the simulation, the temperature is maintained constant by exponentially relaxing, at each time step (2 fs), the average squared atomic velocity using a time constant of 0.1 ps.

Though temperature results in visible movements of atoms, it does not significantly affect grain boundary sliding and migration. Figure 4 (0 ps) is the equilibrium configuration of the Σ $3(1\bar{1}1)$

boundary at 500 K before the shear stress is applied, the interface remains within a range less than one interplanar spacing in y-direction and there is no apparent grain boundary migration occurred across the grain boundary interface. However, under the applied shear, the coupled sling and migration are also observed, i.e., the sliding was taken place first (Figure 4, 1.5 ps) and then the interface migrates one atomic layer in y direction (Figure 4, 2 ps). As the simulation processes further, a more complicate boundary structure appears as the result of increased simulation temperature. Two different twin variants (twinning planes are outlined in Figure 4) have been generated with the help of the thermal vibration. Twin structure is a low energy coherent defect and considered as one of the high temperature deformation mechanisms. In FCC metals, "annealing twins" are often observed at high temperature following plastic deformations [13].

SUMMARY

Three different level modeling of superplasticity has been introduced. A micromechanical model based on grain-level diffusion and dislocation has been extended to characterize the dual-phase superplastic deformation process. The material constants a_l and a_s are evaluated from a single set of experimental data, while threshold stress (σ_*) is temperature dependent but grain size independent. The model is successfully applied to two-phase Ti-6Al-4V and Zn-22Al materials. The flow stresses as a function of temperature and grain size and strain rate sensitivity (m) are predicted for a wide range of strain rates. Atomistic simulations using EAM potentials have been used to predict GB structure, energy, sliding and migration processes in aluminum. Grain boundary energy varies with different grain boundaries. Coupled grain boundary sliding and migration can be promoted by combined applied stress and temperature.

REFERENCES

1. J. Pilling and N. Ridley, *Superplasticity in Crystalline Solids*, The Institute of Metals, London, UK, 1989.
2. K. Murali and N. Chandra, *Acta Metall.*, 43, 1783 (1995).
3. N. Chandra and K. Murali, *Scripta Mater.*, 32, 1429 (1995).
4. N. Chandra, J. Rama and P. Dang, "Application of micromechanical polycrystalline model in the study of threshold effects on superplasticity", *Mater. Sci. Eng. A*, in print, 1997.
5. N. Chandra and P. Dang, *Scripta Metall.*, 36, 1327 (1997).
6. M. F. Ashby and R. A. Verrall, *Acta Metall.*, 21, 149 (1997).
7. H. J. Frost and M. F. Ashby, *Deformation-Mechanism Maps*, Pergamon Press, Oxford,1992.
8. C. H. Hamilton, in *Superplastic Forming*, ed. by S. P. Agrawal, ASM, Metals Park, Ohio, 13 (1995).
10. F. A. Mohamed and T. G. Langdon, *Acta Metall.*, 23, 177 (1975).
9. F. A. Mohamed, M. M. I. Ahmed and T. G. Langdon, *Metall. Trans.*, 8A, 933 (1977).
10. O. D. Sherby and J. Wadsworth, Superplasticity in Metals, Ceramics, and Intermetallics, *MRS Symposium Proceedings*, ed. by M. J. Mayo, M. Kobayashi and J. Wadsworth, Materials Research Society, Pittsburgh, PA, 196, 3 (1990).
11. D. J. Oh and R. A. Johnson, Atomistic Simulations of Material: Beyond Pair Potentials, Edited by V. Vitek and D. J. Srolovitz, Plenum Press, 223, 1989.
12. A. Otsuki and M. Mizuno, Proc. Symo. on Grain Boundary Structure and Related Phenomena, Thans. Japan Inst. Metals, Suppl. 27 (1986) 789.
13. P. A. Thornton and V. J. Colangelo, Fundamental of Engineering Materials, Prentice-Hall, Englewood Cliffs, New Jersey, 1985.

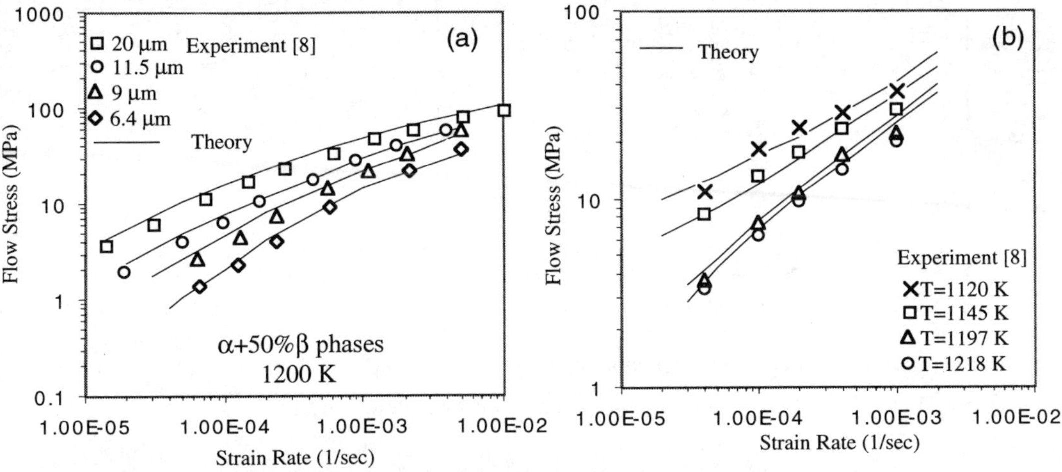

Figure 1: Prediction and experimental comparision of flow behavior of Ti-6Al-4V alloy: (a) grain size effect and (b) temperature effect.

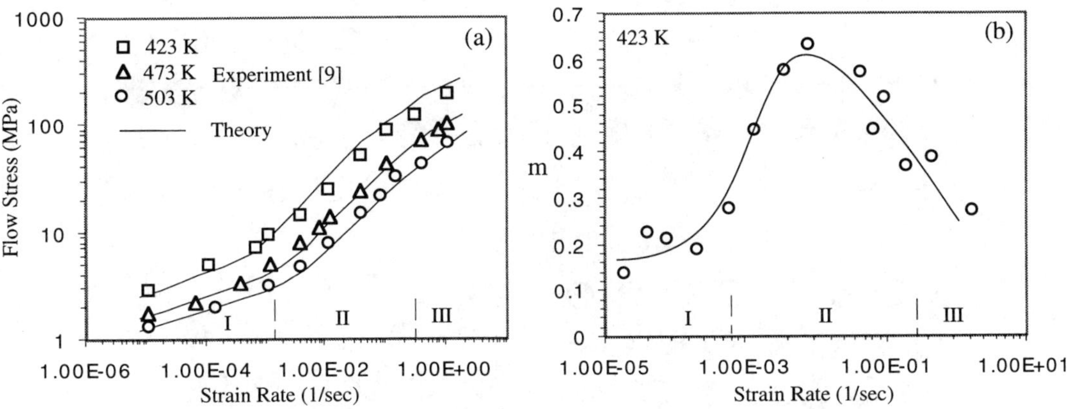

Figure 2: Theoretical prediction of (a) the temperature effect on flow stress-true strain rate behavior and (b) the strain rate sensitivity m of Zn-22Al alloy

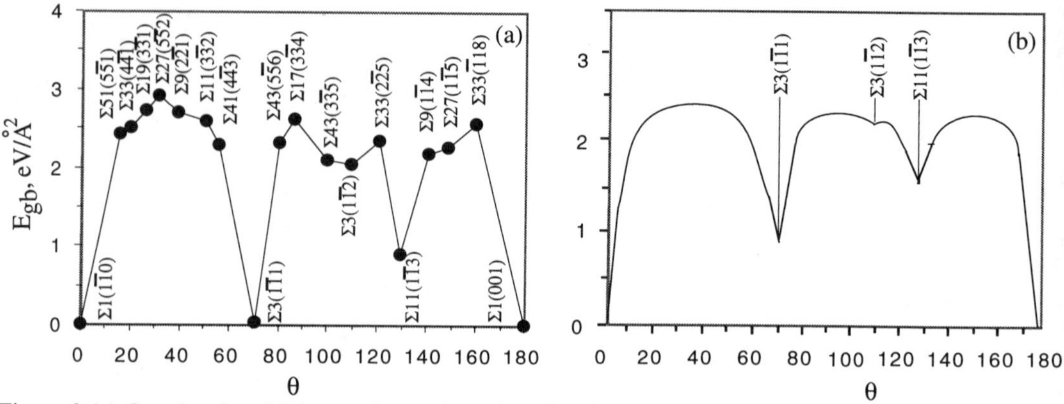

Figure 3 (a) Caculated and (b) experimental results of grain boundary energy, E_{gb} of tilt CSL grain boundaries about [110] of aluminum as a function of misorientation angle, θ.

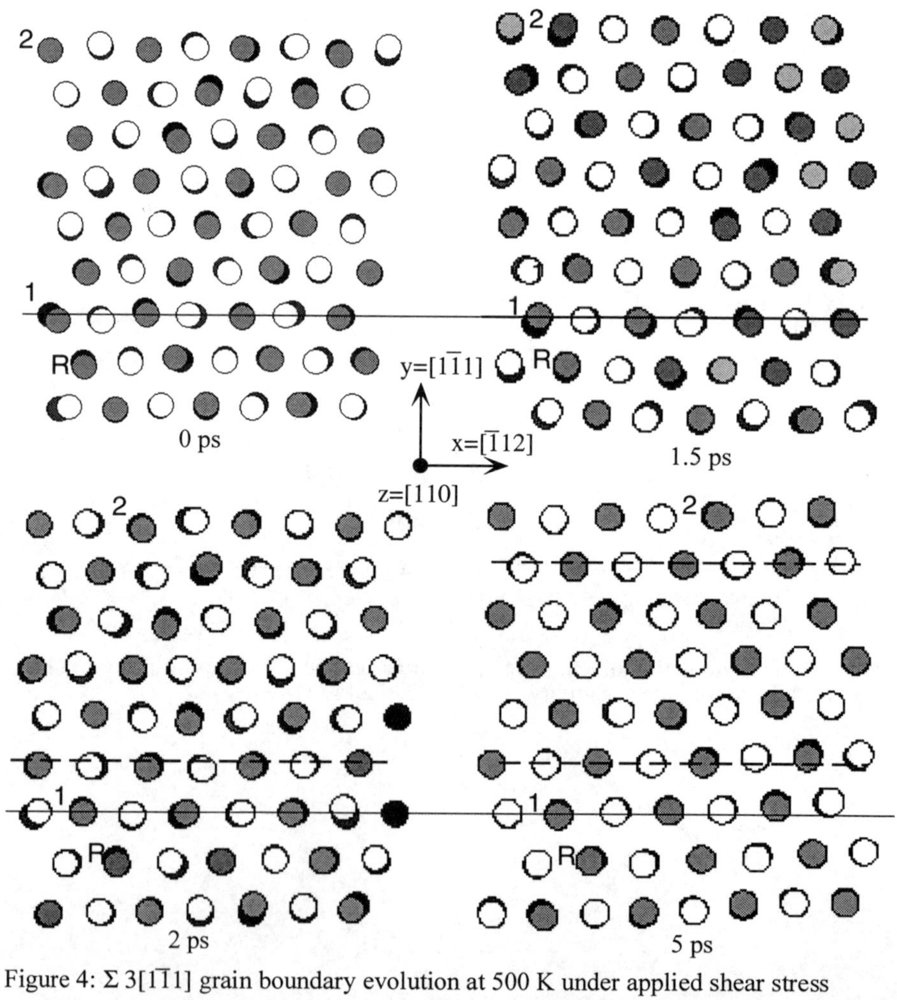

Figure 4: Σ 3[11̄1] grain boundary evolution at 500 K under applied shear stress

Constitutive Models and Experimental Evaluation

High Strain Shear of Lead

M.D.Monti[1], I.R.Shaw[2], W.G.Ferguson[2] and W.H.Robinson[1]

1 Penguin Engineering Limited, P.O.Box 33-093, Petone.
2 Dept of Chemical and Materials Engineering, University of Auckland.

Abstract

The high strain deformation characteristics of pure lead (>99.97%) were assessed for the application of lead as a damping medium for earthquake technologies. Mechanical and metallurgical properties including damping, fatigue and recrystallization of lead, during cyclic deformation were explored, as well as the effects of pressures from 0.1 MPa to 70 MPa on these characteristics. The role of micro-voids, pumping and geometry breakdown, were studied and the contribution of these to failure was assessed.

Lead has been shown to be a high damping capacity material (47 MJ/m^3) suitable for cyclic applications. Imposing a hydrostatic pressure on the samples altered the failure mechanism from a highly ductile yielding to semi-ductile failure via crack propagation.

Modeling the Mechanical Response of
Structural Materials
Edited by E.M. Taleff and R.K. Mahidhara
The Minerals, Metals & Materials Society, 1997

Introduction

Lead has been used to damp the energy associated with earthquake motion since 1970. Lead has been used as a mechanical energy damper in many countries including Japan, USA, China, and Italy. Lead energy dampers have become refined and reliable mechanisms used to successfully protect structures and equipment, the world over.

This research was performed to obtain accurate information regarding the energy absorption capacity of lead during cyclic deformation at room temperature. Lead was tested in tension, cyclic and unidirectional shear, at varying strain rates to produce accurate information regarding its deformation characteristics. Microstructural evidence was sought to clarify the assumed material behaviour.

Theory

The low melting temperature of lead (327° C), and its consequential malleability at room temperature, has made it an important material for technical investigation, modelling and some high deformation applications. The high energy absorbing capacity of lead and its good heat diffusivity, make it ideal for damping applications. Its tendency to recrystallize during deformation, and anneal at room temperature, make it a maintenance-free damping material, with limited susceptibility to fatigue during high cycle plastic deformation.

Recovery and Recrystallization

When lead is deformed at room temperature, the process of annealing, recovery and recrystallization, can begin during deformation. These mechanisms involve motion of point defects, dislocations and subgrain boundaries.

The recrystallization process requires nucleation and growth, and is very different to the recovery process and is extremely time and temperature dependent. The material must be held at a temperature for a specific time to complete the process. Activation energy represents the amount of energy required to initiate nucleation and subsequent recrystallization. This is considered an empirical constant, although it does not remain constant throughout a process, and instead changes constantly as the driving force resulting from cold working, is gradually depleted as the strain energy is invested in realignment of lattice discontinuities. Nucleation and growth occur together and the rate of recrystallization is dependent on the rate at which the nuclei form.

A discontinuity in the stacking order of the close packed <111> plane in a lattice, is a high energy region referred to as a stacking fault. In materials of high stacking fault energy, recovery is most pronounced and does not readily occur in low stacking fault materials[1], such as lead. Dynamic recrystallization is peculiar only to a few metals, in particular low stacking fault energy materials. The low climb rate of dislocations retards the development of sufficient dislocation density and appropriate configurations to drive the recrystallization process.

Mechanisms of Dynamic Recrystallization

Dynamic recrystallization occurs either as a cyclic or stable phenomenon. Initiation is associated with a critical amount of strain and as this strain is attained and reattained, cycles of nucleation and growth occur. Nuclei experience intense growth until the driving force for boundary migration subsides due to increased dislocation density caused by the continuing deformation. The grains nucleate and grow as the grains are deformed. This growth continues until the growing grains impinge upon each other. Another cycle of nucleation will then begin with the nucleation of new grains.

Glover and Sellars[2] observed the critical strain for the onset of recrystallization was primarily dependent on the experimental parameters of temperature and strain rate.

Rapid recrystallization has been observed after deformation has ceased and this was described as metadynamic recrystallization[3]. This mechanism operates at abnormally high speed, as no incubation period is required. This occurs because the high angle boundaries and nuclei required for recrystallization have been formed by the deformation itself. Petrovik *et al*[4] investigated metadynamic recrystallization in austenite during interrupted compression tests. Their findings assimilated this mechanism to static recrystallization, and reinforced the incubation-free process postulated by McQueen[5].

The Effect of Pressure

The common failure mechanism for lead in tension is microvoid coalescence. It is postulated that hydrostatic pressure will reduce the tensile stresses in the specimen neck, thus extending the strain to failure of the material. It has been observed that the relative fracture area failing by void coalescence decreases with pressure, whilst the relative area of shear fracture increases [6]. At a transition temperature the fracture mechanism has been observed to change from void coalescence to fracture via highly localised bands of intense shear deformation [7].

Experimentation, Results and Discussion

Cyclic Testing in Shear

Shear tests were performed to model repeatable, reversible shear strain, to magnitudes of 100%, at a frequency of up to 1 Hz. From the consistent stress-strain loops, damping capacity per unit volume could be calculated. The specimens were then sheared up to 100% for up to 20 cycles. A typical stress-strain diagram is depicted in Figure 1. As the specimen geometry became more sophisticated, the shape and quality of the hysteresis loops improved. Damping was measured as 47 MJ/m^3 per cycle of strain of $\pm$ 1.

Further tests were performed under a range of imposed hydrostatic pressures, producing extended lifetimes before failure. A typical hysteresis record is shown in Figure 2. Following these tests, an attempt was made to constrain the samples with a thin walled rubber tube. No significant improvement was made.

Shear testing consistently produced hysteresis loops indicative of significant energy absorption. As the specimens in this investigation were predominantly unconstrained during testing, production of a uniform shear strain through the whole specimen could only be approximated. Consequently, as the specimens were stretched and compressed the geometry of the specimens quickly deteriorated and the region of deformation in the specimen changed. As this occurred, the damping capacity of the specimen reduced. This instability of the test method was reflected in the minor degeneration of the stress-strain loop.

A finite element analysis of these shear tests indicated that the strain in the hourglass specimens was concentrated in the minimum section of the specimen, with additional tensile and compressive strains in the curved section. As the number of cycles increased cracks began to open up in the curvature and the test became unstable. Samples deformed cyclically under an imposed pressure were subjected to in excess of 100 cycles before failure. However, results indicate that when a few

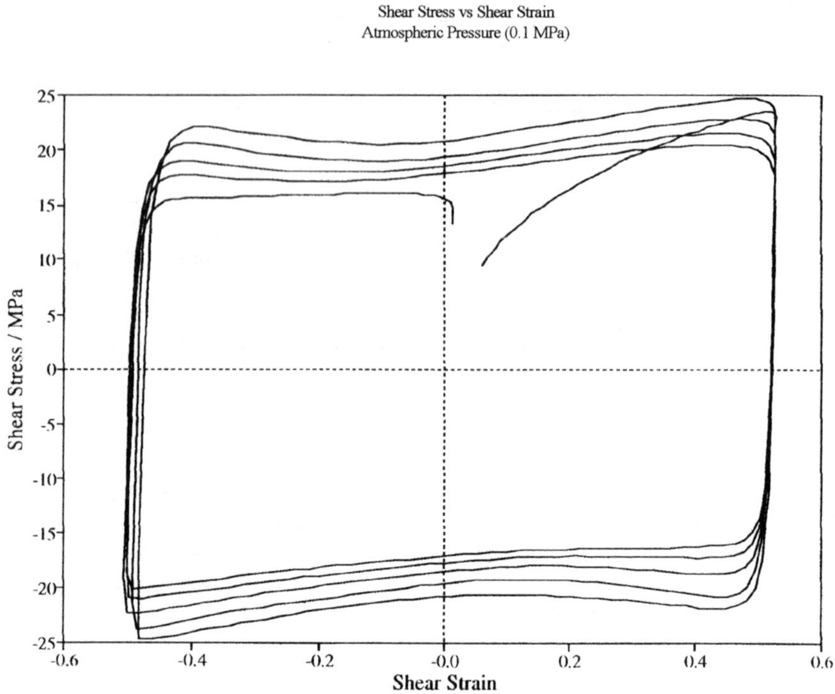

Figure 1. Typical Stress-Strain Hysteresis diagram for lead. Five cycles at ± 10 mm or ± 50% strain.

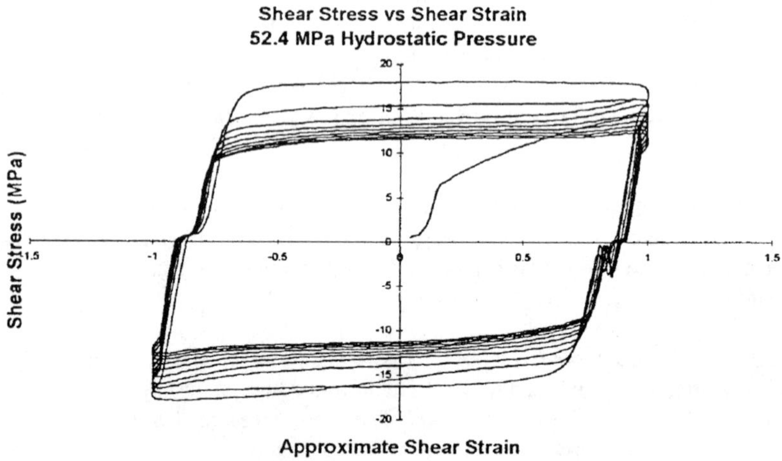

Figure 2. Stress-Strain Hysteresis diagram for lead under hydrostatic pressure. Ten cycles at ± 100% strain.

cycles at a time are applied to the specimens, as is typical of earthquake deformation, the damping capacity of the material remains stable.

The characteristic rectangular loops observed in this investigation indicated that a softening mechanism was occurring. This softening mechanism was thought to be recrystallization. The regular shape of the hysteresis loops suggested the material was maintaining its mechanical properties. It is these characteristics which the successful operation of vibration dampers rely upon.

Single Stroke Shear Testing
Two pairs of shear specimens were tested in unidirectional shear at atmospheric pressure and at 69 MPa, to illustrate the deterioration of the material at strains which induce failure. The resulting stress-strain curves are detailed in Figures 3 and 4. The fracture surface of the specimens sheared under pressure, indicated the failure mechanism was not via microvoid coalescence as is common in extremely ductile materials. Instead it resembled the fracture surface of a semi-ductile material such as steel.

In an experiment conducted previously, a pair of lead shear specimens, constrained in a rubber-steel laminate, were cyclically sheared at a strain of ~+/-10% while a 'hydrostatic' stress of ~60MPa was applied to the to the specimens. After 20,000 cycles the shear force was within 20% of the starting value and on being cut in half the lead was found to be in sound condition. This cyclic shear test was repeated with no pressure being applied to the lead, with the result that after 100 cycles the shear force had dropped to less than 50% of its initial value.

Further work subjected a pair of specimens constrained in a rubber-steel laminate, cyclically sheared to ~+/- 40%, under a pressure of 20 Mpa. After 144,000 cycles the shear force was 10 % greater than the initial shear force. This increase occured in the first few thousand cycles.

Microstructural Evidence for Recrystallization
Microstructural investigation of lead samples was undertaken. The microstructure of both cast and worked material was recorded. When the specimens were re-examined six weeks following the initial investigation, scratches which were a result of the polishing procedure, had recrystallized and grains had grown from these nucleation sites. This can be seen in Figure 5a. As these grains could be seen without re-etching of the material.

A polished lead specimen was scratched with a scalpel in an attempt to observe recrystallization of scratches at room temperature which is known to occur. The specimen required etching after scratching to reveal the grain structure. Consequently a period of up to 5 minutes elapsed before the structure was observed. Figure 5b shows the grains which nucleated in a scratch which varied from 3 to 15 microns deep, 30 seconds after the scratch was made. Some recrystallization can be seen in the region of the specimen immediately adjacent to the scratch and the original much larger grain structure is also visible. The specimen was monitored for six weeks, but no change was detected after the initial 30 second period.

Rounded and distorted annealing twins were observed. According to Jonas [7] this observation suggests dynamic recrystallization. Twins were also present in the deformed structure, which were not deformed. It is likely the undeformed twins are formed in the final stages, or after deformation and the deformed twins, are formed before or during deformation. This evidence collates to indicate that recrystallization had occurred before, during and possibly after deformation. This gives a solid indication that lead recrystallizes at room temperature, statically, dynamically and possibly metadynamically.

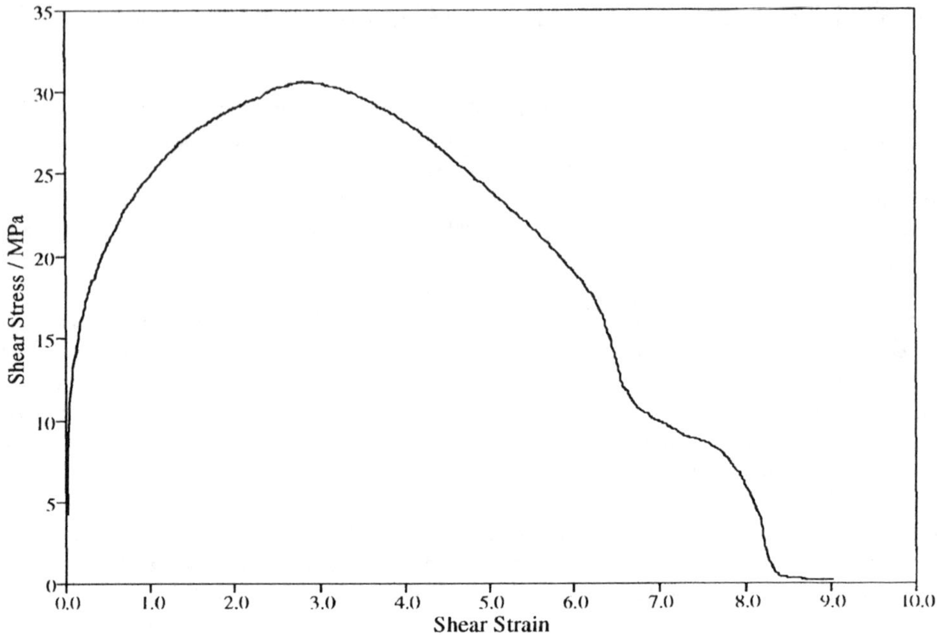

Figure 3. Stress-Strain diagram for lead in single stroke shear. Deformed at a strain rate of 0.1 s^{-1}.

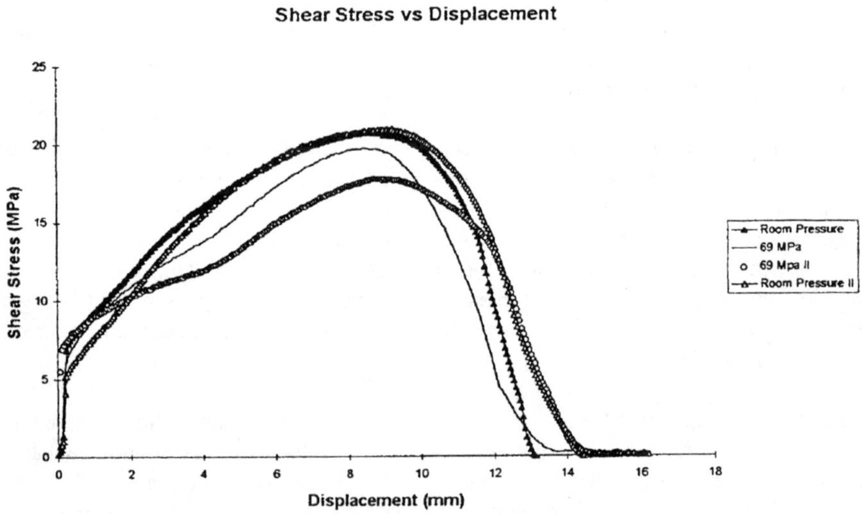

Figure 4. Shear Stress-Displacement for four lead specimens in single stroke shear. Deformed at a strain rate of 0.1 s^{-1}.

a) 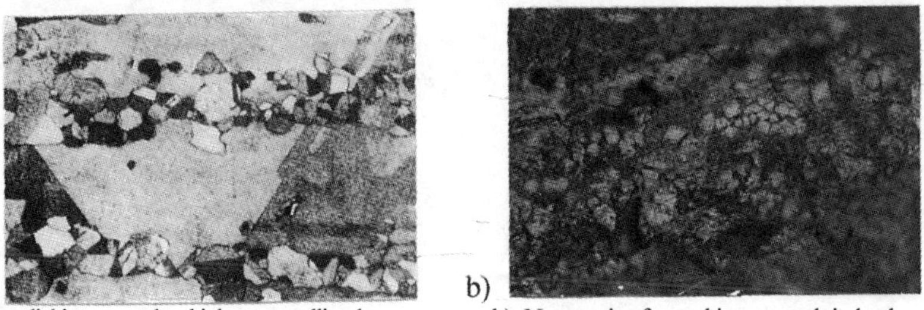 b)

Figure 5. a) A polishing scratch which recrystallized over a period of six weeks. b) New grains formed in a scratch in lead.

Conclusions

- Lead is able to withstand high strain cyclic deformation
  a) During unconstrained cyclic deformation to very high strains, lead samples provided damping consistently within 20 % of the initial cycle. Failure was by cracking induced by geometric breakdown.
  b) Constrained samples under pressure have been subjected to 144,000 cycles, producing consistent damping.
- The damping capacity of lead is 47 MJ/m^3 per cycle of strain of ± 1.
- It has been shown that high purity lead will recrystallize at room temperature, from the characteristic features observed in the stress-strain diagram and the features of the microstructure of lead. From a comparison with results from other investigations, it appears that recrystallization is occurring, statically, dynamically and meta-dynamically.
- Results indicate that more cycles are required to produce failure in a specimen strained under an imposed hydrostatic pressure. Specimens under hydrostatic pressure fail in a semi-ductile manner as opposed to the high ductility failure prevalent in lead.

References.

1. Reed-Hill R.E. and Abbaschian R., Physical Metallurgy Principles, 3rd Edition, PWS-Kent Publishing Company, Boston, 1991, pp 227-271.

2. Glover G. and Sellars C.M., Recovery and Recrystallization During High Temperature Deformation of Iron, *Metallurgical Transactions A*, 4, 1973, pg 765.

3. McQueen H.J. and Bergerson S., Dynamic Recrystallization of copper during hot torsion, *Met. Sci. Journal*, 6, 1972, pp 25 - 29.

4. Petkovic R.A., Luton M.J. and Jonas J.J., Recovery and Recrystallization of Carbon Steel Between Intervals of Hot Working, *Canadian Metallurgical Quarterly*, 14, 2, 1975, pp 137-145.

5. McQueen H.J., *Transactions of the Japanese Institute of Metals,* 9 Supplement, 1968, pp 170-183.

6. French I.E. & Weinrich P.E., The Influence of Hydrostatic Pressure on the Tensile Deformation and Fracture of Copper, Metallurgical Transactions A, Vol 6, 1975, pp 785-791

7. Jonas J.J., Sellars C.M. and Tegart W.J.McG, Strength and Structure Under Hot Working Conditions, Metallurgical Reviews, 14, 130, 1969, pp 1-24.

EFFECT OF TEXTURE ON THE PSEUDOELASTIC RESPONSE OF A SHAPE MEMORY ALLOY

M. Somerday and J.A. Wert

Department of Materials Science and Engineering
University of Virginia
Charlottesville, VA 22903

Abstract

The dependence of the pseudoelastic elongation of a shape memory alloy (SMA) on single crystal orientation with respect to the applied stress or strain state suggests that texture may have an impact on the pseudoelastic behavior of SMA polycrystals. Prior investigations have recognized this orientation effect and also the impact of grain constraint on the pseudoelastic elongation of a polycrystal. The present article systematically analyzes the effect of texture on pseudoelastic behavior of SMAs, demonstrating that the three main considerations are the specific texture component examined, the constraint experienced by sample grains, and the SMA. Model results show that a <110> fiber texture component has little effect on the pseudoelastic behavior of unconstrained and constrained NiTi polycrystals and unconstrained Cu-Zn-Al polycrystals, but has a modest impact on constrained Cu-Zn-Al polycrystals. A <111> fiber texture affects the pseudoelastic behavior of unconstrained NiTi polycrystals slightly but affects unconstrained NiTi and constrained Cu-Zn-Al polycrystals moderately, and unconstrained Cu-Zn-Al polycrystals greatly. The role of texture in determining the pseudoelastic behavior of SMA can be understood by considering the variation of inverse Schmid or Taylor factor with orientation of the uniaxial tensile stress or extension axis.

Modeling the Mechanical Response of
Structural Materials
Edited by E.M. Taleff and R.K. Mahidhara
The Minerals, Metals & Materials Society, 1997

Introduction

Imparting a texture to a polycrystalline material that exhibits anisotropic mechanical behavior in single-crystal form can strongly impact the mechanical properties of the polycrystal. It has been shown [1] that a shape memory alloy (SMA) can demonstrate single-crystal anisotropy with respect to pseudoelastic behavior. Texture can therefore be expected to affect the pseudoelastic strain vs. tensile stress relationship of polycrystalline SMAs. Previous investigators [2,3] have recognized the link between the texture of a SMA and the pseudoelastic elongation. For example, Wang, et al. [2] noted that in Cu-Zn-Al sheet, the pseudoelastic strain increased with increased intensity of the {001}<110> texture component when tensile tested at 45° to the rolling direction. Eucken and Hirsch [3] produced samples of Cu-Sn, Cu-Al-Ni, and NiTi with varying textures and a columnar grain structure to determine the relationship between the crystallographic characteristics of the samples and the mechanical response. They recognized that the pseudoelastic strain depends on the crystallographic orientation of the grains in the sample and used this to correlate the measured texture and pseudoelastic elongation measurements.

These previous studies [2,3] have primarily examined the correlation between texture and pseudoelastic elongation, while acknowledging that the constraint experienced by each grain in a polycrystalline sample will also affect the anisotropy of pseudoelastic strain. Other investigators have explored the effect of grain constraint on the pseudoelastic characteristics of polycrystalline SMAs and have shown a large effect in Cu-based alloys [4-6]. The purpose of the present article is to expand the analysis of SMA polycrystal behavior by combining the effects of texture and constraint into a systematic investigation and discussing these results with respect to previous research. This analysis will evaluate the effects of texture on Cu-Zn-Al and NiTi for both unconstrained and fully constrained grains.

Model

The pseudoelastic strain which characterizes SMAs results from a martensitic transformation. If a SMA is in the austenitic state, a stress may be applied which is sufficient to induce transformation to martensite, a process which produces a shape strain. When the stress is removed, the shape strain is recovered. The overall shape strain can be compactly represented as two strain components—one parallel to and one perpendicular to the habit plane which separates the austenite and martensite phases during transformation. In all technologically significant SMA, the dilatational component of strain is small compared to the shear strain [7]. For this reason, the dilatational component of strain is not accounted for in the present calculations.

Most SMA have a cubic austenite phase. Because of the symmetry associated with a cubic crystal structure, 24 possible habit plane/shear direction combinations, or *transformation systems*, exist. The transformation systems for any SMA can be determined using the phenomenological theory of martensite crystallography (PTMC) [8,9]. Using the transformation system concept permits analogy with plasticity theory. Slip occurs when a critical resolved shear stress (τ_{CRSS}) is reached on a slip plane. Similarly, the transformation of a SMA single crystal from austenite to martensite occurs when a critical resolved transformation stress (τ_t) is attained on a habit plane. The difference between the two processes is that the amplitude of shear strain associated with slip is arbitrary, while the magnitude of shear strain produced on the habit plane during SMA transformation, γ_t, is a specified value and is fixed for a given SMA.

Unconstrained Single Crystal

If an arbitrary stress state is applied to a SMA single crystal, usually only one transformation system must be activated for the crystal to transform to martensite. The favored transformation system can be determined using the generalized Schmid's law [10]. In addition, the work required to transform a unit volume of austenite is a constant. This information can be used to relate τ_t, γ_t, and the applied stress components, σ_{ij}^A, as follows:

$$W_t = \tau_t\gamma_t = \sigma_{ij}^A \varepsilon_{ij}^T \tag{1}$$

where W_t is the work of transformation per unit volume of austenite, and ε_{ij}^T represents the transformation strain components. If τ_t is determined (by experiment) and the transformation system orientations are known, the pseudoelastic strain produced by a given applied stress state and crystal orientation may be calculated.

Constrained Single Crystal

A constrained single crystal is subjected to an applied strain state. For the austenite crystal to transform to martensite, at least five independent transformation systems must be activated simultaneously to satisfy the independent strain components. The transformation systems which are favored are determined using the Bishop-Hill principle of maximum work [11]. The stress states which activate the required number of transformation systems are called *stress corners*, referring to intersections of five or more faces on the six-dimensional transformation surface. As before, because the work to transform a unit volume of austenite is a constant, knowledge of the corner stress state can be used to calculate the tensile stress components necessary to induce transformation:

$$W_t = \sigma_{ij}^{*} \varepsilon_{ij}^A = \tau_t\gamma_t = \sigma_{ij}^T \varepsilon_{ij}^A \tag{2}$$

where ε_{ij}^A are the imposed strain increment components, σ_{ij}^{*} represent the corner stress state components, and σ_{ij}^T represent the tensile stress components.

Unconstrained or Constrained Polycrystal

Sachs [12] and Taylor [13] first modeled plastic deformation of unconstrained and constrained polycrystals, respectively. These models treat a polycrystal as an aggregate of grains, each of which is treated as a single crystal. If the polycrystal is assumed to be unconstrained, the stress state applied to the sample is the stress state experienced by each grain. The stress is considered to be in equilibrium at the grain boundaries, but the strain state is not. For a constrained polycrystal, the strain state applied to the sample is the strain state also applied to each grain. Compatibility is maintained at the grain boundaries, but stress equilibrium is not.

In either model, the sample is assumed to transform grain by grain. The grain most favorably oriented with respect to the applied stress or strain state will transform first. This continues until all grains have transformed, from most to least favorably oriented. For each grain, the inverse Schmid factor (unconstrained model) or Taylor factor (constrained model) and pseudoelastic strain values are calculated and plotted on a cumulative plot. For a uniaxial tensile test, values of inverse Schmid and Taylor factor are related to uniaxial stress by the following relationship:

$$\frac{1}{m_{11}} = \frac{\sigma_{11}^{A}}{\tau_t} = \frac{\gamma_t}{\varepsilon_{11}^{T}} \quad \text{or} \quad M_{11} = \frac{\sigma_{11}^{T}}{\tau_t} = \frac{\gamma_t}{\varepsilon_{11}^{A}} \qquad (3)$$

where $1/m_{11}$ and M_{11} represent the inverse Schmid factor and Taylor factor, respectively. These relationships allow comparison of model results with experimental testing. In the present article, the model results will be presented using inverse Schmid and Taylor factor values. The plots are then normalized stress-strain curves (tensile stress normalized by τ_t), as opposed to stress-strain curves.

<u>Texture</u>

In the present analysis, the polycrystalline model is represented by a set of 2000 grains with random orientations. To represent a specific ideal texture component, those grains out of the set of 2000 grains which have orientations near the specified ideal texture component are "weighted" such that they represent a larger volume fraction than 0.0005. These grains were weighted accordingly to simulate the following volume fractions of a given texture component: 0.5 (50% of the total volume), 0.3, 0.1, and a random texture.

The ideal textures chosen were a <110> fiber texture, a <111> fiber texture, and a {001}<110> rolling texture. These were chosen because previous investigators have reported {001}<110> textures in rolled Cu-Zn-Al sheets [2,14]. Model results were generated for both {001}<110> and <110> fiber textures to evaluate the effect of fixing the <100> direction relative to the specimen coordinate system. The model results for the two cases are very similar. Because of this, only the model results for a <110> fiber texture are included in the present article.

The <111> fiber texture was chosen because this tensile axis orientation represents an upper or lower bound for inverse Schmid factor or Taylor factor for NiTi or Cu-Zn-Al. This ideal texture component is expected to reveal the potential magnitude of texture effects for the NiTi and Cu-Zn-Al SMA.

Results

Model stress-strain curves for <110> fiber texture and <111> fiber texture are presented in Figures 1 and 2, respectively. These are presented as plots of transformation strain vs. either inverse Schmid factor (unconstrained polycrystal model) or Taylor factor (constrained polycrystal model). As mentioned above, model results are presented for 0.5, 0.3, or 0.1 volume fraction of grains of the ideal texture component, and for a random texture. From the plots, it should be evident that the effect of the particular texture component on the mechanical behavior of the SMA is a function of the constraint state and the alloy examined. A more detailed analysis will be presented in the discussion section.

<u><110> Fiber Texture</u>

Increasing the <110> fiber texture component in NiTi does not have a large impact on the mechanical behavior in either the unconstrained or constrained polycrystal case. In each model, the stress-strain curves change little in shape, and total pseudoelastic strain remains nearly constant with increased texture component. The Cu-Zn-Al unconstrained polycrystal model demonstrates the same lack of variation of properties with increased <110> fiber texture.

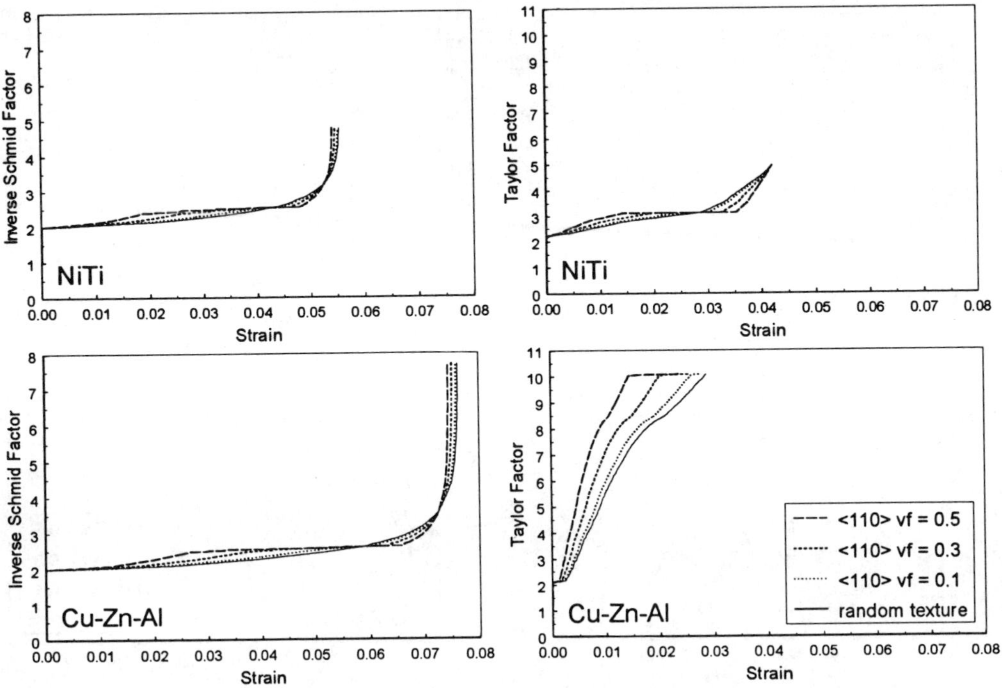

Figure 1. Model results of the effect of <110> fiber texture.

The <110> texture component does affect the mechanical response of the constrained Cu-Zn-Al polycrystal. The randomly textured polycrystal yields a total pseudoelastic strain of 0.03. As the <110> component is increased to a volume fraction of 0.5, the slope of the curve increases, and the total pseudoelastic strain decreases to 0.025. This may appear to be a small decrease in absolute magnitude of strain, but it represents a significant 20% decrease in total pseudoelastic strain.

<u><111> Fiber Texture</u>

The <111> fiber texture component has a more dramatic effect on the behavior of NiTi and Cu-Zn-Al. The unconstrained NiTi polycrystal exhibits the least variation with increased <111> fiber texture. The shape of the curve is nearly constant, and the total amount of pseudoelastic strain varies only slightly. The constrained NiTi polycrystal demonstrates a small increase in pseudoelastic strain with increased <111> fiber texture. The strain rises from 0.042 for a random polycrystal to 0.049 for a textured polycrystal that is 0.5 volume fraction <111> fiber texture. It is interesting to note that as the component of <111> fiber texture is increased, the shape of the curve tends toward that found for the unconstrained polycrystal model.

The unconstrained model of the Cu-Zn-Al polycrystal shows the most dramatic change with an increased <111> fiber texture component. The random texture yields 0.076 strain. When the volume fraction of the <111> texture component is increased to 0.5, the shape of the curve changes markedly as the strain contribution from grains with inverse Schmid factor of 6 or greater is increased. The total pseudoelastic strain drops sharply to 0.052, a decrease in pseudoelastic strain of 32%. In the constrained model of Cu-Zn-Al, the shape of the normalized stress vs. strain curve is affected by the increased <111> fiber texture, and the total pseudoelastic strain produced drops about 15%, from 0.03 to 0.025.

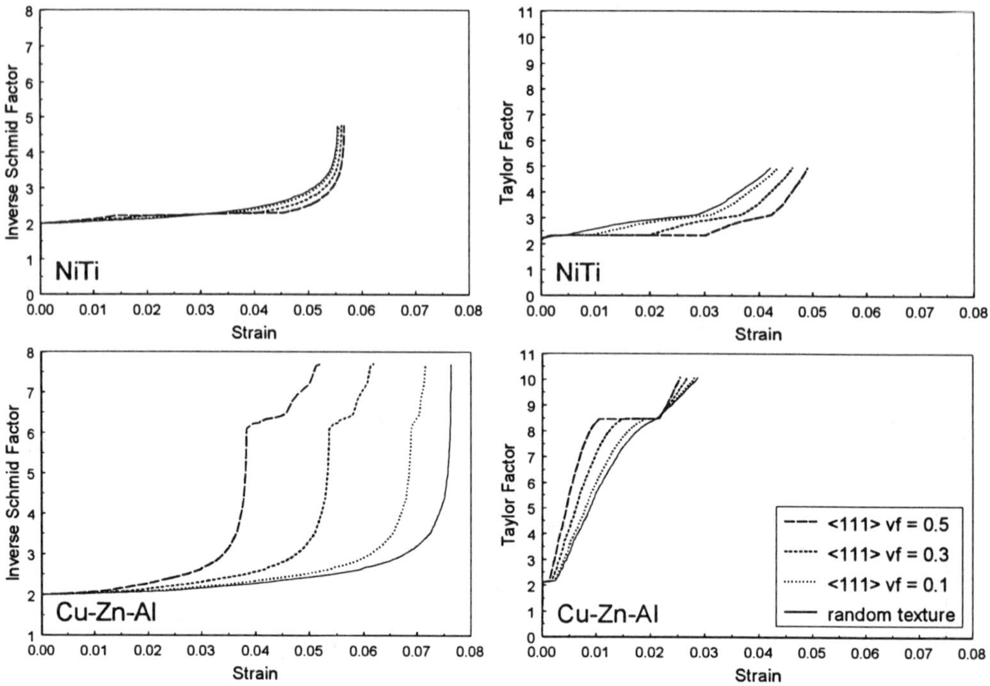

Figure 1. Model results of the effect of <111> fiber texture.

Discussion

Previous studies have shown that the pseudoelastic behavior of SMA tested in uniaxial tension is a function of both the crystallographic orientation of the transformation systems of the alloy [1] and the degree of grain constraint in the sample [6]. As mentioned in the Model section, the inverse Schmid or Taylor factor equivalent provides a relationship between the uniaxial tensile stress and τ_t. The greater the value of inverse Schmid or Taylor factor for one grain orientation, the greater the applied stress needed to transform the grain. Because of the requirement of constant work of transformation per unit volume of austenite, as the tensile stress increases, the uniaxial tensile transformation strain produced for that grain decreases. As a result, contour plots of inverse Schmid factor or Taylor factor with varying tensile axis orientation can provide insight to the expected psuedoelastic response of a SMA. These plots have been constructed and are included in Figure 3; they will be used to clarify the effect, or lack of effect, of texture on SMA transformation behavior.

Model Results

While the minimum values of inverse Schmid factor (unconstrained model) or Taylor factor (constrained model) for NiTi and Cu-Zn-Al is always near 2, Fig. 3 shows that the maximum value is substantially different for the four cases. In addition, the tensile axis orientations associated with the maxima and minima varies with alloy and constraint state. This indicates that the magnitude of a texture effect on SMA transformation behavior depends on: the alloy examined, the constraint state the grains in the sample are subjected to, and the texture. This is clarified by examining the results presented in Figs. 1 and 2, with reference to the contour plots in Fig. 3.

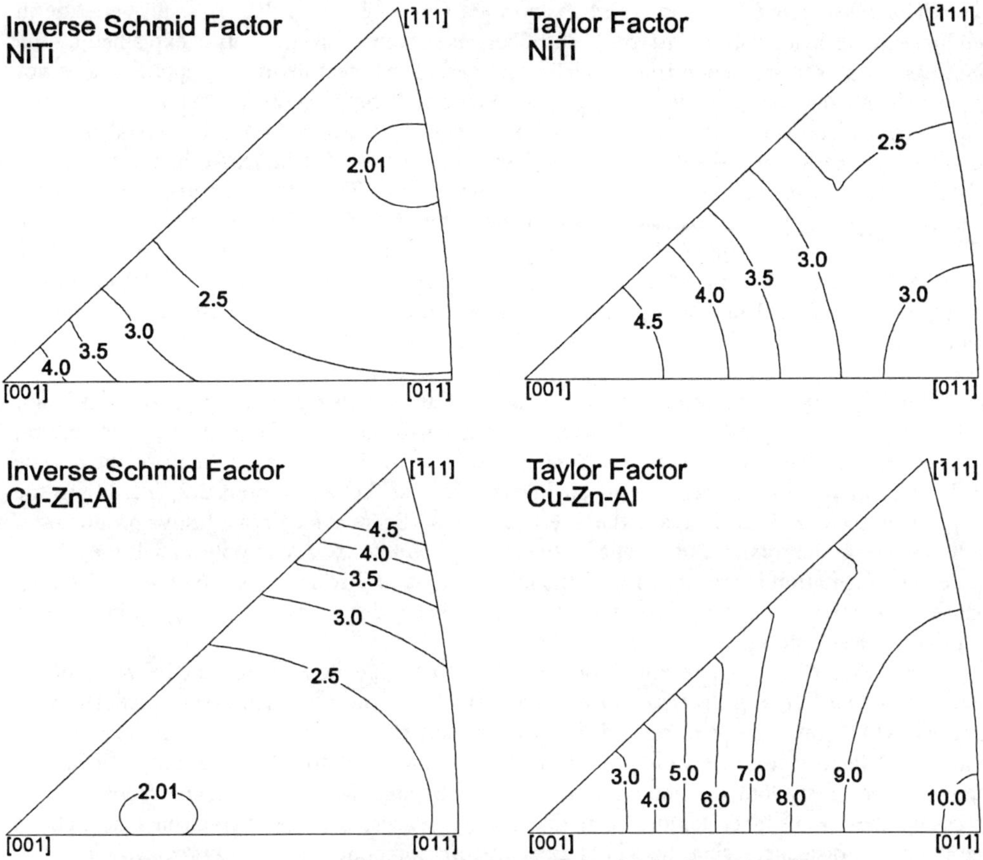

Figure 3. Contour plots of inverse Schmid and Taylor factor.

The range of Schmid factor or Taylor factor and pseudoelastic strain vary between NiTi and Cu-Zn-Al. Because this is true even for the random texture case, this is not a texture effect but instead a result of the different inherent crystallographic characteristics of the two SMA [6]. These differences include: magnitude of γ_t (approximately 0.13 for NiTi and 0.18 for Cu-Zn-Al) which determines the total pseudoelastic strain available from the transformation, and the different orientations of the habit planes and transformation shear, which affect the inverse Schmid and Taylor factor values.

<u><110> Fiber Texture</u>. Fig. 1 shows that the impact of <110> fiber texture on the pseudoelastic behavior of an *unconstrained NiTi* polycrystal is minor. This is illustrated using the contour plot of inverse Schmid factor in Fig. 3. Increasing the <110> fiber texture component reduces the volume fraction of grains with other tensile axis orientations. The value of inverse Schmid factor for a <110> orientation is slightly greater than 2.5, and the inverse Schmid factors associated with other tensile axis orientations range from greater than 4 at <100> to about 2 near <111>. The range of inverse Schmid factor is relatively narrow, and many tensile axis orientations in a random texture have inverse Schmid factor close to 2.5. As a result, as the <110> fiber texture component is increased, the average inverse Schmid factor is relatively invariant, and the normalized stress-strain curve changes only slightly.

The *constrained NiTi* and the *unconstrained Cu-Zn-Al* also exhibit a small variation in pseudoelastic behavior with increased <110> fiber texture component. This is explained by the same reasoning used for unconstrained NiTi—increasing the ideal texture component does not induce a significant change in the average value of inverse Schmid factor or Taylor factor.

The effect of a <110> fiber texture on a *constrained Cu-Zn-Al* polycrystal is more pronounced. For a constrained Cu-Zn-Al polycrystal, a <110> tensile axis orientation has the highest Taylor factor of all possible tensile axis orientations. Two other important features are the large range of Taylor factor values, and the lack of a prominent value of Taylor factor in the standard stereographic triangle. Because of these characteristics, as the <110> fiber texture is increased, the increase in the number of grains with Taylor factor greater than 10 increases, causing an upward shift in average Taylor factor and a concomitant decrease in pseudoelastic strain.

<111> Fiber Texture. The pseudoelastic behavior of an *unconstrained NiTi* polycrystal is not significantly affected by the <111> fiber texture, similar to the case of the <110> fiber texture, because the change in average inverse Schmid factor is small. However, because the <111> tensile axis orientation corresponds to an inverse Schmid factor of about 2.2, increasing the <111> fiber texture does decrease the average inverse Schmid factor modestly, producing a small increase in the total pseudoelastic strain. For the *constrained NiTi* polycrystal, the <111> tensile axis orientation is associated with the lowest value of Taylor factor. As the <111> fiber texture component is increased, the average Taylor factor decreases, producing an increase in total pseudoelastic strain.

Increased <111> fiber texture component has the opposite effect on the behavior of the unconstrained *Cu-Zn-Al polycrystal*. Here, the transformation strain decreases drastically with increased <111> texture component. It is not surprising that this orientation is associated with an inverse Schmid factor of about 7.8 (contours this high would not fit on the plot). Because a large number of random orientations have inverse Schmid factors that range from 2 to 3, increasing the <111> fiber texture component greatly decreases the average inverse Schmid factor. In contrast, increasing the <111> fiber texture component in the *constrained Cu-Zn-Al* case, associated with a mid-range Taylor factor of about 8.5, increases the average Taylor factor, but the effect is not as severe.

Prior Investigations

Wang, Park, and Bunge [2]. These investigators rolled Cu-Zn-Al and Cu-Zn-Al-Ti sheets consisting of only β phase, the ordered BCC phase that displays pseudoelastic behavior, and also processed Cu-Zn-Al and Cu-Zn-Al-Mn wires that contained $\alpha+\beta$. They measured the texture of these samples and used this to calculate pole figure intensities. Using this information and the Taylor model for constrained deformation of a polycrystal, values of pseudoelastic elongation for polycrystalline samples were calculated.

Their model predicted that rolled samples would exhibit larger pseudoelastic strain as a result of a stronger {001}<110> texture component when tested at 45° to the rolling direction. They attributed this trend to the fact that in this orientation, the tensile axis is a <100> type direction, which yields the largest transformation strain for constrained Cu-Zn-Al polycrystals. Other tensile axis orientations did not show definite trends in a sheet with a pronounced {001}<110> texture component. This contrasts with results from the present study which suggest that an enhanced {001}<110> texture component would decrease the total pseudoelastic elongation when the angle of the tensile axis with the rolling direction is 0°. We have no explanation for this apparent discrepancy. However, Wang et al. do not provide information on the effect of other texture components nor an explanation of the specific methods used to calculate the pseudoelastic elongation, so it is difficult to determine why this correlation is not seen in their model results. Furthermore, while these authors acknowledge that the choice of either the Sachs or Taylor model for calculations of SMA pseudoelastic elongation is based on the constraint state of the sample, they apparently did not recognize the effect of grain size on sample constraint. Since they did not report the sample grain size for either the sheets or wires, it is impossible to assess the state of grain constraint and, consequently, to understand whether their choice of the Taylor model for their calculations is appropriate for their samples.

For the wires investigated by Wang et al. [2], the pole density of <100>, <110>, and <111> orientations is measured from samples sectioned normal to the wire axis at each stage of drawing. The calculated pseudoelastic elongation decreases with increasing reduction for both alloys. Correspondingly, for the Cu-Zn-Al wire, the <110> and <111> pole figure densities tend to increase with increasing reduction, while the <100> pole figure density is nearly constant. The increased pole densities of <110> and <111> tensile axis orientations, both unfavorable in constrained Cu-Zn-Al, may cause a decrease in the total pseudoelastic strain. The calculated values of pseudoelastic elongation at each stage of deformation are less for Cu-Zn-Al-Mn wires compared to the Cu-Zn-Al wires. However, no definite trend is seen in any of the three reported pole figure densities with increasing reduction for the Cu-Zn-Al-Mn wires. Further details about the sample texture and the method used to calculate the pseudoelastic elongation are not provided, so it is difficult to discuss these results in more detail relative to the current analysis of texture and grain size effects in SMAs.

Eucken and Hirsch [3]. Melt-spun samples of NiTi and Cu-Al-Ni were produced, each with the same texture and a columnar grain structure, to analyze the connection between texture and mechanical behavior. These investigators recognized two salient points: (1) pseudoelastic strain is a function of grain orientation and (2) sample configuration and grain structure determine the applicability of either the unconstrained or constrained polycrystal model.

Eucken and Hirsch noted that in Cu-Al-Ni samples with a "very strong" texture of <110> and <100> orientations aligned parallel to the tensile axis, the sample yielded about 6.5% tensile strain, almost all of which was fully recovered. They attributed this increase in recoverable strain to the texture induced in the sample, since <110> and <100> directions are favorable tensile axes for transformation in most Cu-based alloys. However, a portion of this recoverable strain is probably attributable to a detwinning phenomenon. Previous PTMC-based models of unconstrained SMA single crystals [15] have shown that the maximum uniaxial tensile strain that Cu-Al-Ni samples can produce is about 4.6%. Two models of unconstrained single crystals which account for detwinning—the PTMC + detwinning model, and the lattice deformation model [15]—predict strains as high as 5.9% and greater than 6%, respectively.

The high level of pseudoelastic strain observed by Eucken and Hirsch indicates that detwinning almost certainly occurred. An important point is that both detwinning models predict that the most favorable tensile axis orientation for transformation is <110>, instead of an orientation closer to <100> for the PTMC model without detwinning. As a result, we deduce that detwinning may have an important role in establishing the effect of an ideal texture component on pseudoelastic strain.

The value of uniaxial tensile pseudoelastic strain reported by Eucken and Hirsch for NiTi samples is about 4%. This number falls within the range of expected pseudoelastic strains for <110> and <100> tensile axis orientations for both PTMC calculations and calculations including detwinning. The tensile axes associated with maximum and minimum inverse Schmid factor are approximately the same whether or not detwinning occurs. Therefore, in NiTi, the presence of detwinning can be expected to increase slightly the total pseudoelastic strain, but the effect of texture on pseudoelastic behavior should be similar regardless of whether detwinning occurs.

Conclusions

1. Previous investigators have recognized the orientation and constraint dependence of SMA pseudoelastic elongation. The present work extends the prior analyses by systematically demonstrating that SMA pseudoelastic behavior depends upon the strengthened texture components, the state of grain constraint in the sample, and the alloy examined.

2. Model results show that the <110> fiber texture has a moderate impact on the pseudoelastic behavior of constrained Cu-Zn-Al polycrystals. The unconstrained Cu-Zn-Al, unconstrained NiTi, and constrained NiTi polycrystals demonstrate little change in pseudoelastic behavior by strengthening the <110> fiber texture component.

3. The model demonstrates that a <111> fiber texture has a pronounced effect on the pseudoelastic behavior of unconstrained Cu-Zn-Al. The pseudoelastic behavior of unconstrained Cu-Zn-Al and constrained NiTi polycrystals show a minor dependence on <111> fiber texture, but the unconstrained NiTi polycrystal is virtually unaffected by strengthening the <111> fiber texture component.

4. The impact of specific texture components on the pseudoelastic behavior of SMA in uniaxial tension can be rationalized from knowledge of the magnitude and distribution of inverse Schmid (unconstrained model) or Taylor factor (constrained model) as functions of stress or extension axis orientation.

Acknowledgments

This work is sponsored by the National Science Foundation under grant no. DMR-9924487.

REFERENCES

1. M. Somerday, R. J. Comstock, Jr., and J. A. Wert, "A Systematic Analysis of Transformation Stress Anisotropy in Shape Memory Alloys," Phil. Mag. A, 75 (5) (1997), 1193-1207.

2. C. Wang, N. J. Park, and H. J. Bunge, "Prediction of Pseudoelastic Elongation in Cu-Zn-Al Shape Memory Alloy," Rare Metals, 14 (2) (1995), 131-136.

3. S. Eucken and J. Hirsch, "The Effect of Textures on Shape Memory Behavior," Mat. Sci. Forum, 56-58 (1990), 487-492.

4. I. Dvorak and E. B. Hawbolt, "Transformational Elasticity in a Polycrystalline Cu-Zn-Sn Alloy," Metall. Trans., 6A (1975), 95-99.

5. G. N. Sure and L. C. Brown, "The Mechanical Properties of Grain Refined β-CuAlNi Strain-Memory Alloys," Metall. Trans., 15A (1984), 1613-1621.

6. M. Somerday, R. J. Comstock, Jr., and J. A. Wert, "Effect of Grain Size on the Observed Pseudoelastic Behavior of a Cu-Zn-Al Shape Memory Alloy," Metall. Mater. Trans., in press.

7. Hiroyasu Funakubo, ed., Shape Memory Alloys (New York, NY: Gordon and Breach Science Publishers, 1987), 23.

8. J. S. Bowles and J. K. Mackenzie, "The Crystallography of Martensite Transformations I," Acta Metall., 2 (1954), 129-137.

9. D. S. Lieberman, M. S. Wechsler, and T. A. Read, "Cubic to Orthorhombic Diffusionless Phase Change—Experimental and Theoretical Studies of AuCd," J. Applied Phys., 26 (1953), 473-484.

10. U. F. Kocks, "The Relation Between Polycrystal Deformation and Single-Crystal Deformation," Metall. Trans., 1 (1970), 1121-1143.

11. J. F. W. Bishop and R. Hill, " A Theory of the Plastic Distortion o f a Polycrystalline Aggregate Under Combined Stresses," Phil. Mag., 42 (1951), 414-427.

12. G. Sachs, Z. Verein Deut. Ing., 72 (1928), 734-736.

13. G. I. Taylor, "Plastic Strain in Metals," J. Inst. Metals, 62 (1938), 307-324.

14. N. J. Park and H. J. Bunge, "Determination of Texture in CuZnAl Shape Memory Alloys in the High Temperature Austenitic Phase," Mat. Sci. Forum, 157-162 (1994), 167-174.

15. T. E. Buchheit and J. A. Wert, "Predicting the Orientation-Dependent Stress-Induced Transformation and Detwinning Response of Shape Memory Alloy Single Crystals," Metall. Mater. Trans., 27A (1996), 269-279.

The Electroplastic Effect in Cermets

R. J. Monfredo Gee, C. Persad, V. K. Sarin*

Institute for Advanced Technology, University of Texas at Austin, Austin, Texas
*Department of Manufacturing Engineering, Boston University, Boston, Massachusetts

Abstract

The crack propagation pattern and fracture toughness of carbide-based cermets subjected to high density current pulses were examined. The effect of high current density pulses, known as the electroplastic effect, on the plastic deformation behavior of metals and alloys is well-known. To study this effect in cermets, fracture toughness values were calculated using a modified Evans-Charles formula, data from microhardness tests and measurements of crack lengths resulting from Vickers indentation. Prior to experimentation, measurements were HK=1636 ± 24 kg/mm^2 and K_{IC} = 21.25 ± 0.67 kg/mm$^{3/2}$ for average hardness and average fracture toughness, respectively. The influence of high density current pulses (800-1600 amps/mm^2 for 50-150 μsec duration) on the extent of crack propagation was examined. Samples that were pulsed during Vickers indentation were observed to produce longer crack lengths than unpulsed samples. Indented samples that were pulsed prior to cyclic compression loading produced different crack extension patterns than unpulsed samples. Mechanisms influencing crack propagation behavior in these cermets during and after pulsing are discussed.

Modeling the Mechanical Response of
Structural Materials
Edited by E.M. Taleff and R.K. Mahidhara
The Minerals, Metals & Materials Society, 1997

Introduction

The low fracture toughness of cermets has limited their application in many industries where the high hardness, high corrosion resistance, and high temperature performance of cermets is needed, but where brittle failure could be catastrophic [1, 2]. Improvements in cermet fracture toughness have commonly reduced their attractive hardness and wear resistance characteristics [3-5].

The objective of this research was to investigate the effect of high density current pulses on crack propagation in cermets. The purpose was to determine whether improvements in fracture toughness could be achieved without compromising hardness values. High density current pulses (approximately 1000 amps/mm^2 for 50-150 microseconds) have been shown to enhance plasticity of pure metals and alloys through reduction in crack initiation and crack propagation [6]. This enhancement is known as the electroplastic effect [7]. An electroplastic effect in cermets has not been examined. This research sought to investigate the magnitude of an electroplastic effect on cermet fracture toughness values through evaluation of crack propagation behavior.

Experimental Approach

Overview

The results of two different experimental trials are reported. In the first set of electroplasticity experiments, current pulses were applied to cermet samples while they were loaded by a pyramidal Vickers indenter. In the second set of electroplastic experiments, cyclic compression loading was performed on previously indented samples. These samples were in two conditions: one set was pulsed prior to loading, and the other was not.

Fracture toughness for all samples was calculated from sample microhardness and measurements of crack lengths and diagonals from Vickers indentation using a modified Evans-Charles formula [8, 9]. Fracture toughness and crack propagation behavior was also determined in the absence of current pulses.

The electroplastic response of cermets was evaluated by measurements of propagated crack lengths and sample microhardness before and after pulsing, and by examination of crack paths by scanning electron microscopy.

Material Characterization

As received cermet samples were supplied by a commercial producer. The cermets were prepared by liquid-phase sintering at 1590°C for one hour, followed by hot isostatic pressing at 1400°C for forty-five minutes. Sample composition is reported in Table I. The calculated density of the received samples was 6.95 g/cm^3.

Table I. Weight percent composition and electrical resistivities of components in examined cermet.

Cermet Component	Weight Percent	Resistivity (μΩ-cm)
Co	8.6	6.24
Ni	4.9	9.5
WC	19	22
(Ta,Nb)C	10.3	35-50
TiN	15.1	50.0
TiC	35	51-110
Mo_2C	7.1	96

Rectangular, prismatic samples with dimensions of 2.5 x 5.0 x 15 mm were received. Both 5.0 x 15.0 mm faces on each sample were ground and polished using diminishing abrasive sizes to 1 μm diamond paste following standard metallographic procedures(see Figure 1). Final sample thickness of 2.0 mm was found to be sufficient to minimize the opposing free-surface interactions during fracture toughness indentations [10]. Samples were removed from phenolic polishing mounts before experimentation .

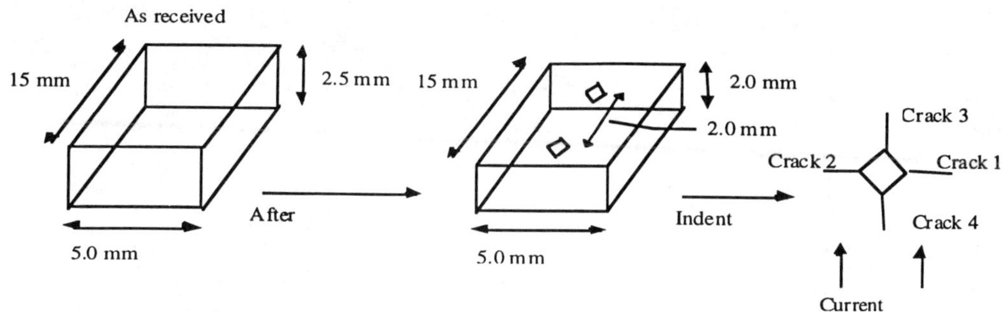

Figure 1. Schematic of Vickers indent with orientation of current pulse. Cracks 3 and 4 were parallel to the 15mm length sides of the sample, and parallel to the direction of the electric current.

<u>Experimentation</u>

The following modified Evans-Charles formula was used to calculate fracture toughness [8, 9]:

$$\text{Fracture toughness, } K_{IC} = \frac{0.113\ HK\ (D_{AV})}{(1 + C_{TOTAL}/2{*}D_{AV})^{\wedge\ 3/2}} \qquad (1)$$

where HK is the Knoop microhardness, D_{AV} is the average of the two diagonals formed from a Vickers diamond indentor, and C_{TOTAL} is the sum of the four propagated crack lengths measured from the indenter tip to the crack tip (see Figure 1).

Fracture toughness indents were generated using a pyramidal Vickers diamond indentor mounted on a Rockwell hardness tester under a 30 kg load [9]. Cracks and diagonals were measured at 500x magnification using an optical microscope. As indicated in Figure 1, indents were centered width-wise on the sample, and spaced approximately 2 mm from each other and the top and bottom edges to avoid stress interactions. Tested samples were stored in a desiccator to minimize effects of the environment on final crack lengths.

Hardness indents were made with a Knoop diamond indentor mounted on a Tukon Microhardness tester using a load of 0.8 kg. Hardness was calculated according to the formula $HK = 14.229\ P / d^2$, where **P** is the applied load (0.8 kg) and **d** the length of the long diagonal. Initial average sample hardness was calculated based upon 5-6 indents measured on one polished side of the sample. Vickers indentations were made on the reverse side, and subsequent hardness measurement indents were evenly spaced between those indents.

High density current pulses were delivered by a circuit of charged capacitors (see Figure 2). Discharge signals were collected and stored on a Tektronix digital multimeter/oscilloscope. The current pulse passing through the sample was calculated from the peak voltage of the critically damped signal, based on the formula I = V/R. A current sensing resistor, R= 9.989 E-4 ohms, was in parallel with the oscilloscope, across which the signal was read. Current density, the magnitude of the peak current passing through the cross sectional area of the sample, ranged from 800 to 1600 amps/mm^2. Pulses were 50-150 microseconds in duration. Approximately 35-50 seconds was needed to recharge the capacitor between delivery of multiple pulses. A typical pulse train is shown in Figure 3.

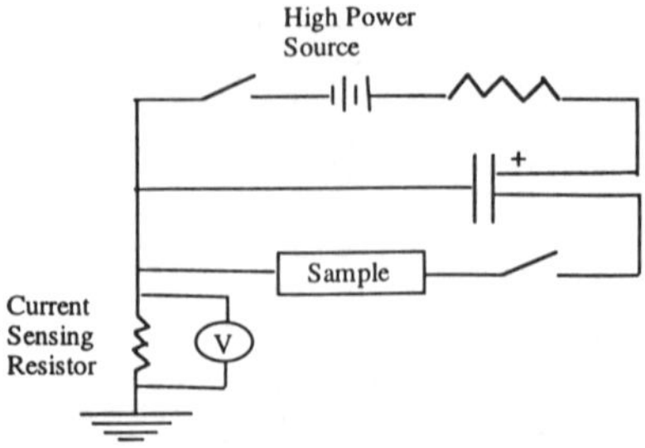

Figure 2. Schematic design of the pulsing circuit used for electroplasticity experiments.

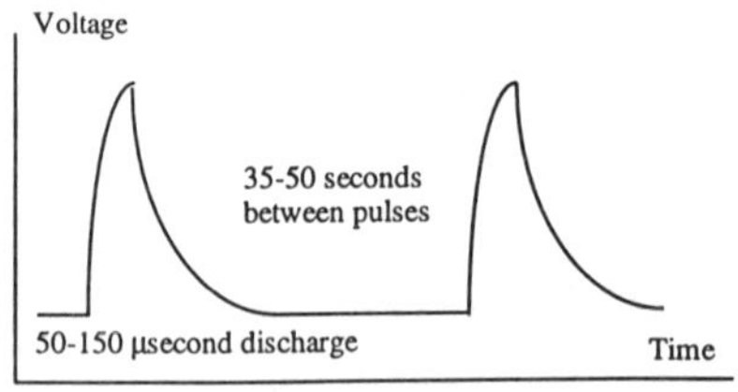

Figure 3. Schematic of pulse train.

Electroplastic Experiments

1. Pulsing during indentation

The current pulse circuit was connected to the cermet samples as shown in Figure 2. During the pulsing, the sample was mechanically loaded by the Vickers indentor under an 0.8 kg load. A train of six pulses per indent was typical. The nominal current density produced in the sample by these pulses was 600-800 amps/mm^2. A five indent sequence was used: one indent made prior to pulsing, the train of six pulses applied to indents two through four. The final (fifth) indent was created with no additional pulsing. Surface orientation of the Vickers indents on the samples was normal to the current pulse.

For comparison, cermet samples were indented in the absence of pulsing: three indents prior to any pulsing treatment, and three indents following a train of six pulses.

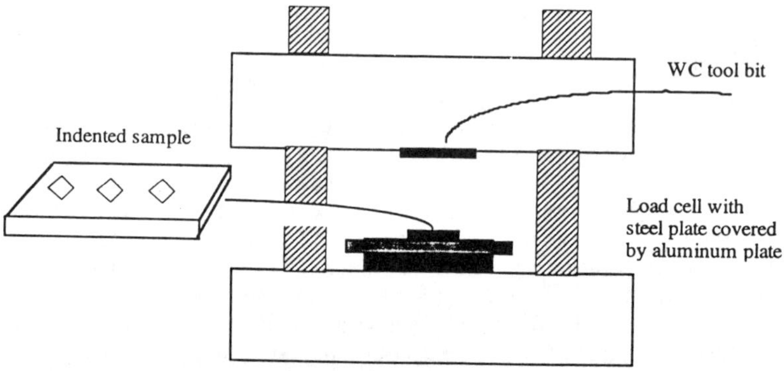

Figure 4. Schematic design of compression loading experiment, using a pneumatically loaded Wabash compression device.

2. Cyclic compression loading

Cyclic compression loading was performed on a Wabash hydraulic press (see Figure 4). Prior to loading, samples received Vickers indents under 15 kg loads. Orientation of the Vickers indents on the samples was normal to the compression load force. Samples received one of three possible treatments: one round of six pulses, two rounds of six pulses with a reverse of current polarity between rounds, or no pulses. Compressive loading cycles consisted of ten repetitious loads delivered over sixty seconds at 15-18 kpsi, and repeated once. This loading procedure was observed to promote crack extension in the samples without causing catastrophic failure.

Table II is a summary of the specimen designations and treatments to which they were subjected.

Table II. Summary of cermet sample preparation and treatment.

Electroplastic Pulsing During Indentation		Cyclic Compression Loading		
Sample 1A	**Sample 1B**	**Sample 2A**	**Sample 2B**	**Sample 2C**
Test for sample hardness	Test for sample hardness	Test for sample hardness	Test for sample hardness	Test for sample hardness
Three indents for fracture toughness calculation; measurement of cracks and diagonal lengths	One indent for fracture toughness (30 kg load)	Five indents (15kg load) for fracture toughness. Crack lengths measured.	Five indents (15kg load) for fracture toughness. Crack lengths measured.	Five indents (15kg load) for fracture toughness. Crack lengths measured.
Six pulses (average current density 1300 amps/mm2)	Indent generated during train of six pulses (average current density, 600-800 amps/mm2)	(No current pulses)	Train of six pulses, current density 1300-1600 amps/mm^2	Train of six pulses, current density 1300-1600 amps/mm^2
Three indents for fracture toughness calculation; measurement of cracks and diagonal lengths	Repeat procedure twice (for total of three indents)	Cyclic compression loading	Cyclic compression loading	Polarity of pulses reversed.
Test for sample hardness	Final indent for fracture toughness (in absence of pulses)	Crack lengths re-measured	Crack lengths re-measured	Train of six pulses, current density 1300-1600 amps/mm^2.
	Measurement of cracks and diagonal lengths for all five indents			Cyclic compression loading
	Test for sample hardness			Crack lengths re-measured

Results

Material Characteristics

The cermet samples were found to have an average measured hardness of HK_{800} = 1636 ± 24 kg/mm^2 and an average fracture toughness, K_{IC} = 21.25 ± 0.67 kg/mm$^{3/2}$ prior to experimentation. The experimentation procedures on the five samples are summarized in Table II.

The crack path through an unpulsed, indented cermet is shown in the SEM micrograph in Figure 5. As can be observed from the phase contrast in Figure 5, the material consists of contiguous carbide and nitride grains ranging in size from 0.5 to 2.5 µm in uniform distribution. The dark grains are TiC cores surrounded by lighter gray solid solutions of (W, Ti)C, while the light gray core is TiN surrounded by (W, Ti)N solid solution. [11] Varying compositions of solid solutions exist resulting from the fabrication process. The crack path is seen to propagate around the smaller grains, and through the larger, more brittle grains.

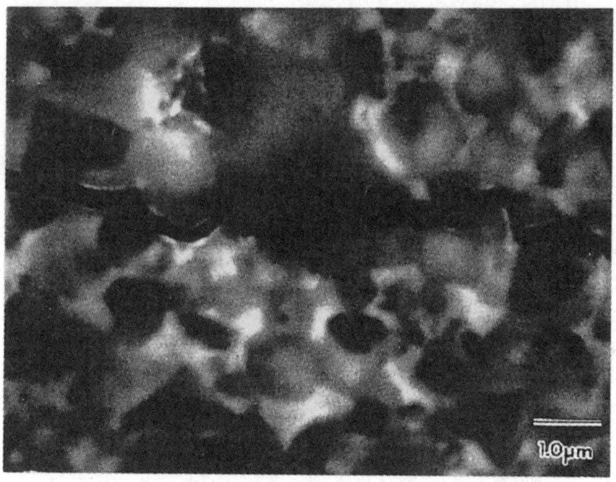

Figure 5. SEM micrograph of polished surface of unpulsed cermet sample showing crack running across sample.

Pulsing during indentation

As can be seen in Figure 6a and 6b, crack lengths increased when generated during current pulses for cermet 1B, compared to cermet 1A. The data indicates an increase in total crack length in sample 1B to a value of 680 µm compared to the maximum observed value in sample 1A of 655 μm. A cumulative effect of pulsing on crack length is also apparent in cermet 1B, observed as a monotonic increase in the total propagated crack lengths from an initial 628 μm to the final length of 680 μm. Calculated fracture toughness values for these samples are indicated in Figure 7a and 7b. There is a concurrent decrease in the fracture toughness of cermet 1B with the lengthening of propagated cracks.

The effect of plastic flow for varying lengths of time under the indentor was examined. It was found that increasing the duration of the indentation time from the standard five seconds to 180+ seconds enlarged the pyramidal indentation, but had no effect on the total crack length.

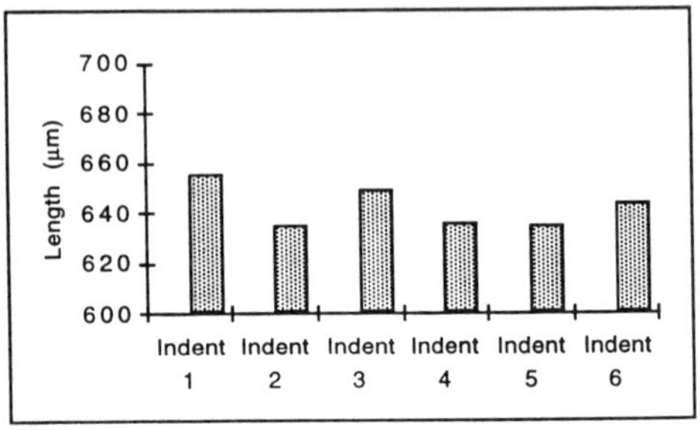

Figure 6a. Total crack length for cermet 1A, Vickers indents:
1-3 prior to pulsing, indents 4-6 after pulsing.

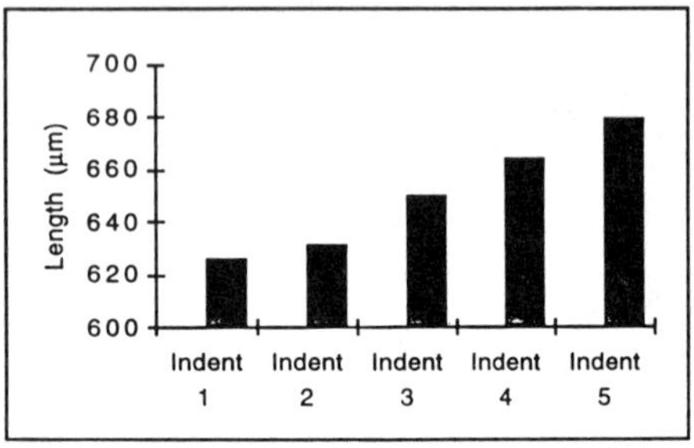

Figure 6b. Total crack length for cermet 1B, Vickers indents:
1 prior to pulsing, indents 2-4 during pulsing, indent 5 after pulsing.

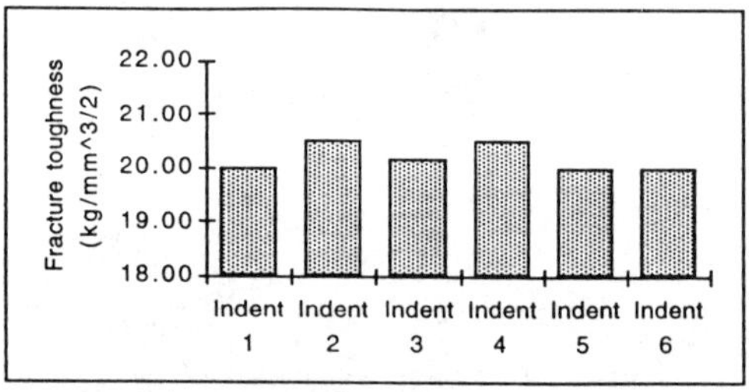

Figure 7a. Fracture toughness measurements for Vickers indents in cermet 1A:
indents 1-3 prior to pulsing, indents 4-6 after pulsing.

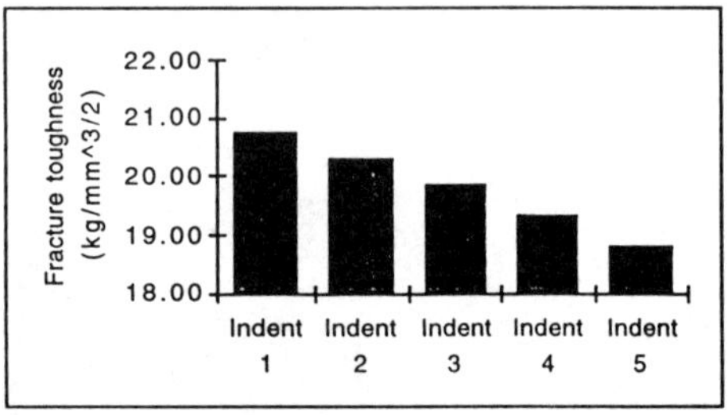

Figure 7b. Fracture toughness for cermet 1B, Vickers indents:
1 prior to pulsing, indents 2-4 during pulsing, indent 5 after pulsing.

<u>Cyclic compression loading</u>

In the second series of experiments, three samples received Vickers indentation, pulsing and compression treatments as summarized in Table II. Cyclic compression loading was performed on samples 2A, 2B, and 2C individually.

Substantial growth in total crack length was observed in the unpulsed cermet 2A from an average total crack length of 420 μm , to well over 1000μm (see Figure 8). In several instances, cracks propagating from one indent extended into crack paths of neighboring indents.

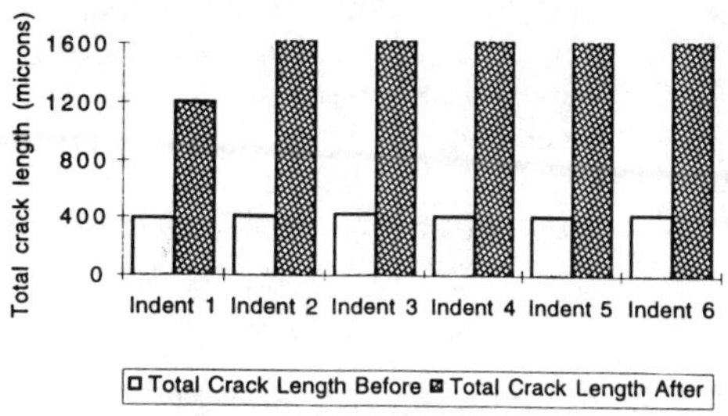

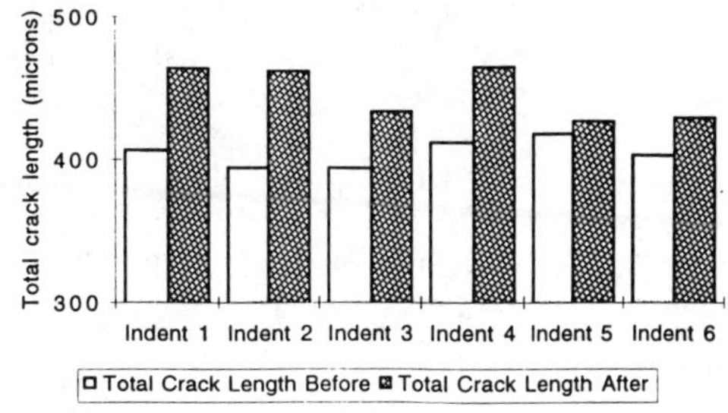

Figure 8. Crack lengths before and after cyclic compression loading in absence of current pulses on cermet 2A.

Figure 9. Crack lengths before and after cyclic compression loading on cermet 2B. Indents received one train of pulses.

More moderate growth was observed in the pulsed samples, 2B and 2C. As indicated in Figures 9 and 10, increases are found from a total crack length average of 406 μm to 448 μm in cermet 2B, and from a total length average of 399 μm to 450 μm in cermet 2C.

Physical distance of the indent from the current pulse input source is reflected by the amount of crack extension, demonstrating an impact of pulse polarity, seen in indents 5 and 6 in Figure 9 , and in the indents at both ends of the sample in Figure 10. Also of interest is the comparison of length increases in cracks parallel to the current pulse direction, compared to those normal to the pulse direction. As shown in Figure 11, in a breakdown of individual crack lengths for each indent in sample 2C, Cracks 1 and 2, perpendicular to the direction of the current pulse, are proportionally longer than Cracks 3 and 4, parallel to the pulse. Possible explanations for this behavior are offered in the discussion section.

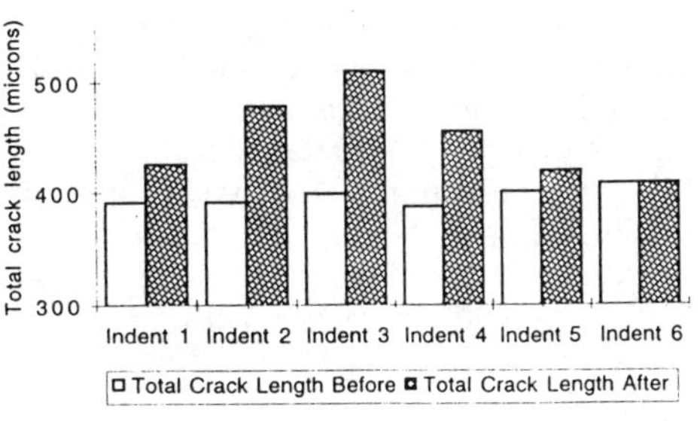

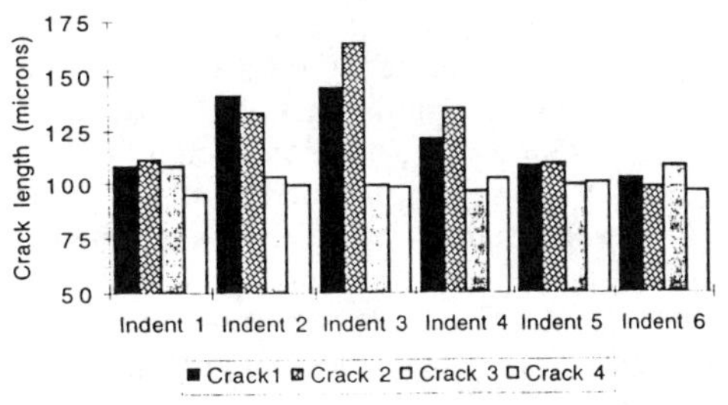

Figure 10. Crack lengths before and after compression on cermet 2C. Indents received two trains of six pulses with reverse of polarity between trains.

Figure 11. Individual crack lengths for cermet 2C, following cyclic load compression testing. Cracks 1 and 2 were perpendicular to the current pulse direction.

Discussion

The experiments sought to measure the electroplastic effect in cermets through two different performance measures: the effect of current pulsing during Vickers-induced crack propagation, and the effect of current pulsing on cracks subsequently subjected to compression loading.

It is well-known that in a resistive network current flows in the path of lowest resistance, which in cermets would be through the more conductive cobalt-nickel binder, around the carbide and nitride ceramic particles (resistivity values of the components of the cermet are listed in Table I). Consequently, current distribution through these samples was not uniform. Higher concentrations of current are predicted around the non-conductive voids or cracks, and around the less conductive ceramics. This concentration of current has been shown in other research to generate highly localized Joule heating [15, 16]. This heating relieves the mechanical stresses at the crack tip, and can increase the radius of the crack tip, reducing the Griffith crack propagation forces [17]. The net effect is that the rate of crack propagation is slowed or halted, depending upon the applied forces.

This is the effect observed in the second set of experiments where existing cracks were subjected to pulses and then to compression loading. Cracks that were pulsed appear to have been effectively blunted, and extension of existing cracks during cyclic loading experiments was less than those cracks that received no current pulses prior to loading. Further supporting this argument is the observation that this effect was even more pronounced with cracks oriented in the direction of the current pulse.

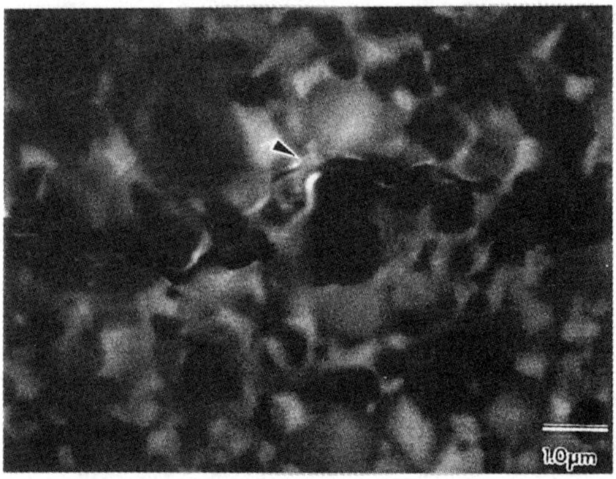

Figure 12. SEM micrograph of polished surface of cermet with crack propagated during pulsing of cermet sample running across surface.

Recently, Kotoul [18] predicted a partition betwen carbide- and binder-phase related fracture. It was shown that a multiligament zone formed by ductile binder ligaments behind the crack tip strongly affects the cleavage process in the WC matrix. This is due to a decrease in both the crack tip opening displacement, and the plastic zone size ahead of a blunting crack tip. A high current pulse acting on such a crack tip will tend to enlarge the plastic zone size due to localized Joule heating in the vicinity of the crack tip. At the same time, the ductile tearing of the binder ligaments will be accelerated by the flow of current through the ligaments, analogous to explosion of a microwire [19]. This tearing effect is predicted to be highly dependent on the current flow path as cracks running parallel to the current flow will experience a crack closure florce due to the pinch effect. This type of pinch effect has been observed and described in the early electroplasticity work of Triotskii [20].

This described phenomenological model may explain the crack behavior observed in the first set of experiments. These samples were not cyclically loaded after pulsing. In this case, localized heating is likely to have relieved some of the applied stresses at the crack tip. However, the flow of current through the sample creates a strong electromagnetic effect. Current flow along the surfaces of the cracks (the 'skin' of the inner material surface) creates a magnetic force perpendicular to the flow of the current. Depending upon the orientation of the cracks with current flow, the forces can either propel the surfaces away from each other, or force closure (the pinch effect). In theory, the stress on the crack tips in the cermets from indentation loading would have been enhanced internally by crack-opening magnetic forces generated by the flow of current, causing the correspondingly longer cracks.

Observation of the crack path of a pulsed sample by SEM introduces another aspect of crack propagation to consider with respect to the materials examined. Crack propagation in cermets is governed by the size of the ceramic particles--the larger the particle, the more brittle, the more susceptible to fracture under stress [21]. In contrast, the ductile binder slows crack propagation through greater energy consumption during elastic-plastic deformation in the zone ahead of the moving crack. Large ceramic grains in a cermet contribute to the properties of hardness, wear resistance and strength, but are more susceptible to transgranular fracture. Smaller grains are tougher, and serve to deflect a propagating crack path back into the binder [22]. As is visible in Figure 5, in the unpulsed sample, the crack path propagates through the larger ceramic grains, but is deflected by the smaller grains back to the ductile binder phase. In the pulsed sample, as shown in Figure 12, it was found that larger (i.e. greater than 1.5 μm) grains also served to deflect the crack path.

Poech et al [23] and Bhaumik [24] discuss the competing fracture-promoting mechanisms in different cermet systems. Poech looks at models designed to predict the mechanical behavior of hard metals and other particle composites. The case can be made that both the plastic deformation and fracture of the composites can be adequately described using the in situ properties: the known stress-strain behavior of bulk cobalt alloys of proper composition and grain size, provided that due regard is given to the plastic constraint exerted by the rigid carbide grains and to the consequential triaxiality of the stress in the binder phase which results from the plastic constraint of the rigid carbide grains. Bhaumik deals with the fracture characteristics of cemented carbides with modifications in both carbide and binder phases . The results show that the basic fracture processes are similar in all the compositions, namely decohesion of the particle/particle, particle/binder interfaces and cleavage of hard particles. However, the predominant local fracture mode changes from alloy to alloy due to the change in alloy chemistry. In alloy WC-TiN-Co, decohesion of particle/particle boundary followed by transgranular fracture is the predominant local fracture mode, whereas debonding of WC/binder interface and transgranular fracture of TiN are more prevalent in WC-TiN-Mo_2C-Co(Ni) cemented carbide.

Conclusion

The results suggest the complex fracture behavior of multiphase cermets is modified by an electroplastic effect. The documented standard fracture description for this class of cermets includes features associated with the ductile metal matrix; the hard ceramic particles, with the metal-ceramic interface; and with the plastic constraint of the deforming ductile phase sandwiched between the hard particles. The competing electroplastic mechanisms of crack-tip blunting, the large, local electromagnetic forces for crack closure or expansion, and the differing fracture response of the several metallic and ceramic phases combine to produce distinctive crack paths.

Crack propagation behavior in cermets was examined in samples subjected to high density electric current pulses. Crack propagation in cermets during current pulses was found to produce longer crack lengths than in the absence of pulsing. Samples with controlled crack propagation that were pulsed prior to cyclic compression loading showed less growth in crack lengths compared to samples that were not pulsed. Crack tip blunting through localized Joule heating in pulsed samples is believed responsible for crack behavior under subsequent cyclic compression loading. This effect can be overwhelmed by local electromagnetic forces generated on the crack surfaces during pulsing. Depending upon crack orientation relative to the current vector, the crack can be pinched closed, or forced open, the former being the case for cracks aligned with the current vector, and the latter for cracks normal to the current vector.

Bibliography

1. V. K. Sarin, "Cemented Carbide Cutting Tools," *Advances in Powder Technology*, ed. Gilbert Y. Chin, American Society for Metals: Metals Park, Ohio, 1981, pp. 253-288.

2. S. T. Buljan, "Cermet Cutting Tool Materials: State-of-the-Art and Development Opportunities," internal paper, GTE Laboratories Incorporated, Waltham, Massachusetts, 1991.

3. J. Gurland, "New Scientific Approaches to the Development of Tool Materials," *International Materials Review*, vol. 33, no. 3, 1988, pp. 151-166.

4. B. A. Movchan, "Dimensional-structural relationships of the strength of two-phase polycrystalline inorganic materials," *Materials Science and Engineering*, A138 (1991) pp. 109-121.

5. N. Ingelstrom and H. Nordberg, "Fracture Toughness of Cemented Tungsten Carbides," *Engineering Fracture Mechanics*, vol. 6, no. 3, 1974, pp. 597-607.

6. H. Conrad, A.F. Sprecher, W.D. Cao, X.P. Lu, "Electroplasticity--The Effect of Electricity on the Mechanical Properties of Metals," *JOM*, September 1990, pp. 23-33.

7. O. A. Troitskii and V.I. Likhtman, *Dokladi Akad. Nauk*, USSR, vol. 148, 1963, pp. 324-332.

8. A. G.. Evans, E. A. Charles, "Fracture Toughness Determinations by Indentation," *Journal of the American Ceramic Society*, vol. 59, nos. 7-8, 1976, pp. 179-182.

9. J. G.. Baldoni, "Indentation Fracture Toughness Measurement: Recommended Procedures," Internal Document TM 83-562, GTE Laboratories Incorporated, Waltham, Massachusetts, 1983.

10. G. L. Kehl, *The Principles of Metallographic Laboratory Practice*, New York: McGraw Hill, 1949.

11. S. Kang, "Some Issues in Ti(C, N)-WC-TaC Cermets," *Science of Hard Materials*, ed. V. K. Sarin, vol. 5, Elsevier Publishing Company, New York, 1995, pp. 306-312.

12. *Properties and Selection of Materials*, ASM Metals Handbook, 8th edition, vol. 1, American Society for Metals, Metals Park, Ohio, 1971.

13. L. E. Toth, *Transition Metal Carbides and Nitrides*, Academic Press, New York, 1971.

14. E. K. Storms, *The Refractory Carbides*, Academic Press, New York, 1967.

15. V. Parton and J. Tani, "On the retardation of a crack by means of passing a current pulse," *International Journal of Applied Electromagnetics in Materials*, vol. 5, 1994, p. 279-290.

16. V. M. Finkel', Y. I. Golovin, and A. A. Sletkov, "Disintegration of a crack tip with a strong electromagnetic field," *Soviet Physics Dokl.*, vol. 22, no. 11, 1977, pp. 683-685.

17. B. Lawn, *Fracture in Brittle Solids*, Cambridge University Press, New York, 1993.

18. M. Kotoul, "On the shielding effect of a multiligament zone of a crack in WC-Co," *Acta Materialia*, vol. 45, no. 8, 1997, pp. 3363-3376.

19. T. J. Tucker, R. P. Toth, "EBW1: A computer code for the prediction of the behavior of electrical circuits containing exploding wire elements," Interal Document SAND-75-0041, Sandia Laboratories, Albuqueque, New Mexico, 1975.

20. O. A. Troitskii, "Characteristics of the plastic deformation of a metal during passage of an electric current," *Problemi Prochnosti*, vol. 7, no. 7, 1975, pp. 14-19. Translated in Strength of Materials, 1976.

21. D. B. Marshall, W. L. Morris, B. N, Cox, M. S. Dadkhah, "Toughening mechanisms in cemented carbides," *Journal of the American Ceramic Society*, vol. 73, no. 10, 1990, pp. 2930-2943.

22. L. S. Sigl, H. F. Fischmeister, "On the Fracture Toughness of Cemented Carbides," *Acta Metallurgica*, vol. 36, no. 4, 1988, pp. 887-897.

23. M. H. Poech, H. Fischmeister, R. Spiegler, "Assessment of the in situ properties of the cobalt phase in WC-Co hard metals," *Journal of Hard Materials*, vol. 2, no. 3-4, 1991, pp. 197-205.

24. S. K. Bhaumik, G. S. Upadhyaya, M. L. Vaidya, "SEM study of the fracture behavior of WC-TiN-Co and WC-TiN-Mo2C-Co(Ni) cemented carbides," Praktische Metallographie, vol. 29, no. 7, 1992, pp. 366-373.

MODELING HYPERVELOCITY IMPACT CRATERING IN COPPER AND ALUMINUM

Stella A. Quinones, L. E. Murr, Alicia Ayala, and E. Ferreyra T.

Department of Metallurgical and Materials Engineering
The University of Texas at El Paso, El Paso, Texas 79968

Abstract

A 2-D computer code (Autodyn 3.0) has been utilized in modeling the residual hardness contours in crater half sections produced by impacting 3.2 mm diameter 1100 aluminum spheres into 1.3 cm thickness targets of polycyrstalline copper; over a projectile velocity range from 1 to 6 km/s. Utilizing both Lagrangian and Eulerian processors with a Johnson-Cook constitutive relationship, and appropriate failure criteria, experimental hardness and yield strength measurements have been qualitatively modeled along with residual crater geometries, rim jetting and fracture, and target spallation. The validated computer calculations in the 1 to 6 km/s range have allowed predictive calculations of cratering in the hypervelocity impact range above 8 km/s where laboratory gun experiments are untenable. Target softening and especially the development of a wide zone of dynamic recrystallization at the crater wall are observed at hypervelocities in both copper and aluminum targets using TEM, and this feature has also been modeled in computer extrapolations to impact velocities of 12 and 24 km/s.

Modeling the Mechanical Response of
Structural Materials
Edited by E.M. Taleff and R.K. Mahidhara
The Minerals, Metals & Materials Society, 1997

Introduction

There is an obvious advantage in utilizing computer simulations in examining hypervelocity impact in metal targets: laboratory guns are often expensive to utilize and can conveniently fire small projectiles (~3.18 mm diameter) up to the minimum orbital impact velocity (~8 km/s). Consequently, efforts to examine hypervelocity impacts in low-Earth orbit (LEO) or deep space (u_0> 10 km/s) must rely solely upon computer simulations. Murr and Rivas (1) have recently attempted to predict cratering features on spacecraft surfaces based upon simulated crater geometries; with little success. Harrison, et al. (2) have also simulated simple crater geometries in aluminum targets over a range of steel projectile impact velocities ranging from 2 to 8 km/s, but specific crater shapes, including rim fracture, etc. were not simulated. These are important features in the study of hypervelocity impact phenomena since Murr, et al. (3) have already demonstrated that a considerable fraction of the debris created in LEO may accrue as a consequence of ejection of both projectile and target fragments during impact. Furthermore, an important design consideration for impacts on space structures over long time periods may not only involve crater formation, proximity, and debris formation, but also surface hardening and the interaction of crater-related, residual stress or hardening contours which can contribute to long term erosion and related surface degradation profiles.

The key to effective impact simulations, and especially extrapolations to practical space regimes well beyond laboratory capabilities must involve the validation (or calibration) of hydrocodes and their particular referentials, over a range of laboratory experiments which can provide a basis for impact extrapolations into the hypervelocity regimes. This paper provides some examples of such validations for laboratory impacts of 1100 aluminum projectiles into copper targets over a range of velocities (from 1 to 6 km/s). This paper also begins to examine some low-velocity (0.5 to 1.5 km/s) anomalies recently observed by Bernhard and Hörz (4) for soda-lime glass projectiles impacting 1100 aluminum targets by investigating the impact of ferritic stainless steel projectiles into both copper and 1100 aluminum targets in the laboratory, and comparing and validating computed simulations for these observations.

Experimental and Analytical Considerations

The range of experimental impact systems to be considered in this presentation is illustrated in the impact data reproduced in Fig. 1(a) for 1100 aluminum projectiles impacting copper targets (5), soda-lime glass impacting copper targets (6), soda-lime glass projectiles impacting 1100 aluminum targets (4), and ferritic stainless steel projectiles impacting copper targets (7). Figure 1(b) illustrates the crater geometry and projectile impact (with velocity u_0) implicit in Fig. 1(a), and provides a reference schematic for the measurements to be reported in this study. All of the projectiles represented in Fig. 1(a) had a diameter (dp) of 3.18 mm. These spherical projectiles were launched from a powder gun with a 7 mm diameter bore barrel for velocities below 3 km/s, and a light gas gun was employed for impact velocities above 3 km/s. Data points indicated by open diamonds for crater cross-section analysis in Fig. 1(a), corresponding to soda-lime glass projectiles impacting 1100 aluminum targets, indicate craters examined by optical and transmission electron microscopy along with cross-sections from the data points shown for other impact systems. The electron microscopy procedures, etc. have been described in detail elsewhere (3, 5-8). In addition, microhardness measurements were made along the impact axis beyond the crater wall for all samples, and complete, residual microhardness contour maps were constructed for the 1100 aluminum projectiles impacting copper targets using a Vickers indenter in a Shimadzu HMV-2000 hardness tester.

AUTODYN 2-D (version 3.0) hydrocode software was employed in calculations and simulations to be reported herein. This is an interactive, integrated hydrocode available on a PC-compatible disk (9, 10), and based on early code work of Wilkins (11) among others. A range of processors, including Lagrangian and Eulerian processors, is available and can be fully coupled. Use of the Lagrangian processor without any fracture criterion requires a selection of an erosion algorithm which is simply a numerical technique for avoiding grid tangling (10). A variety of strength and fracture models are available along with matrices for model parameters for many materials based on Hugoniot and other data. Successful modeling in this study for the 1100 aluminum projectile impacting copper targets utilized a Johnson-Cook constitutive

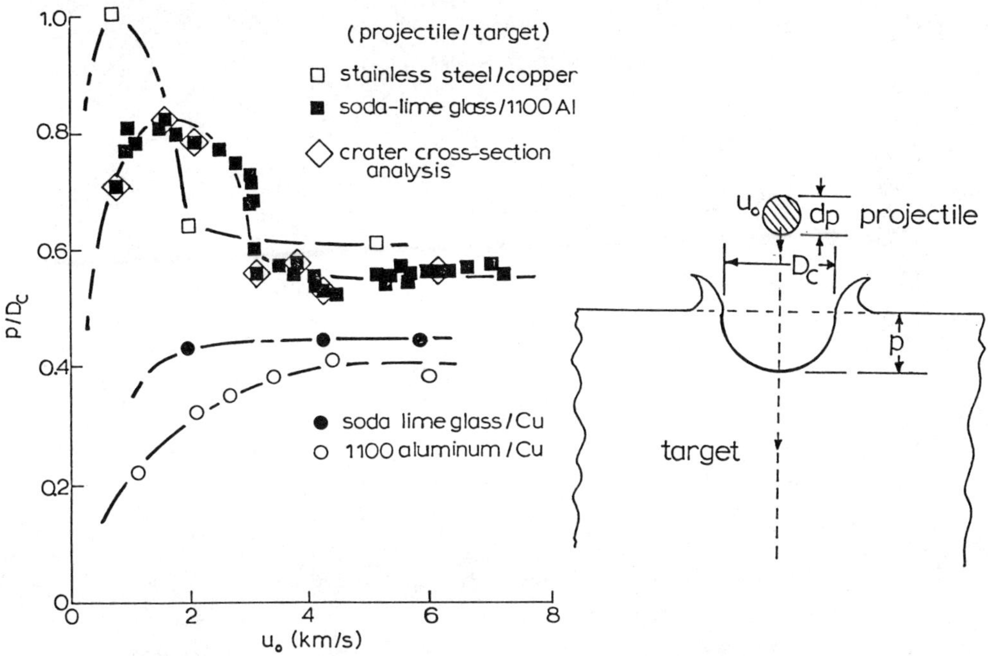

Figure 1 - (a) Crater depth/crater diameter (p/D_c) ratio versus projectile impact velocity (u_0) for a number of projectile/target material systems. (b) Schematic showing ideal crater geometry and associated parameters. The impact axis is shown dotted along u_0 and into the target.

relationship in characterizing the target and a von Mises equation for the projectile strength model (12). Yield strengths were based upon steady-state values obtained from hardness measurements; where $\sigma = H/3$; H is the Vickers hardness (in GPa or Mbar). Computed, residual stress contours were also compared or converted to residual hardness contour maps utilizing this simple linear relationship.

Results and Discussion

Experimental

It is probably a worthwhile approach to first develop some comparisons between the impact systems illustrated initially in Fig. 1(a). Figure 2 shows such comparisons for residual hardness measurements along the impact axes (shown dotted in Fig. 1(b)) corresponding to a range of impact velocities. While these plots require careful perusal, there are several pervasive issues. One is a very subtle softening noted for each target at higher velocities of impact. The other is the hardness decline and the nature of this decline into the target. Both of these features arise because of residual microstructures and microstructural evolution associated with the cratering process, and are illustrated only very generally in the transmission electron micrographs shown for comparison in Fig. 3. Figure 3(a) and (c) show dynamically recrystallized grain structure very close to (<0.5 mm) the crater walls for the highest impact velocities for an 1100 aluminum projectile into copper and a soda-lime glass projectile into 1100 aluminum respectively. This accounts for the softening near the crater wall, and it appears that this dynamically recrystallized zone will increase in size with increasing impact velocity. Furthermore, as shown in Fig. 3(b), there is a zone of microbands in the copper targets which also widens with impact velocity (5). Microbands do not appear in the 1100 aluminum target microstructure, and the microstructure beyond the dynamic recrystallization zone consists of

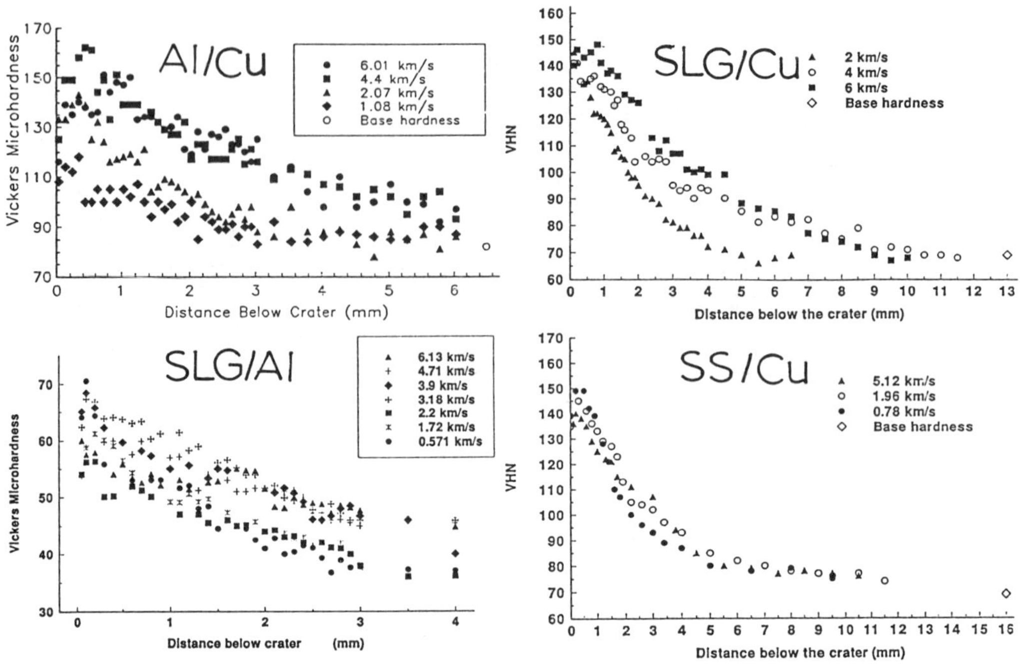

Figure 2 - Residual hardness versus distance form the crater bottom along the impact axis (Fig. 1(b)); corresponding to the four systems illustrated in Fig. 1(a). Impact velocities are noted. SLG is soda-lime glass; SS is ferritic stainless steel.

dislocation cells whose size increases with distance from the crater wall.

Simulations, Modeling, and Computer Predictions

One of the important issues in this study involved the ability to simulate the residual target hardness not only along the impact axis, but as contours of hardness or stress corresponding to zones of microstructure illustrated very generally in Fig. 3. Figure 4 illustrates this issue. Figure 4(a) and (b) show crater half sections in copper targets created by 1100 aluminum projectiles impacting at velocities of 1.08 and 6.01 km/s respectively. Notable features include the projectile component residing in the crater at 1.08 km/s (arrow) and the spall at the back of the target at 6.01 km/s. Figure 4(c) and (d) show corresponding hardness contour maps measured for the targets in Fig. 4(a) and (b) above. Figure 4(e) and (f) show corresponding Lagrangian simulations for residual stress contours while Fig. 4(g) and (h) show the corresponding Eulerian simulations (which include a fracture criterion with the Eulerian processor). While both the Lagrangian and Eulerian simulations tend to emulate general features of the experimentally determined, residual hardness contours, the numerical distinctions and the ability to accurately model these features are more quantitatively illustrated in the data - both measured and calculated - in Fig. 5. Here we see that the agreement deviates more strongly for both processors as the impact velocity increases. It even appears that the Lagrangian calculations are closer to the experimental data. But inspecting the actual crater simulations (Fig. 4(a) to (h)) it can be seen that the Eulerian calculations (with fracture) more closely model the actual crater shapes in Fig. 4(a) and (b). This was also demonstrated on comparing measured and computer calculated p/D_c values with impact velocity, u_o which tended to convey the notion that both Lagrangian and Eulerian processors provide reasonably accurate crater geometry calculations, but the actual shapes and shape features - such as rim fracture and ejection of fragments and spallation (implicit in the cross-section of Fig. 4(b)) were neither modeled nor projected by Lagrangian calculations. This feature is convincingly illustrated in Fig. 6 which also shows the increase in the width of the dynamically recrystallized region adjacent to the crater wall as the impact velocity increases. In addition, these computer projections also show that perforation will occur in the 1.3 cm thick copper target at $u_o = 24$ km/s.

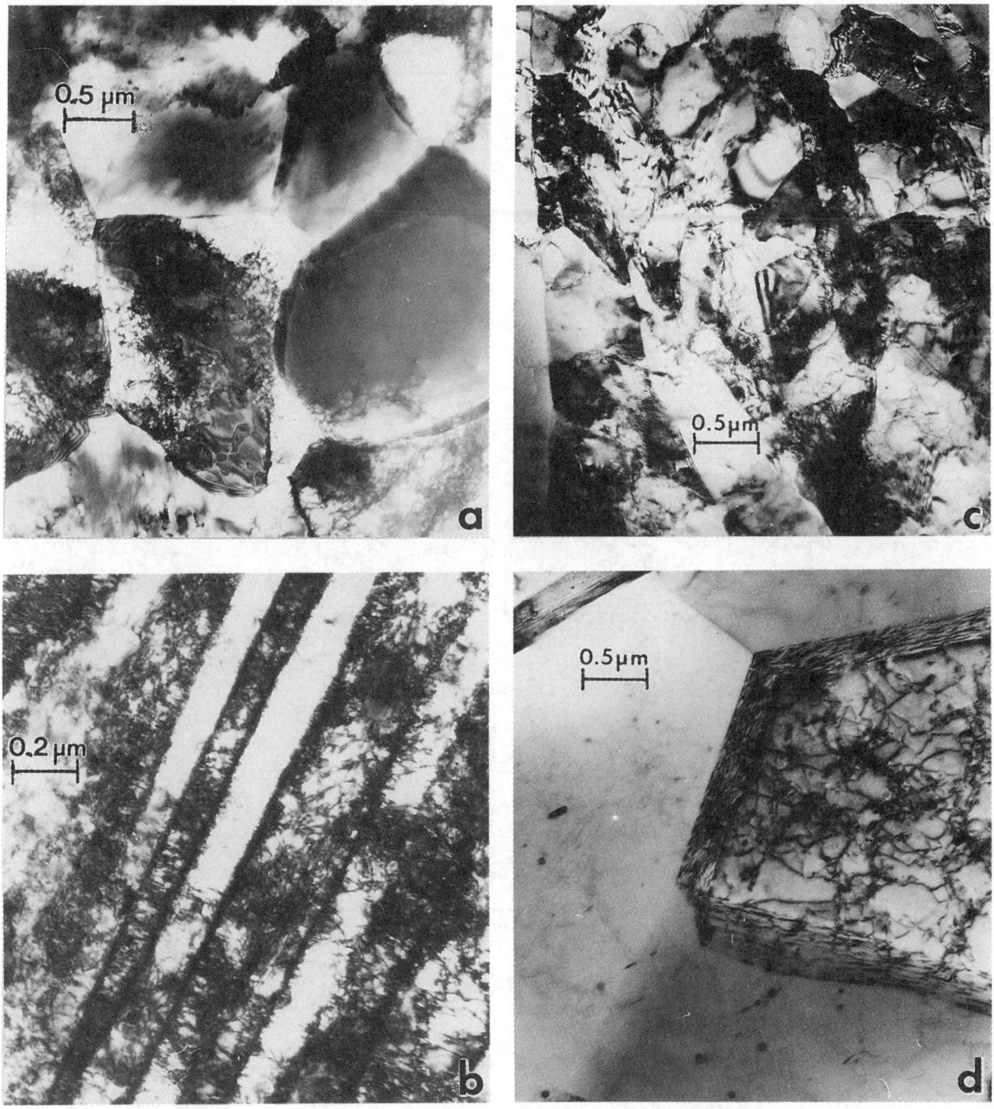

Figure 3 - TEM (bright-field) images showing prominent crater-related microstructures. (a) Dynamic recrystallization very near the crater wall in copper crater formed by 1100 aluminum projectile impacting at 6.01 km/s. (b) Microbands and very dense dislocation cells about 2 mm from the crater wall in (a). (c) Dynamic recrystallization near 1100 aluminum crater for 6.13 km/s soda-lime glass projectile. (d) 1100 aluminum target base microstructure for comparison with (c).

Because of space limitations, it is not possible to expand this work further. However, only preliminary, additional simulations have been achieved for stainless steel projectiles impacting copper. We have been unable to make any sensible simulations for the soda-lime glass/1100 aluminum system because of a lack of Hugoniot data for glass projectiles and a sensible strength model. It should be apparent of course that the extended penetration at low velocities illustrated in Fig. 1(a) for stainless steel/copper and soda-lime glass/1100 aluminum in contrast to the other systems seems to depend strongly upon the projectile strength and $(\rho_p/\rho_t)^{1/2}$, where ρ_p and ρ_t

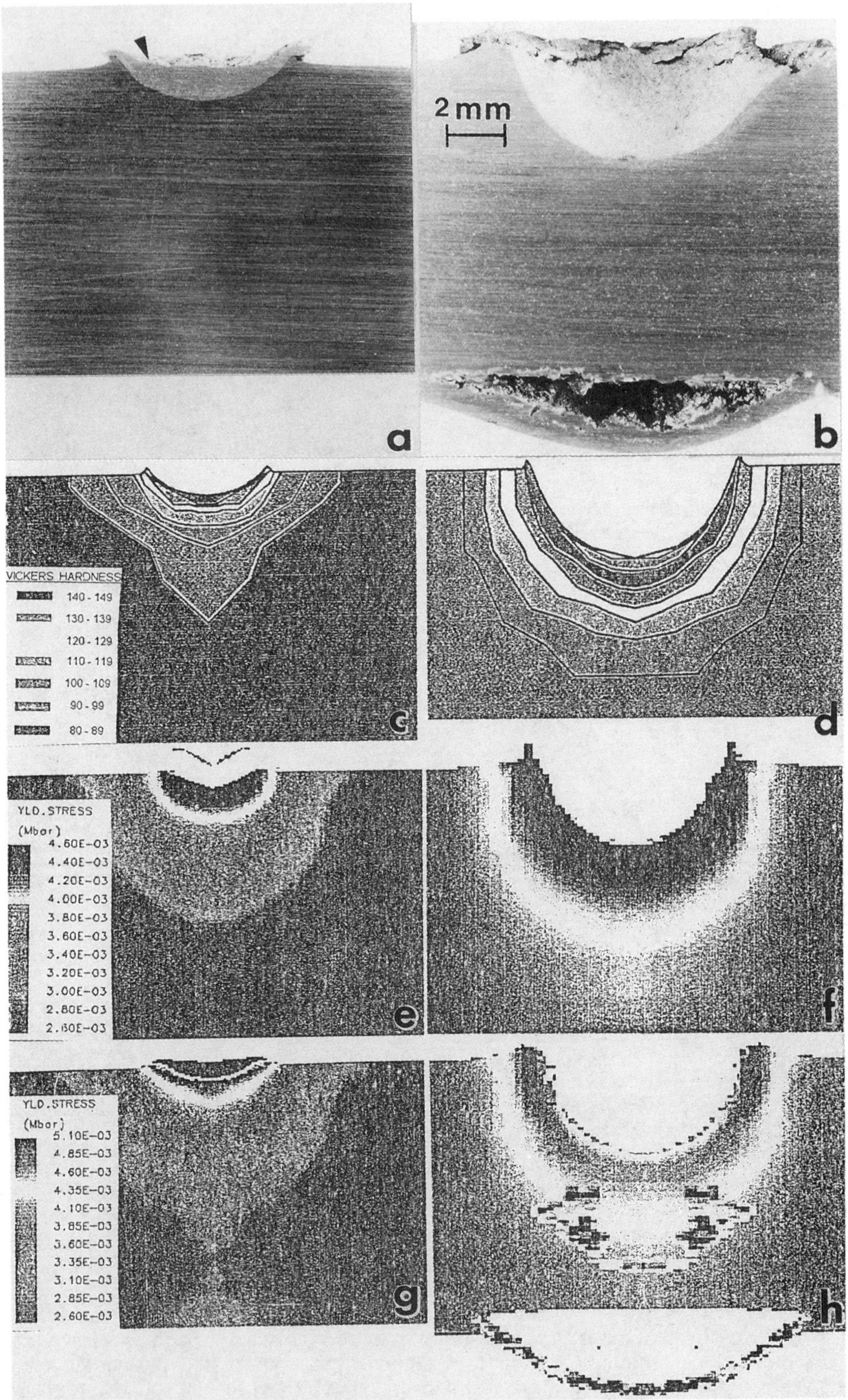

Figure 4 - 1100 Al/Cu System. (a) 1.08 km/s crater cross-section. (b) 6.01 km/s cross-section. Note projectile in (a) (arrow) and spall in (b). (c) and (d) show corresponding, residual hardness contour measurements. (e) and (f) show Lagrangian simulations for corresponding stress contours. (g) and (h) show Eulerian (with fracture) simulations for corresponding stress contours.

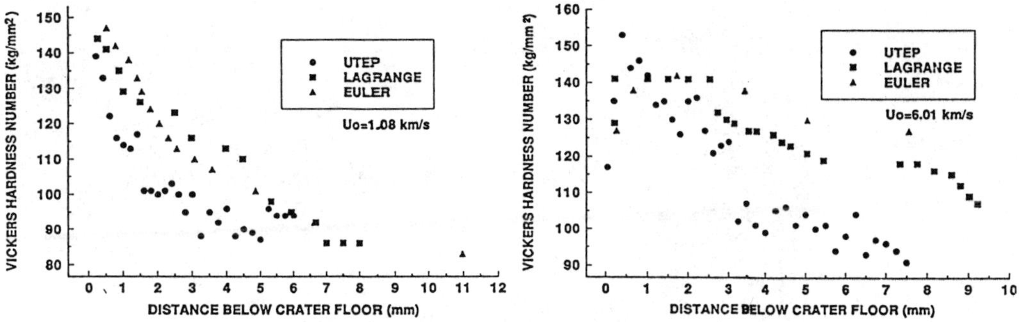

Figure 5 - Comparison of residual, axial hardness measurements (UTEP) with values from Lagrangian and Eulerian processor simulations of residual stress as in Fig. 4.

are the projectile and target densities respectively. However, the details of these low-velocity anomalies will only be apparent after careful interrogation of the computer processor alternatives, the application of appropriate strength models, and the generation of valid simulations, i.e., emulation of experimental crater observations, including specific shape and related features.

Summary and Conclusions

We have examined a number of impact crater systems including 1100 Al/Cu, ferritic stainless steel/Cu, soda-lime glass/Cu, and soda-lime glass/1100 Al. All of these systems are characterized by a zone of softening at the crater wall as a consequence of dynamic recrystallization which increases with impact velocity. Computer simulations at extrapolated impact velocities of 12 and 24 km/s illustrated this feature for the 1100 Al/Cu system. In addition, computer simulations in this system were able to produce residual stress contours which qualitatively emulated experimentally determined hardness contours in crater cross-sections. Eulerian processor calculations utilizing a fracture criterion provided more accurate crater shape simulations because crater rim fracture and ejection as well as spallation in 1.3 cm thick targets were also accurately simulated. While the results of this study are exemplary and preliminary, they point up the potential for impact crater simulations at hypervelocities unattainable in the laboratory.

Acknowledgments

Research supported by NASA-Johnson Space Center Grant NAG-9-481 and NASA-NGT-70343. We are especially grateful to Bence Gerber of Century Dynamics, Inc., San Ramon, CA for AUTODYN 2-D software support. Dr. F. Hörz of NASA-JSC provided archived crater samples and also performed some specific laboratory gun shots. Portions of this work were also supported by a Defense Logistics Agency Grant DN-009.

References

1. L. E. Murr and J. M. Rivas, "Measuring Hypervelocity Impact Velocity from Micrometeroid Crater Geometry," Int. J. Impact Engng. 15 (6) (1994) 785-795.

2. W. Harrison, et al., "Experimental Data and Hydrocode Calculations for Hypervelocity Impacts of Stainless Steel into Aluminum in the 2-8 km/s Range," ibid, 17 (1995) 363-374.

3. L. E. Murr, et al., "Electron Microscope Observations of Impact Crater Debris Amongst Contaminating Particulates in Materials Surfaces Exposed in Low-Earth Orbit, J. Mater. Sci., 28 (1993) 4553-4561.

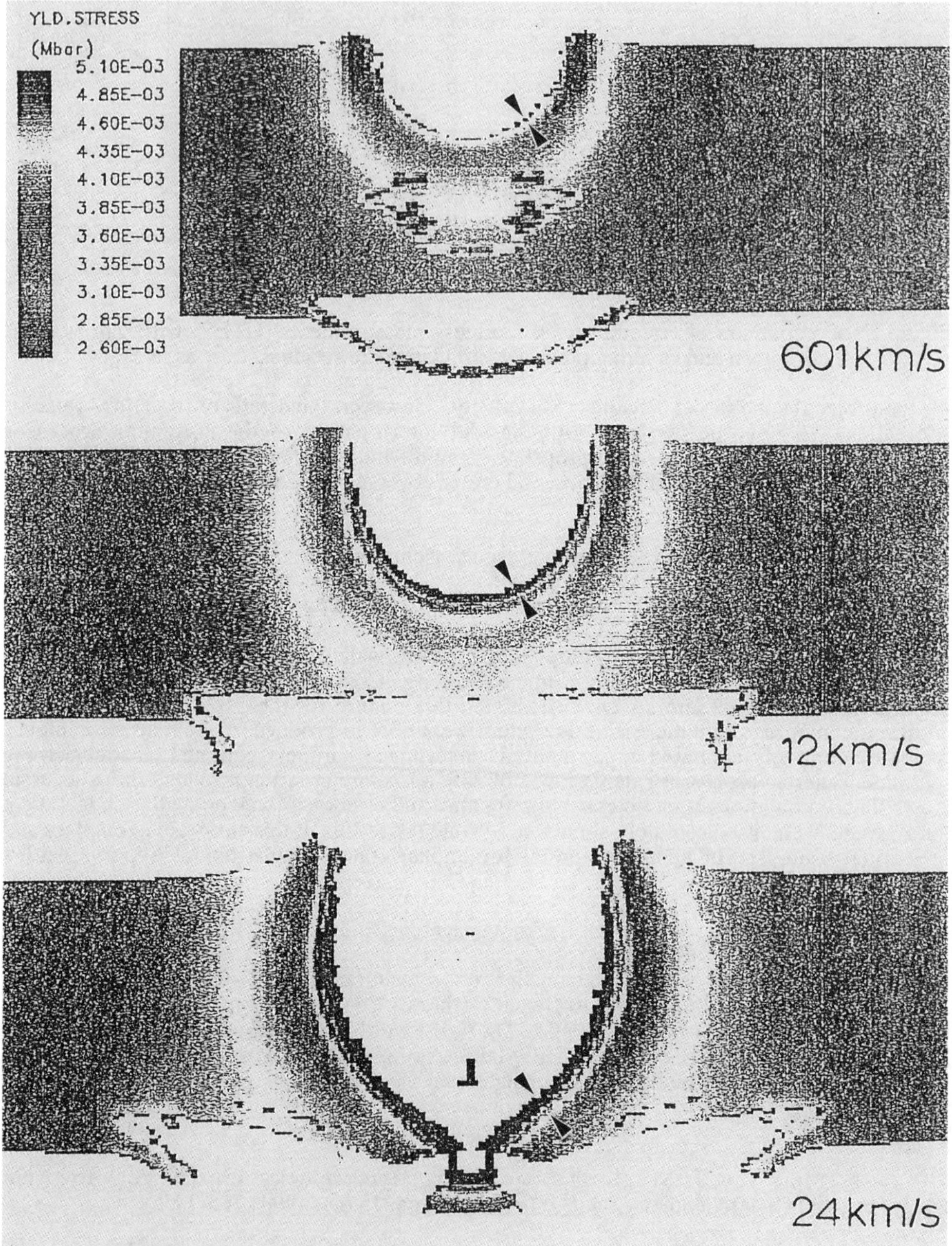

Figure 6 - Eulerian processor (with fracture) simulations for the 6.01 km/s crater (Fig. 4(b)) and extrapolated impact velocities of 12 km/s (a) and 24 km/s (b). Note the dynamic recrystallization zones between arrows at the crater wall.

4. R. P. Bernhard and F. Hörz, "Craters in Aluminum 1100 by Soda-Lime Glass Spheres at 1 to 7 km/s", Int. J. Impact Engng. 17 (1995) 69-80.

5. S. A. Quinones, et al., "Microstructural Evolution Associated with Hypervelocity Impact Craters in OFHC Copper," Chap. 36 in Metallurgical and Materials Applications of Shock-Wave and High-Strain-Rate Phenomena, L. E. Murr, K. P. Staudhammer, and M. A. Meyers (Eds.), Elsevier Science B.V., Amsterdam, 1995, pp. 293-302.

6. E. Ferreyra T., et al., "Effect of Initial Microstructure on High Velocity and Hypervelocity Impact Cratering and Crater-Related Microstructures in Thick Copper Targets: Part 1: Soda-Lime Glass Projectiles," J. Mater. Sci., 32 (1997) 2573-2585.

7. L. E. Murr, et al., "Effect of Initial Microstructure on High Velocity and Hypervelocity Impact Crating and Crater-Related Microstructures in Thick Copper Targets: Part 2: Ferritic Stainless Steel Projectiles," ibid;, 3143-3156.

8. L. E. Murr, Electron and Ion Microscopy and Microanalysis: Principles and Applications, 2nd Edition, Marcel Dekker, Inc., New York, 1991.

9. N. K. Birnbaum, et al., "Autodyn - An Interactive, Non-Linear Dynamic Analysis Program for Microcomputers Through Supercomputers," 9th Int. Conf. on Structural Mechanics in Reactor Technology, Lausanne, August, 1987.

10. C. J. Hayhurst, et al., "Modelling of Microparticle Hypervelocity Oblique Impacts on Thick Targets," Int. J. Impact Engng., 17 (995) 375-386.

11. M. Wilkins, "Calculation of Elastic-Plastic Flow," in Methods of Computational Physics, B. Alder, S. Fernback, and M. Rotenberg (Eds.), Academic Press, New York, 1964, pp. 211-263.

12. M. A. Meyers, Dynamic Behavior of Materials, J. Wiley & Sons, Inc., New York, 1994.

Creep Deformation

A NEW MODEL FOR CREEP OF DISPERSION-STRENGTHENED METALS

BASED ON STRESS CHANGE TEST RESULTS

Stuart E. Broyles
W. L. Gore and Associates, Inc.
P.O. Box 500, Flagstaff, Arizona 86002-0500

Jeffery C. Gibeling
Division of Materials Science and Engineering
University of California, Davis, California 95616

Abstract

The results of constant structure, stress reduction creep experiments of dispersion-strengthened copper and aluminum are described. The data reveal that the creep transients are similar in appearance to those for solid solution alloys rather than pure metals. This observation illustrates clearly the important contribution of mobile dislocation density to the creep of these materials. A new model is developed to describe creep of dispersion strengthened metals based on the kinetics of thermally activated dislocation glide, and incorporating the influence of applied stress on dislocation density. The thermal activation event is assumed to involve detachment of the dislocation line from the particle-matrix interface. The steady state and constant structure creep behavior of dispersion-strengthened materials is characterized by an athermal flow stress, $\hat{\sigma}$, representing the microstructural strength, and a Helmholtz free energy of activation, ΔF. The constant structure results are used to determine values for these two parameters. As expected, the values of these parameters are in the range appropriate for dislocation motion controlled by strong obstacles. The model is shown to provide a satisfactory description of the steady state and constant structure creep behavior of dispersion-strengthened copper and aluminum alloys. In contrast to other approaches, this model does not rely on threshold stresses or non-physical activation energies for creep.

Modeling the Mechanical Response of
Structural Materials
Edited by E.M. Taleff and R.K. Mahidhara
The Minerals, Metals & Materials Society, 1997

Introduction

Creep of Dispersion-Strengthened Metals

Dispersion strengthening (DS) emerged as a practical method to improve the elevated temperature creep resistance in metals in the late 1960's through the 1970's. Early studies of the mechanisms of dispersion strengthening focused on the behavior of γ' and oxide dispersion-strengthened nickel-based superalloys [1], while later work on fundamental issues utilized model alloys of copper containing SiO_2 and Al_2O_3 particles [2-5]. In the mid-1980's, advances in rapid solidification technology and related powder metallurgy methods for producing particle-strengthened aluminum aerospace alloys spawned renewed interest in understanding the mechanisms of dispersion strengthening [8-10]. Most recently, several manufacturers have developed dispersion-strengthened copper alloys for actively cooled structures, such as fusion reactor first walls and leading edges of aerospace vehicles. These alloys, which are essentially pure copper matrices with a relatively low volume fraction of dispersoids and are much more uniform in composition and structure than the early model alloys, provide a new opportunity to study the mechanisms of dispersion strengthening at elevated temperatures.

Over the intervening years, several authors have summarized the experimental observations and the theoretical understanding that has been gained from these studies. The reviews of Blum and Reppich [6], McLean [7], Arzt [8], Gibeling and Nix [11] and Mukherjee and Mishra [12], among others, provide helpful perspectives. One experimental observation that is common to all studies of dispersion strengthened metals is that application of the conventional Dorn equation,

$$\dot{\varepsilon} = A\left(\frac{\sigma}{E}\right)^n \exp\left(\frac{-Q_c}{k_B T}\right) \tag{1}$$

to describe steady state creep data commonly requires stress exponents, n, in the range of 10-100. These values are significantly greater than the value of 3 that is typical of solid solution strengthened alloys and the 4-8 that is commonly observed for pure metals. In addition, the activation energy for creep, Q_c, is significantly greater than the activation energy for lattice self-diffusion that normally characterizes the temperature dependence of creep in pure metals and solid solution alloys. These extremely large values of n and Q_c have proven difficult to rationalize on the basis of any theoretical descriptions of dislocation motion past unshearable, incoherent dispersoids.

A number of authors have analyzed their data by incorporating a threshold stress, σ_o, that is presumed to oppose the motion of dislocations. Thus, Eq. (1) is modified to give

$$\dot{\varepsilon} = A\left(\frac{\sigma-\sigma_o}{E}\right)^n \exp\left(\frac{-Q_c}{k_B T}\right). \tag{2}$$

In order to analyze the data, n is assumed to be either 5 if the dislocation structure is expected to evolve during deformation or 8 if the particles are expected to constrain the substructure. The latter condition is usually termed “constant structure” and the exponent of 8 is chosen because that value has been observed when the data for stress reduction tests of pure aluminum are analyzed. However, recent work on a variety of pure metals has demonstrated that this value is only approximately correct, and that creep data obtained under constant structure conditions following stress reduction tests are more appropriately characterized by an exponential law rather than a power law [13-16]. Experimentally, this approach is justified by the observation that the strain rates in dispersion-strengthened metals often become asymptotically low at low stresses. While the correlation of the data is often good, the specific values and temperature dependencies of the threshold stress have been difficult to justify from a mechanistic perspective, and the microstructural significance of the threshold stress is not often clear [6].

As an alternative to the threshold stress concept, Arzt and coworkers have developed a model in which creep in dispersion-strengthened metals is controlled by the attractive interactions between particles and dislocations [8, 17-19]. Rather than considering the rate limiting process to be the climb of dislocations as they approach particles, the critical step in the Rösler and Arzt model is taken to be the thermally activated detachment of dislocations from the particle-matrix interfaces as the dislocations attempt to depart in their glide plane [18, 19]. This treatment is supported by TEM observations of dislocations that are restrained from leaving the particle-matrix interface after climbing over the dispersoids in a variety of materials. In this model, a portion of the strain energy of the dislocation is relaxed at the particle-matrix interface. In order for the dislocation to leave the interface, thermal activation is needed. Rösler and Arzt assume this step involves the movement of vacancies to the particle-matrix interface by lattice diffusion, although the need for diffusion is not fully explained. The resulting expression for the creep rate in a material governed by this dislocation-particle interaction process is given by

$$\dot{\varepsilon} = \dot{\varepsilon}_o \exp\left[-\frac{\mu b^2 r}{k_B T}(1-k)^{3/2}\left(1-\frac{\sigma}{\sigma_d}\right)^{3/2} \right] \tag{3}$$

where $\dot{\varepsilon}_o$ is a reference strain rate given by $3D_V \lambda r / b$ [19]. In these equations, μ is the shear modulus, λ is the interparticle spacing, r is the average particle radius, k is an interaction parameter and σ_d is the athermal detachment stress. The interaction parameter provides a measure of the extent to which the dislocation line energy is relaxed at the particle-matrix interface. As the value for k increases, the interaction becomes weaker, reaching an upper limit for detachment control of k = 0.94 The athermal detachment stress, σ_d, is related to the interaction parameter according to the expression

$$\sigma_d = \sigma_{Or}\sqrt{1-k^2} \tag{4}$$

where σ_{Or} is the Orowan stress calculated from the particle size and spacing.

Analysis of steady state creep data according to the model of Rösler and Arzt depends sensitively on the difference between the experimentally measured apparent activation energy for creep, Q_c, and the activation energy for vacancy diffusion, which is equivalent to the activation energy for creep in the pure matrix [19]. Values of the interaction parameter and the detachment stress must be determined by the quality of the fit of the rate equation to the creep data. One significant shortcoming of the model is that the dislocation densities are assumed to be constant and the values required to give an adequate fit are always unrealistically high. The assumption of constant dislocation density is in contradiction to the observations of Lloyd and Martin, who reported that the dislocation density varies as σ^2 in dispersion-strengthened copper [20].

Nevertheless, creep data for a variety of dispersion-strengthened materials have been successfully analyzed by Rösler and Arzt, as well as others [19,21-23].

General Description of Time Dependent Deformation

One of the principal shortcomings of previous descriptions of the creep behavior of dispersion-strengthened metals, including those summarized above, is that these theoretical treatments neglect the development of dislocation density and arrangement during creep. As a consequence, these analyses are limited to describing steady state behavior and it is not possible to distinguish between the formation of the dislocation structure that is at dynamic equilibrium, and characteristics of deformation within the developed structure. A more general description requires consideration of the evolution of the dislocation structure and a separate description of the kinetics of deformation within a fixed structure. This distinction requires two sets of rate equations. A kinetic law that describes the rate at which deformation occurs at an imposed stress and temperature condition within a given, constant structure can be written as

$$\dot{\varepsilon} = f\left(\sigma, T, \hat{\sigma}_1, \hat{\sigma}_2, ..., \hat{\sigma}_i\right) \tag{5}$$

where σ is the applied stress, T is the temperature, and the structure parameters, $\hat{\sigma}_i$, describe the strength of the existing or developed dislocation structure [13,24]. In dispersion strengthened materials, the effect of the dispersoids on dislocation motion would be included as an additional structure parameter, and may dominate the dislocation interaction terms. The dislocation structure parameters are be expected to change during deformation due to the competing effects of recovery and strain hardening. The rate of change of the structure parameters is given by a structure evolution law in the general form:

$$\frac{\partial \hat{\sigma}_i}{\partial t} = g_i\left(\sigma, T, \hat{\sigma}_1, \hat{\sigma}_2 ... \hat{\sigma}_i\right) \tag{6}$$

Kocks, Argon and Ashby provided a general treatment of the thermodynamics and kinetics of slip within a fixed structure [25]. In this general description, the obstacle in a slip plane may be attractive or repulsive, and may represent a dislocation-dislocation interaction, a particle-dislocation interaction or an interaction of another type. The dislocation is assisted in overcoming the obstacle by mechanical work supplied by the applied stress and by thermal activation. This gives rise to a stress dependence of the free enthalpy of thermal activation, ΔG, which can be described as

$$\Delta G = \Delta F\left[1 - \left(\frac{\sigma}{\hat{\sigma}}\right)^p\right]^q \tag{7}$$

where $\Delta F = g_o \mu b^3$ is the Helmholtz free energy, σ is the applied stress and $\hat{\sigma}$ is a single structure parameter, the mechanical threshold, which represents the maximum obstacle glide resistance for a fixed structure [13,24,25]. In other words, $\hat{\sigma}$ is the flow stress required for a dislocation to overcome an obstacle in the absence of any thermal energy. The exponents p and q describe the obstacle shape and distribution. Typical values for p and q are in the range of $0 \leq p \leq 1$ and $1 \leq q \leq 2$. For regular and random distribution of obstacles, p is usually given values of 1 and 2/3, respectively. For strong obstacles, the value of q is typically 3/2. For weak obstacles, q is generally equal to 1.

Based on the above formulation, the creep rate is given by

$$\dot{\varepsilon} = \dot{\varepsilon}_0 \exp\left[\frac{-\Delta F}{k_B T}\left(1-\left(\frac{\sigma}{\hat{\sigma}}\right)^p\right)^q\right] \quad (8)$$

where $\dot{\varepsilon}_0$ is a pre-exponential term that is only weakly dependent on stress as given by $\dot{\varepsilon}_0 = \rho_m b\nu$. Here, ρ_m is the mobile dislocation density and ν is the thermal attempt frequency assumed equal to the vibration frequency of 10^{13} s^{-1} for all temperature of interest.

Because Eq. (8) describes the kinetics of deformation under conditions of constant structure (fixed mobile dislocation density), it is necessary to explore this relationship under conditions where the dislocation structure does not vary with applied stress. One approach to this problem is based on stress reduction tests, which have been used to characterize the creep mechanisms in various pure FCC metals and alloys. During the short time following a reduction in stress, very little change in the dislocation structure occurs. Thus, the creep rate measured after a stress reduction characterizes deformation under conditions of "constant structure", where the structure that is presumed to remain constant is that which controls the deformation process. In this way it is possible to study separately the kinetics of deformation and the structure evolution characteristics.

Nakayama and Gibeling [13-15] and Broyles and Gibeling [26] have performed stress reduction testing to examine the kinetics of dislocation glide in a variety of FCC materials. For these materials, this work has shown that setting p = q = 1 gives a good correlation to the data. Nakayama and Gibeling also outlined procedures which enable the Helmholtz free energy and the structure parameter to be determined from constant structure creep data [13]. When applied to Al, Cu, Ni and Ni-Co alloys, this method reveals that the structure parameter scales linearly with the applied stress and that the primary influence of stacking fault energy is on structure evolution, as reflected in the differences in $\hat{\sigma}/\mu$ [15]. We note that the results of these investigations are consistent with those obtained by many other investigators, as reviewed recently by Biberger and Gibeling [26]. However, it appears that the method of stress change testing and kinetic analysis has not been previously applied to study creep mechanisms in dispersion-strengthened materials.

The goal of the present study is to develop a new model of the creep behavior of dispersion-strengthened metals that describes both steady state and transient deformation and is based on thermally activated dislocation motion. The development of this model is informed by the results of stress reduction tests of dispersion-strengthened copper and dispersion-strengthened aluminum that characterize the deformation response under constant structure conditions. In addition, a method of analyzing the steady state and constant structure data is developed in order to quantify the model parameters. The resulting calculations are shown to provide a good description of the steady state and constant structure deformation behavior.

Experimental

Materials and Specimen Preparation

The dispersion strengthened copper alloy studied in this investigation, GlidCop Al-15 (C15715), was supplied by SCM Metal Products, Inc. in the form of an extruded rectangular bar. This alloy of copper containing 0.15 wt% aluminum was produced by a proprietary powder process during which the aluminum was oxidized. The resulting dispersoid content is 0.7 volume%. In addition, 150 ppm boron was added to minimize the amount of oxygen remaining in solution. The microstructure of this materials has been characterized in some detail by previous investigators [27-29]. The extrusion process creates an elongated grain structure approximately 2 μm × 2 μm × 40 μm. Large (0.2 to 1.0 μm) cubic α-Al_2O_3 particles are observed near grain boundaries. However, the creep resistance is apparently controlled a uniform distribution of extremely fine (10 nm diameter) γ-Al_2O_3 within the grain interiors.

The Al-8.5 wt% Fe-1.3 wt% V-1.7 wt% Si alloy (designated FVS-0812) was prepared by the rapid solidification and powder consolidation techniques developed at Allied Signal Metals and Ceramics Laboratory. The extruded alloy is hot rolled to final thickness. This alloy is strengthened by a dispersion of intermetallic $Al_{13}(Fe,V)_3Si$ particles. The dispersoid size in this material is in the range 40-80 nm, which is slightly larger than the γ-Al_2O_3 particles in the copper alloy. This alloy is also characterized by very fine, equiaxed grains approximately 500 nm in diameter. Specimens were annealed in nitrogen for one hour at 673K to relieve any internal stresses remaining from the machining process.

Stress Reduction Testing Procedures

The equipment and experimental methods developed by Nakayama and Gibeling [13-15] were used to study the creep response following stress reductions. During the course of this investigation, several significant improvements were made to the testing system as described in detail elsewhere [16, 28]. The data for dispersion strengthened aluminum were acquired prior to these modifications while the tests of DS copper were conducted using the modified system. Briefly, a load cell was added to the creep system so that a direct measurement of force could be used to more accurately detect the instant of the stress change. In addition, the resolutions of time and strain have been enhanced significantly compared to earlier measurements. The LVDT was recalibrated to more closely match a full-scale range of the digital voltmeters, resulting in an improvement in the strain resolution from 1×10^{-6} to 8×10^{-8}. The time resolution of the strain and load data was also enhanced by increasing the reading rates of the voltmeters from 20 readings per second to a maximum of 50 readings per second. In addition, the accuracy of the time readings was significantly improved (from 50 ms to 1μs) by replacing the internal computer clock with a high-resolution counter/timer card. This change was critical in reducing error in the strain rate calculations and permitted a change from calculating strain rates from a manual fit of the strain *vs.* time graphs to an automated calculation based on a ten-point least squares linear fit of the strain *vs.* time data.

All creep tests of the dispersion-strengthened copper were conducted at a temperature of 973K in a vacuum of 5×10^{-4} Pa. Tensile creep tests of the aluminum alloy were conducted at a test temperature of 573K at test pressures equal to or lower than 4×10^{-3} Pa. All stress reduction experiments were performed after the specimens had attained a steady state creep rate.

Results

Constant Structure Creep of Dispersion-Strengthened Metals

In order to describe the results of the stress reduction experiments, we adopt the operational notation proposed by Biberger and Gibeling [26]. This approach is illustrated in Figure 1, which shows schematically the form of the creep transients expected after stress reduction tests of pure metals (and Class M alloys) and solid solution-strengthened (Class A) alloys. In both classes of material, region C represents the constant structure behavior of interest in the present study. This

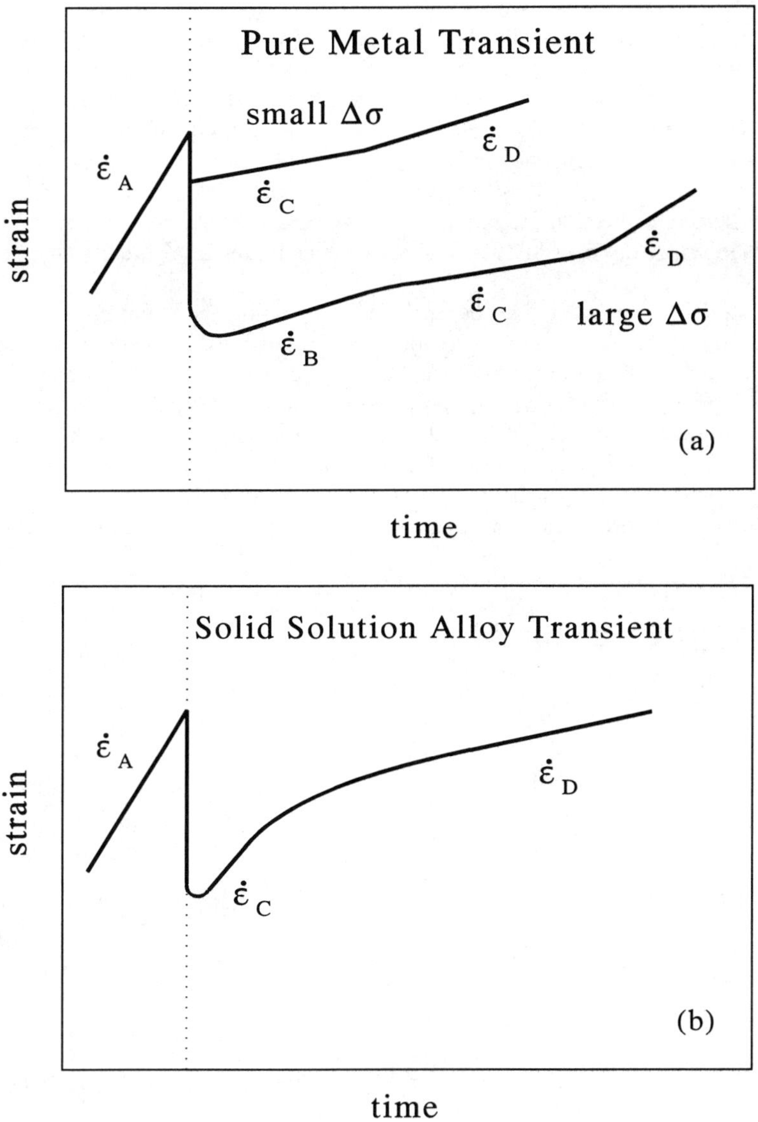

Figure 1: Schematic representation of stress reduction creep transients for (a) pure metals and (b) solid solution strengthened alloys.

figure illustrates the normal observation that the creep rates observed after a stress reduction are less than the eventual steady state rate (region D) for the pure metal case. This situation arises arises because the dislocation substructure must evolve from a hard structure characteristic of the initial applied stress to a softer structure appropriate to the reduced stress. However, in the case of solid solution strengthened alloys, the creep rate immediately following a stress reduction (region C) is higher than that observed at the new steady state (region D). We note that for pure metals, all creep rates following the stress reduction are lower than the eventual steady state rate (region D). This observation reflects the fact that the mobile dislocation structure density cannot adjust immediately to the value appropriate to the reduced stress due to the pinning effects of the solute atoms [30].

An example of the transients exhibited by the dispersion-strengthened alloys studied in the present investigation is shown in Figure 2. All stress reduction tests conducted during this investigation resulted in transients of this form. The most important observation from this result is that the shape of the transient resembles that observed for solid solution strengthened alloys rather than that obtained when pure metals are tested. From this important observation, we conclude that mobile dislocation density must play an important role in our understanding of the creep mechanisms in dispersion strengthened metals and that the particles may stabilize the mobile dislocation structure in a way that is similar to the influence of solute atoms.

The constant structure creep data can be analyzed by plotting the strain rates in region C for various initial and reduced stresses in a double logarithmic manner as shown in Figure 3. As expected from the form of the transients, the constant structure creep rates are higher than the eventual steady state rates, so a line through these values lies above the steady state line. This observation is in direct contrast to the results obtained for pure metals, for which the constant structure line is below the steady state line [13-16]. The results for the dispersion-strengthened copper and dispersion strengthened aluminum are qualitatively similar, as illustrated in this figure. While it is clear form these plots that a power law of the form given in Eq. (1) only

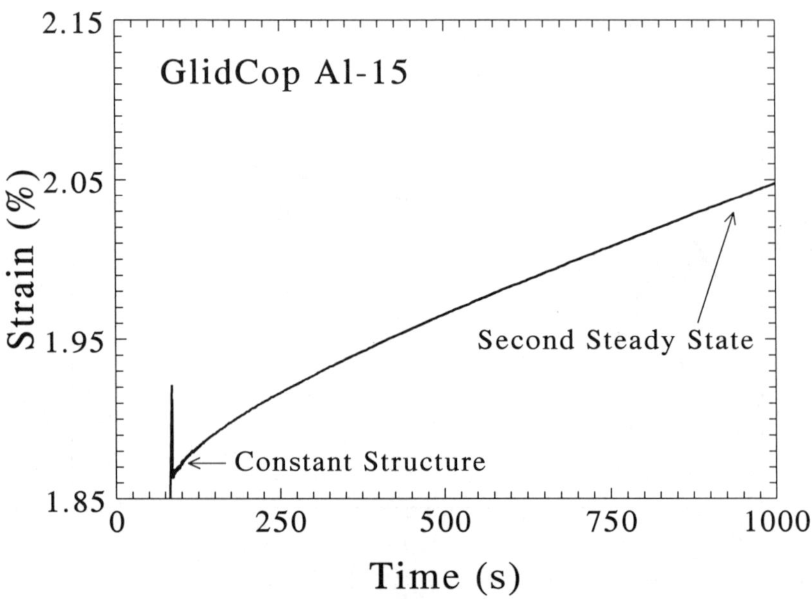

Figure 2: Typical stress reduction creep transient for dispersion strengthened copper.

approximately describes the constant structure results, it is informative to examine the stress exponents that are obtained from the slopes obtained from a linear regression analysis of the data in Figure 3. For both materials, the steady state stress exponents (n_{ss}) are greater than 5, as expected for dispersion-strengthened metals. The constant structure exponents (n_{cs}) are less than n_{ss} by values of 4 to 6, depending on the material and the test condition.

Specimens of both materials were cooled under the applied creep load (in tests without stress reductions) to preserve the dislocation structure for TEM observation. A typical observation for GlidCop Al-15 is presented in Figure 4. This figure reveals the presence of a relatively high density of dislocations interacting with individual particles. No subgrain structure is apparent in this figure, nor was one observed in the DS aluminum alloy. These observations confirm that creep in these materials, with fine grain sizes and fine particle structures, is controlled by dislocation-particle interactions and not by the characteristics of a dislocation substructure. Further, these microstructural observations are consistent with the results of the stress reduction experiments, which do not exhibit characteristics of subgrain controlled creep as is observed in pure metals.

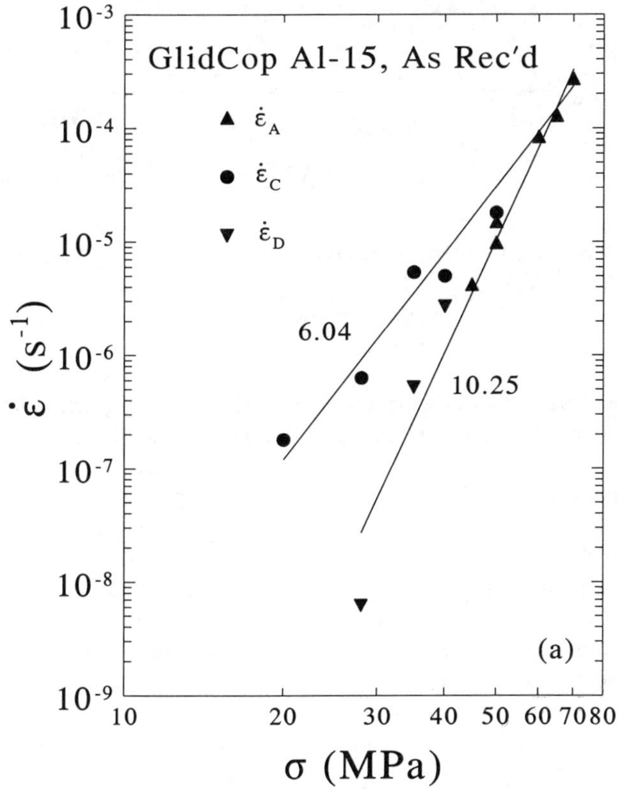

Figure 3a:Constant structure and steady state creep data for GlidCop Al-15 at 973K.

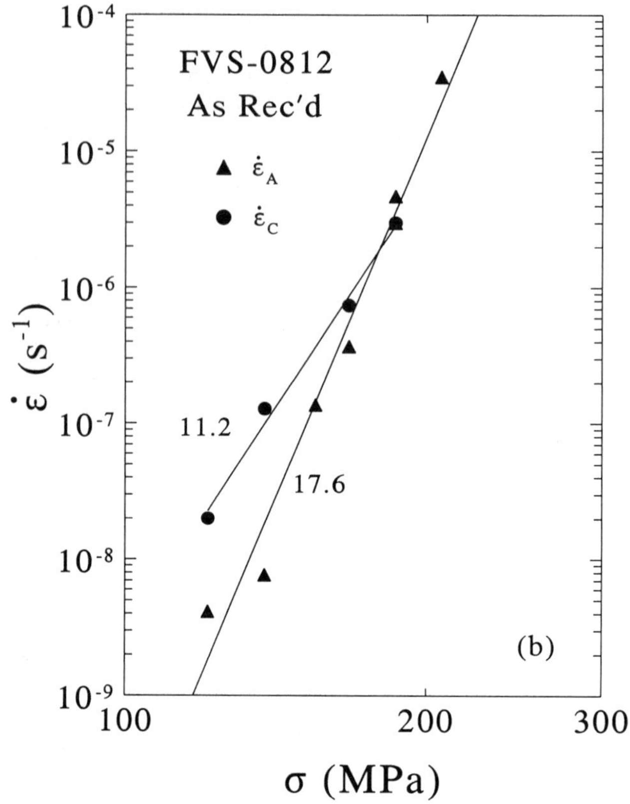

Figure 3b:Constant structure and steady state creep data for FVS-0812 at 573K.

Analysis and Discussion

As noted above, several important observations from the stress change experiments may be used to guide the development of a new model of creep in dispersion-strengthened metals. First, the observation of creep transients demonstrates that the existing models are inadequate in that they predict only steady state behavior based on the rate that a single dislocation passes a dispersoid. None include terms that represent development of a dislocation structure that would remain invariant following a change of stress. A fundamental description of creep in DS metals must include an ability to predict these transients. Further, the form of the strain-time record following stress reductions is similar to that observed in solid solution strengthened alloys. This observations suggests that the mobile dislocation density must be taken into account and that the particles restrict the response of these mobile dislocations to the stress change in much the same manner as the solute atoms. This similarity to the behavior of solid solution strengthened metals is also borne out when the constant structure strain rates are examined as a function of the applied stress.

In order to develop a new model of creep that properly describes both the steady state and constant structure behavior of dispersion strengthened metals, we assume that the detachment of dislocations from particles is the rate controlling process and that it occurs with the aid of

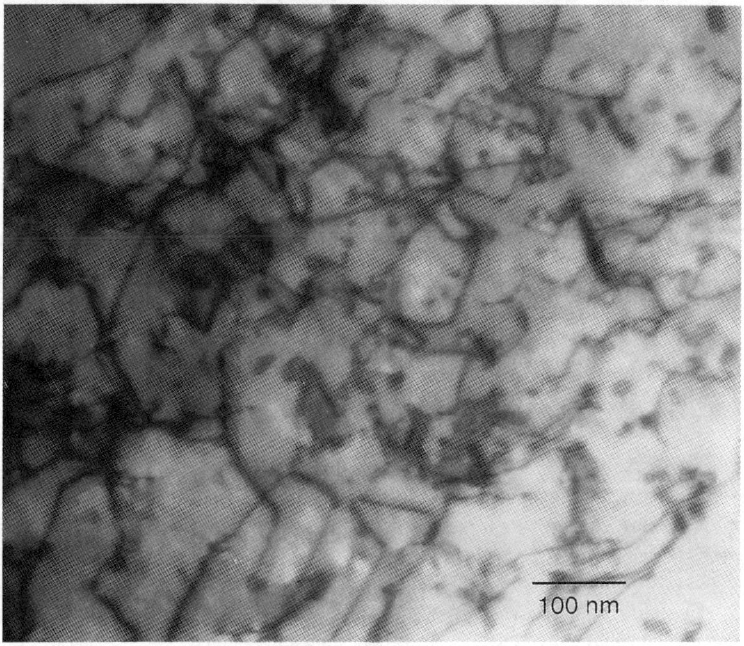

Figure 4: TEM micrograph of GlidCop Al-15 tested at 40 MPa showing high dislocation densities and dislocation-particle interactions.

thermal activation. Thus, we begin with an expression for creep rate derived from the general kinetic law given in Eq. (8). This equation is clearly similar to that written by Rösler and Arzt (Eq. (3), but for the moment is left in a more general form. For simplicity of analysis and physical appropriateness, we assume a regular distribution of obstacles (p=1). Further, Kocks, Argon and Ashby have suggested that q=3/2 gives a reasonable representation of the shape profile of strong obstacles [25], and we assume this value in the present case. We note that this choice also gives a creep equation that is similar to that of Rösler and Arzt, which includes an exponent of 3/2. With these choices, Eq. (8) can be written as

$$\dot{\varepsilon} = \dot{\varepsilon}_o \exp\left[\frac{-\Delta F}{k_B T}\left(1 - \frac{\sigma}{\hat{\sigma}}\right)^{3/2}\right] \tag{9}$$

Several investigators have observed that the mobile dislocation density in DS metals varies as the square of the applied stress. In particular, Lloyd and Martin reported that this relationship described their observations for single crystals of copper containing Al_2O_3 particles [20]. Accordingly, we assume the dislocation density follows the Taylor relationship:

$$\rho_m = \alpha\left(\frac{\sigma}{\mu b}\right)^2 \tag{10}$$

As a result, Eq. (9) describing the steady state creep rate can be written as

$$\dot{\varepsilon} = A\left(\frac{\sigma}{\mu}\right)^2 \nu \cdot \exp\left[\frac{-\Delta F}{k_B T}\left(1-\frac{\sigma}{\hat{\sigma}}\right)^{3/2}\right]. \tag{11}$$

where A is a constant. Although the form of Eq. (11) is very similar to the treatment of Rösler and Arzt (Eq. (3)), the Helmholtz energy of activation is left in general form and is not equated to the activation energy for lattice diffusion. In this representation, the structure parameter, $\hat{\sigma}$, has the same meaning as the detachment stress, σ_d, both of which describe the mechanical strength of the particle structure and are expected to be independent of the applied stress. This situation differs form the case for pure metals, in which the structure parameter represents the strength of the dislocation subgrain structure that forms during creep, hence depends on the applied stress and evolves with creep strain.

The shape of the strain-time curves indicates that the mobile dislocation density does not change when the applied stress is reduced. Referring to Eq. (11), this observation suggests that constant structure creep in dispersion-strengthened metals can be described by letting the term inside the exponential include the reduced stress, since this part of the equation does not contain structure dependent terms. However, based on the dislocation density arguments presented above, the pre-exponential would be proportional to the initial applied stress. Then, if the data are represented as a power law over a limited range of stress and strain rate, the constant structure stress exponent (n_{cs}) is related to the steady state exponent (n_{ss}) as $n_{cs} = n_{ss}-2$.

We envision the creep microstructure to include an array of dislocations interacting with the particles and gliding forward. The high stresses will give rise to a small dislocation spacing that is less than the average particle spacing. This picture is supported by the observations of Liu and Cowley who noted arrays of closely spaced dislocations interacting with HfC particles in a Hf-Re alloy [31]. When the stress is reduced, these arrays seek to adjust the dislocation spacing to re-equilibrate the interdislocation forces. However, the dislocations are prevented from gliding backward by the particles. In this way, the arrays of gliding dislocations interacting with particles are like pile-ups that cannot run back to adjust the interdislocation spacing. The forces causing reverse glide are much less than those remaining in the forward direction, hence the probability (and rate) of reverse glide is extremely low. The interdislocation forces thus give rise to dislocation self forces that add to the forces associated with the reduced applied stress. We represent this extra component of forward force by a self stress term, σ_s, giving the final relationship for constant structure creep as

$$\dot{\varepsilon}_C = A\left(\frac{\sigma_{initial}}{\mu}\right)^2 \nu \cdot \exp\left[\frac{-\Delta F}{k_B T}\left(1-\frac{\sigma_{reduced}+\sigma_s}{\hat{\sigma}}\right)^{3/2}\right]. \tag{12}$$

The value of σ_s cannot be calculated or measured directly. Rather, this value is determined by iterative calculation such that the difference between n_{cs} and n_{ss} is 2. n_{cs} is determined from the slope of a double logarithmic plot of constant structure strain rate *vs.* effective reduced stress, $\sigma_{reduced}+\sigma_s$. The values of σ_s for the materials tested in the present study are given in Table I.

The constant structure data can also be used to identify values of ΔF and $\hat{\sigma}$. Taking logarithms of both sides of Eq. (12) gives

$$\ln \dot{\varepsilon}_C = \ln\left[A\left(\frac{\sigma_{initial}}{\mu}\right)^2 \nu\right] - \frac{\Delta F}{k_B T}\left(1 - \frac{\sigma_{reduced} + \sigma_s}{\hat{\sigma}}\right)^{3/2} \tag{13}$$

which can be differentiated with respect to the reduced applied stress to give

$$\left(\frac{d(\ln \dot{\varepsilon}_C)}{d\sigma_{reduced}}\right)^2 = \left(\frac{3}{2}\cdot\frac{-\Delta F}{k_B T}\right)^2 \left(\frac{1}{\hat{\sigma}}\right)^2 \left(1 - \frac{\sigma_{reduced} + \sigma_s}{\hat{\sigma}}\right). \tag{14}$$

This expression represents a linear relationship between the term on the left of the equality and the reduced applied stress. Operationally, the term on the left hand side of Eq. (14) is obtained by smoothing the constant structure creep data using a 2nd order polynomial curve fit. A linear regression of the data plotted in the form suggested by Eq. (14) yields values ΔF and $\hat{\sigma}$. The results of these calculations are presented in Table I for the two dispersion-strengthened materials of interest. We note that for both alloys, ΔF and g_o are approximately half the values of that observed in the corresponding base alloy and $\hat{\sigma}$ is substantially larger than in the base alloy [13-16].

Table I Model Parameters from Constant Structure Creep Experiments of DS Metals

	Al-15	FVS-0812
$\sigma_{initial}$ (MPa)	70	186.2
σ_s (MPa)	11	59.3
$\hat{\sigma}$ (MPa)	99.8	270.8
ΔF (kJ/mole)	118.5	138.9
g_o	0.385	0.481
k	0.927	0.964

The model developed here does not include an explicit parameter representing the nature of the dislocation-particle interaction. Rather, these details are included in the value of ΔF. However, it is possible to determine the interaction parameter, k, in the Rösler and Arzt model from the present results by noting that

$$\Delta F = \mu b^2 r (1-k)^{3/2} \tag{15}$$

based on a comparison of Eq. (3) and Eq. (11). This equation can be rearranged to solve for k, and the results are listed in Table I. We note that these values are slightly higher than the values obtained by application of the Rösler and Arzt model [29,32], reflecting the fact that mobile dislocation density is included explicitly in the present treatment.

The above analysis provides a direct method to calculate all of the interaction parameters needed to describe steady state in the two materials of interest. As steady state, the dislocation arrays are in equilibrium and Eq. (11) can be used to describe the data. Thus, the interdislocation forces should be equal in the forward and reverse directions, and the data should be described by Eq. (11). The constant A can be determined by performing a nonlinear regression fit of this equation to the steady state creep data. The results for dispersion strengthened copper and aluminum are shown in Figure 5, which includes steady state rates measured at both the initial stress ($\dot{\varepsilon}_A$) and the reduced stress ($\dot{\varepsilon}_D$). We note that the procedure developed in this investigation provides a satisfactory fit of the data. Of course, it is important to recognize that the constant A has been adjusted to give a best fit. Nevertheless, the present method of obtaining the necessary parameters is more rigorously quantitative than the method outline by Rösler and Arzt, demonstrating the importance of considering the mobile dislocation density and the value of the constant structure measurements in evaluating the model parameters.

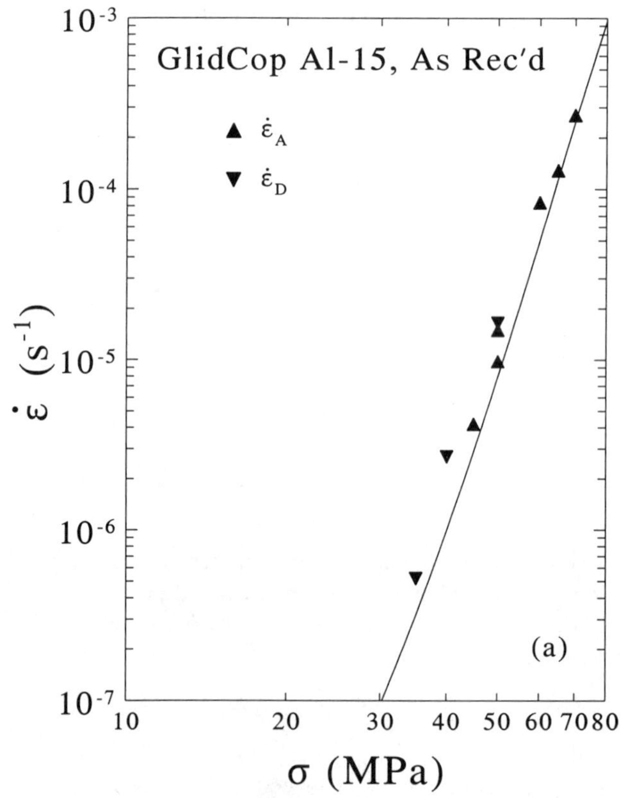

Figure 5a: Curve fits to steady state data for GlidCop Al-15 based on parameters calculated from the constant structure creep data.

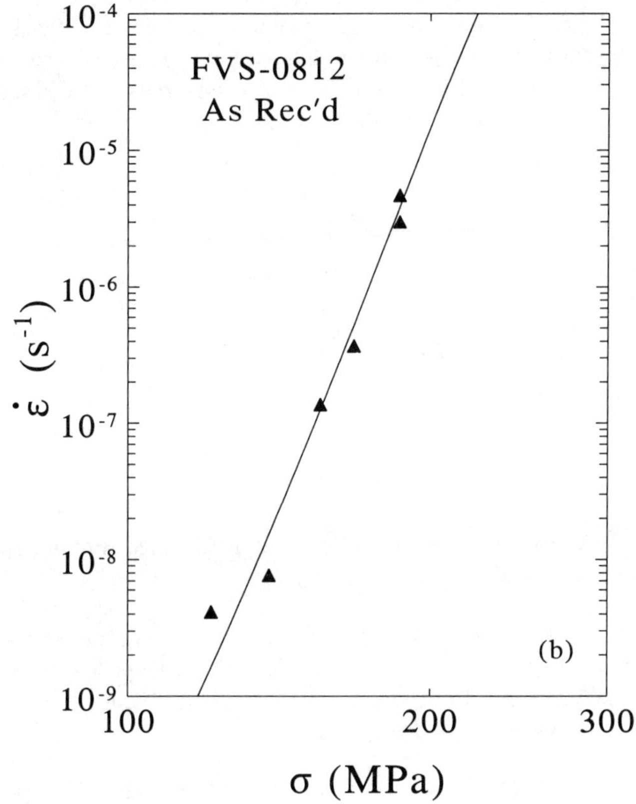

Figure 5b: Curve fits to steady state data for FVS-0812 based on parameters calculated from the constant structure creep data.

Conclusions

Stress reduction creep tests were conducted on dispersion-strengthened copper and aluminum alloys. The strain-time transients recorded in these tests resemble those observed in tests of solid solution strengthened alloys, demonstrating that mobile dislocation density controls the creep behavior of dispersion strengthened metals following stress reductions. The creep transients are similar in form to those observed in stress change tests for solid solution-strengthened alloys, suggesting that dislocation density plays a key role in the behavior of both materials. Previous models of the creep of dispersion-strengthened metals do not account for the development of the mobile dislocation density and are not able to predict transient creep behavior. In order to describe these new results, a new model based on thermally activated detachment of dislocations from particles has been proposed. The model incorporates an effective stress that is greater than the applied stress during constant structure creep as a consequence of the fact that the dislocation density is greater than that expected at steady state for the reduced stress. The constant structure creep data are analyzed to determine all of the necessary model parameters, resulting in a quantitative description of the steady state and constant structure behavior of these materials.

This approach has been successfully applied to GlidCop Al-15 and FVS-0812 alloys in the as-received condition. The FVS-0812 alloy has a higher volume fraction of particles, larger average particle size and smaller particle spacing than the Al-15 alloy. This is reflected in the values of the activation parameters ΔF and $\hat{\sigma}$ determined for the two materials, both of which are greater for the FVS-0812 material than for the Al-15 alloy.

Acknowledgments

This paper is based on work supported by the National Science Foundation under grant DMR 92-08549. The authors wish to thank Mr. Jack D. Troxell of SCM Metal Products, Inc., Research Triangle Park, NC for kindly supplying the copper material used in this investigation and the Alloys Department, Allied Signal, Inc., Morristown, NJ for providing the dispersion-strengthened aluminum.

References

1. G. S. Ansell and J. Weertman, "Creep of a Dispersion-Hardened Aluminum Alloy," Trans. Metall. Soc. AIME 215 (1959), 838.

2. F. J. Humphreys, P. B. Hirsch and D. Gould, "The Effect of Temperature on the Mechanical Properties and Microstructure of Single Crystals of Copper Containing Dispersed Oxide Particles," Proc. Second Intl. Conf. on the Strength of Metals and Alloys, (Metals Park, OH: American Society for Metals, 1970), 550-554.

3. R. S. W. Shewfelt and L. M. Brown, "High-temperature Strength of Dispersion-hardened Single Crystals: I. Experimental Results," Philos. Mag. 30 (1974), 1135-1145.

4. R. S. W. Shewfelt and L. M. Brown, "High-temperature Strength of Dispersion-hardened Single Crystals: II. Theory," Philos. Mag. 35 (1977), 945-962.

5. G. J. Lloyd, R. J. McElroy and J. W. Martin, " Some Creep Studies on Internally Oxidized Copper-Aluminum Crystals," Proc. Third Intl. Conf. Strength of Metals and Alloys, (London: Institute of Metals, 1973), 185-189.

6. W. Blum and B. Reppich, "Creep of Particle-Strengthened Alloys,"Creep Behavior of Crystalline Solids, B. Wilshire and R. W. Evans, eds., (Swansea, UK: Pineridge Press, 1985), 83-135.

7. M. McLean, "On the Threshold Stress for Dislocation Creep in Particle Strengthened Alloys," Acta Metall. Mater. 33 (1985), 545-556.

8. E. Arzt, "Creep of Dispersion Strengthened Materials: A Critical Assessment ," Res. Mechanica 31 (1991), 399-453.

9. D. L. Yaney and W. D. Nix, "Elevated Temperature Deformation Behavior of an Al-8.4 wt Pct Fe-3.6 wt Pct Ce Alloy," Metall. Trans. 18A (1987), 893-902.

10. J. A. Hawk, J. K. Briggs and H. G. F. Wilsdorf, "Creep in Mechanically Alloyed Dispersion Strengthened Alloys," Advances in Powder Metallurgy, Vol. 3, edited by T. G. Gasbarre and W. F. Jandeska, Jr. (Princeton, NJ: Metal Powder Industries Foundation, 1989), 285-299.

11. J. C. Gibeling and W. D. Nix, "The Description of Elevated Temperature Deformation in Terms of Threshold Stresses and Back Stresses: A Review," Mater. Sci. Engr. 45 (1980) 123-135.

12. A. K. Mukherjee and R. S. Mishra, "Creep Mechanisms in Dispersion Strengthened Materials," The Johannes Weertman Symposium, edited by R. J. Arsenault, D. Cole, T. Gross, G. Kostorz, P. K. Liaw, S. Parameswaran and H. Sizek (Warrendale, PA: The Minerals, Metals and Materials Society, 1996), 119-126.

13. G. S. Nakayama and J. C. Gibeling, "Constant Substructure Creep of Aluminum Following Stress Reductions," Acta Metall. 38 (1990), 2023-2030.

14. G. S. Nakayama and J. C. Gibeling, "Creep of Copper Under Constant Structure Conditions," Scripta Metall. 24 (1990), 2031-2035.

15. G. S. Nakayama and J. C. Gibeling, "Experimental Investigation of the Kinetic Law for High Temperature Deformation," Modeling the Deformation of Crystalline Solids, edited by T. C. Lowe, *et al.* (Warrendale, PA: The Minerals, Metals and Materials Society, 1991), 589-600.

16. S. E. Broyles and J. C. Gibeling, "Constant Structure Creep of Copper at 973K," Scripta Metall. 33 (1995), 767-772.

17. J. Rösler and E. Arzt, "The Kinetics of Dislocation Climb Over Hard Particles - I. Climb Without Attractive Particle-Dislocation Interaction," Acta Metall. Mater. 36 (1988), 1043-1051.

18. E. Arzt and J. Rösler, "The Kinetics of Dislocation Climb Over Hard Particles - II. Effects of an Attractive Particle-Dislocation Interaction," Acta Metall. Mater. 36 (1988), 1053-1060.

19. J. Rösler and E. Arzt, "A New Model-Based Creep Equation for Dispersion Strengthened Materials," Acta Metall. Mater. 38 (1990), 671-683.

20. G. J. Lloyd and J. W. Martin, "The Effect of Dislocation Density on the Creep of Dispersion-strengthened Copper Crystals," Mater. Sci. Engr. 46 (1980), 1-13.

21. M. S. Nagorka, G. E. Lucas and C. G. Levi, "Novel Oxide-Dispersion-Strengthened Copper Alloys from Rapidly Solidified Precursors: Part 2, Creep Behavior," Metall. and Mater. Trans. A 26A (1995), 873-881.

22. R. S. Mishra, A. G. Paradkar and K. N. Rao, "Steady State Creep Behavior of a Rapidly Solidified and Further Processed Al-5 wt% Ti Alloy," Acta Metall. Mater. 41 (1993), 2243-2251.

23. A. Orlová and J. Čadek, "On Rösler and Arzt's New Model of Creep in Dispersion Strengthened Alloys," Acta Metall. Mater. 40 (1992), 1865-1871.

24. W. D. Nix, J. C. Gibeling and D. A. Hughes, "Time-Dependent Deformation of Metals," Metall. Trans. 16A (1985), 2215-2224.

25. U. F. Kocks, A. S. Argon and M. F. Ashby, "Thermodynamics and Kinetics of Slip," (Oxford: Pergamon Press, 1975).

26. M. Biberger and J. C. Gibeling, "Analysis of Creep Transients in Pure Metals Following Stress Changes," Acta Metall. 43 (1995), 3247-3260.

27. F. Ernst, P. Pirouz and A. H. Heuer, "HRTEM Study of a Cu/Al_2O_3 Interface," Philos. Mag. 63 (1991), 259-277.

28. S. E. Broyles and J. C. Gibeling, "A New Model for Creep of Dispersion Strengthened Copper Based on Stress Change Experiments," submitted to Acta Materialia.

29. S. E. Broyles, K. R. Anderson, J. R. Groza and J. C. Gibeling, "Creep Deformation in Dispersion-strengthened Copper," Metall. and Mater. Trans. 27A (1996), 1217-1227.

30. M. J. Mills, J. C. Gibeling and W. D. Nix, "A Dislocation Loop Model for Creep of Solid Solutions Based on the Steady State and Transient Creep Properties of Al-5.5 at%Mg," Acta Metall. 33 (1985), 1503-1514.

31. M. Liu and J. Cowley, "Particle/Dislocation Interactions in Dispersion Strengthened Tungsten Alloy at Ultrahigh temperatures," Scripta Metall. Mater. 28 (1993), 307-312.

32. R. J. Lewis, "The Creep and Aging Response of a Rapidly Solidified Al-Fe-V-Si Alloy" (M.S. Thesis, University of California, Davis, 1990).

EFFECT OF THRESHOLD STRESS PROCESSES ON CREEP BEHAVIOR

Farghalli A. Mohamed

Department of Chemical and Biochemical Engineering and Materials Science
University of California, Irvine, CA 92697

ABSTRACT

The operation of a threshold stress process signifies that the high temperature mechanical behavior of the material is not driven by the total applied stress but rather by an effective stress. By incorporating a threshold stress for creep, which depends strongly on temperature, in basic deformation processes proposed for metals and simple alloys, it is demonstrated that the creep behavior of powder metallurgy Al alloys and PM discontinuous SiC-Al composites can be predicted from information and analyses documented for solid- solution alloys.

Modeling the Mechanical Response of
Structural Materials
Edited by E.M. Taleff and R.K. Mahidhara
The Minerals, Metals & Materials Society, 1997

INTRODUCTION

Under conditions of steady-state deformation, the dependence of creep rate, $\dot{\gamma}$, on the applied stress, τ, at low stresses and large grain sizes may generally be represented by an expression of the form [1-3]:

$$\dot{\gamma} = B\left(\frac{\tau}{G}\right)^n \exp\left(-\frac{Q_c}{RT}\right) \quad (1)$$

where B is a constant (sensitive to microstructure), G is the shear modulus, n is the stress exponent, Q_c is the activation energy for creep, R is the gas constant, and T is the absolute temperature. Equation 1 is referred to as power law creep. The validity of Equation 1 has been established for a wide range of materials when the equation is expressed in the following normalized form (the Dorn equation) [2, 3]:

$$\dot{\gamma} = A\frac{DGb}{kT}\left(\frac{\tau}{G}\right)^n \text{ with } D = D_o \exp\left(-\frac{Q}{RT}\right) \quad (2)$$

where A is a dimensionless constant, D is the diffusion coefficient that characterizes the creep process, b is the Burgers vector, k is Boltzmann's constant, D_o is the frequency factor for diffusion, and Q is the activation energy for the diffusion process that controls the creep behavior.

One of the basic characteristics of the creep behavior represented by Equations 1 and 2 is that all the applied stress is responsible for the measured creep rate. However, a number of experimental observations supported by theoretical analyses have suggested that the low-stress creep behavior of some materials may be best described by a modified creep power law which incorporates a threshold stress, τ_o, and which may given by:

$$\dot{\gamma} = A\frac{D_oGb}{kT}\left(\frac{\tau - \tau_o}{G}\right)^n \exp\left(-\frac{Q}{RT}\right) \quad (3)$$

The presence of a threshold stress, τ_o, signifies that an effective stress, rather than the applied stress, is responsible for the measured creep rate. The concept of a threshold stress for creep has recently been adopted to explain the creep characteristics of two types of materials: dispersion strengthened (DS) alloys [4, 5], and micrograin superplastic alloys superplastic ($d <$ 10 μm) [6]. For DS alloys, the interpretation of plastic flow in terms of a threshold stress was centered around some form of interaction between moving dislocation and nanometer-scale dispersion particles [7-10]. For superplastic alloys, the origin of the threshold stress was attributed to strong segregation of impurities at grain boundaries and their interaction with boundary dislocation [11-13], i.e. a threshold stress for boundary sliding exists.

The purpose of this paper is two-fold: (a) to review recent experimental data on the creep the behavior of powder metallurgy Al alloys, and (b) to demonstrate that the creep behavior of these alloys can be predicted by introducing a temperature-dependent threshold stress in the rate controlling equations which describe the creep behavior of solid-solution alloys.

ANALYSIS AND DISCUSSION

Powder metallurgy Al alloys

Powder metallurgy (PM) Al alloys, such as PM 6061 Al and PM 2124 Al, have been used as matrices in the development of discontinuous SiC-Al composites [14]. It has been recognized that creep data on these alloys are essential not only to the characterization of the creep strength of discontinuous SiC-Al composites in terms of deformation processes but also to the assessment of the potential of discontinuous SiC-Al composites for use in high temperature applications. For example, a comparison between the creep behavior of a discontinuous SiC-Al composite and

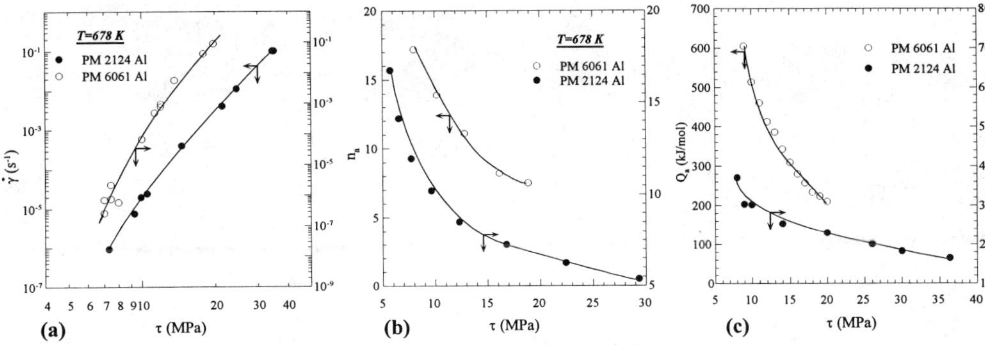

Figure 1. *(a) Apparent creep behavior; (b) apparent stress exponent; and (c) apparent activation energy.*

its unreinforced matrix Al alloy under similar experimental conditions may determine whether the incorporation of SiC particulates or whiskers results in a strengthening effect over the entire stress range used.

Creep Behavior of powder metallurgy Al alloys. In recent investigations [15 ,16], the creep behavior of PM 6061 Al and PM 2124 Al was studied at several temperatures and over six orders of magnitude of strain rate. The results of the investigations have shown [15, 16] that the apparent creep behavior of PM 6061 Al and PM 2124 (Figure 1a) neither resembles that of one of the two Classes of solid solution alloys (Class I or Class II) [1, 2, 17] nor exhibits a transition from Class II to Class I behavior with increasing the applied stress [17-23]. First, the apparent stress exponents for creep in both alloys are higher than those reported for Class I alloys (alloy Class; $n = 3$) and Class II alloys (metal Class; $n = 5$). Second, the apparent stress exponent is not constant but increases with decreasing applied stress (Figure 1b). Third, the apparent activation energy for creep in PM 6061 Al (the major solute element is Mg) and PM 2124 Al (the major elements are Cu and Mg), is not only variable but also much higher (Figure 1c; for PM 6061 Al, $Q_a = 200\text{-}500$ kJ/mole) than that reported for Al-Mg alloys [3, 18, 19] and Al-Cu alloys [21, 22].

The preceding findings indicate that the creep behavior of PM Al alloys is similar to that of dispersion strengthened (DS) alloys [4, 5] with regard to (a) the high values of the stress exponent, (b) the increase in the stress exponent with decreasing the applied stress, and (c) the high values of the activation energy for creep. This similarity suggests that the anomalous stress dependence of creep rate in PM Al alloys, like that in DS alloys and superplastic alloys, can be interpreted in terms of a threshold stress and that the creep behavior can be represented by Equation 3 (modified power law creep).

The above suggestion was examined in detail elsewhere [14-16]. The results of such an examination have indicated that a threshold stress for creep in PM 6061 Al and PM 2124 Al is present; that the creep behavior of both alloys obeys Equation 3; and that the threshold stress for creep is characterized by a temperature dependence which is much stronger than that attributable to the shear modulus and which can be represented by an expression of the form [14-16]:

$$\frac{\tau_o}{G} = B_o \exp\left(\frac{Q_o}{RT}\right) \tag{4}$$

where B_o is a constant, and Q_o is an energy term. While the above equation cannot be accounted for by available threshold stress models [7-10], consideration of creep data on some dispersion-strengthened (DS) alloys, PM 6061 Al, PM 2124 Al, and two PM SiC-Al composites suggests [16, 24] that such an equation may represent general behavior.

Creep substructure in PM Al Alloys. Creep microstructure in PM 2124 Al PM 6061 Al was studied [16, 24] by means of transmission electron microscopy (TEM) and the data obtained have shown the presence of two main features: (a) intermetallic dispersoids and precipitate

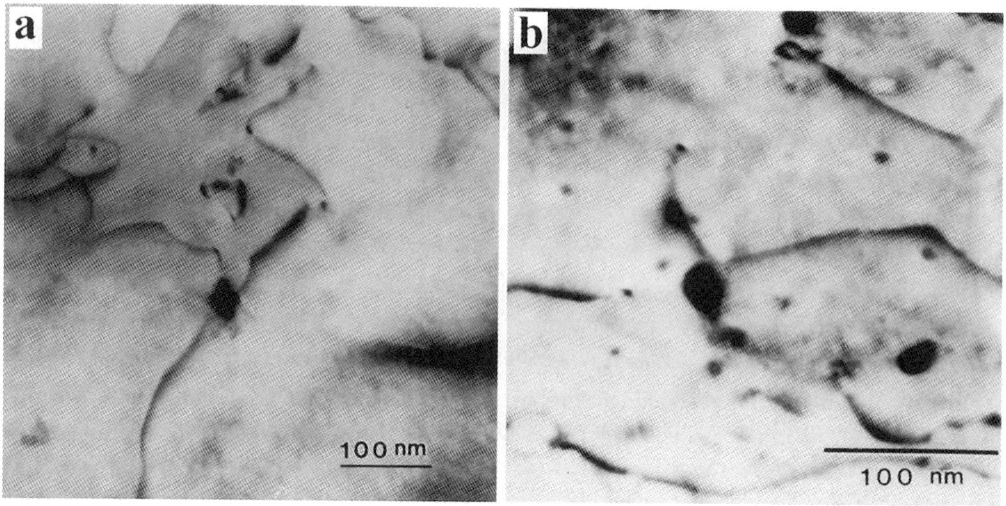

Figure 2. *Typical TEM observations. (a) PM 6061 Al, (b) PM 2124 Al.*

particles that exhibit no significant interaction with moving dislocations, and (b) a distribution of nanometer-scale dispersion particles that interact extensively with dislocations; for example, dislocation attachment to the particles was frequently noted (Figures 2a and 2b). While no effort was made to identify the exact compositions of the intermetallic dispersoids, energy dispersive spectroscopy (EDS) spectra obtained from intermetallic dispersoids both indicate that these dispersoids contain Cu, Mn, and Cu in case of PM 2124 Al and Al, Mg, Cu, Fe, and Si in case of PM 6061 Al.

Consideration of the morphology of the ultra-fine dispersion particles observed in PM 2124 Al and PM 6061 Al along with the processing history of both alloys leads to the conclusion that these particles represent oxides particles that are introduced in the microstructure as a result of processing the alloys by powder metallurgy. It has been suggested [14-16] that the interaction between these oxide particles and dislocations provides the most likely source for the threshold stress for creep in PM 6061 Al and PM 2124 Al, and that the strong temperature dependence of the threshold stress, as described by Equation 4, may be a reflection of an interaction between impurities which are able to segregate at incoherent particles and dislocations which are captured at the detachment side of the particles.

Having reviewed creep behavior and microstructure in PM 2124 and 6061 Al, attention is now focused on the second objective of the paper: the prediction of the creep characteristics of these alloys by using information and analysis documented for the creep behavior of solid-solution alloys.

Creep behavior of solid solution alloys

Significant progress has been made in the interpretation of the creep of solid-solutions alloys [1, 2, 17-23, 25, 26], including Al-based solid-solutions alloys. Such progress is reflected in the following important results including the identification of three regions of deformation [18-23, 25, 26] (region I, region II, and region III) along with two creep transitions in the plot of steady-state creep against applied stress on a logarithmic scale (Figure 3). Information on the creep characteristics of these regions and the rate-controlling equations are provided in Table 1. Also, the criteria describing the two creep transitions are given in Table 2.

Creep behavior of powder metallurgy alloys vs. that of solid solution alloys

It is suggested here that the creep behavior of powder metallurgy Al alloys represents a modified form of the creep behavior of Al-based solid solution alloys, and that such a

TABLE 1. Regions of Deformation in Solid-Solution Alloys.

Region	*Description*	*Rate controlling equation*[*]	*Comments*
I	• Low-stress region • Class II (metal class) behavior • Climb control • Well-developed subgrains	$\dot{\gamma}_I = A_I \frac{D_c G b}{kT}\left(\frac{\tau}{G}\right)^5$ (1) $A_I = B_I\left(\frac{\Gamma}{Gb}\right)^3$	For dilute Al alloys: $Q_c \sim Q_D$ in Al. For solid-solution alloys containing dispersion particles, $\tau = \tau_e$
II	• Intermediate-stress region • Class I (alloy class) behavior • Viscous glide control • Subgrain formation is not extensive	General case $\dot{\gamma}_{II} = \frac{4G}{\Sigma A}\left(\frac{\tau}{G}\right)^3$ (2) Special case (Cottrell-Jawson) $\dot{\gamma}_{II} = 4\left(\frac{kT}{ec^{1/2}Gb^3}\right)\frac{D_g G b}{kT}\left(\frac{\tau}{G}\right)^3$ (3)	For dilute Al alloys: $Q_c \sim Q_s$. For solid-solution alloys containing dispersion particles, $\tau = \tau_e$
III	• High-stress region • May represent a region in which breakaway of dislocations from solute atom atmospheres occurs	Limited data over less than two orders of magnitude of strain rate suggests a creep power law with n ~ 5	For Al-Cu alloys and Al-21%Zn: $Q_c > Q_D$.

* A_I and B_I are dimensionless constants, D_c is the climb diffusivity, G is the shear modulus, b is the Burgers vector, k is Boltzmann's constant, T is the absolute temperature, Γ is the stacking fault energy, ΣA is the sum of the interaction parameters for viscous drag processes, e is the atom misfit parameter, c is the solute concentration, D_g is the glide diffusivity, Q_c is the activation energy for creep, Q_D is the activation energy for self diffusion, and Q_s is the activation energy for solute diffusion.

modification in creep behavior arises from the presence of a threshold stress for creep in the former alloys. This suggestion is based on the following main consideration. Powder metallurgy Al alloys contain several solute elements such as Mg, Cu, Si, Fe, and Mn. Although, according to microstructural data on PM 2124 Al [16, 24] and PM 6061 Al, these element interact with Al or with each other forming intermetallic dispersoids and precipitate particles, it is expected that, in the temperature range used for high temperature deformation, part of some of these element would exist in solid-solution with Al.

The aforementioned suggestion leads to two implications: (a) the creep of powder metallurgy Al alloys and that of Al-based solid solution alloys are governed by the same deformation processes which are mentioned in Tables 1 and 2, and (b) the rate controlling

TABLE 2. Creep Transitions in Solid-Solution Alloys.

Transition	*Description*	*Rate controlling equation*	*Comments*
Region I ↓ *Region II*	Transition from climb to viscous glide with increasing stress.	General case $\frac{kT}{D_g b \Sigma A} = C\left(\frac{\Gamma}{Gb}\right)^3 \frac{D_c}{D_g}\left(\frac{\tau}{G}\right)_t^2$ (4) Special case (Cottrell-Jawson) $\left(\frac{kT}{ec^{1/2}Gb^3}\right)^2 = C\left(\frac{\Gamma}{Gb}\right)^3 \frac{D_c}{D_g}\left(\frac{\tau}{G}\right)_t^2$ (5)	For solid-solution alloys containing dispersion particles, $\tau = \tau_e$
Region II ↓ *Region III*	Represents the breakaway of dislocations from solute atom atmospheres.	$\left(\frac{\tau}{G}\right)_t = 0.05ce^2\left(\frac{Gb^3}{kT}\right)$ (6)	For solid-solution alloys containing dispersion particles, $\tau = \tau_e$

equations proposed to describe the behavior of solid-solution alloys in the deformation regions shown in Figure 3 can also be used to account for the creep behavior of powder metallurgy Al alloys provided that the applied stress in these equations (Tables 1 and 2) is replaced by an effective stress, τ_e ($\tau_e = \tau - \tau_o$); the threshold stress in the above equations is given by Equation 4.

Predictions for creep behavior in PM Al alloys

On the basis of the preceding discussion, it is possible to predict several cases for steady-state creep behavior in PM Al alloys. Primary among these cases are the following:

Case I. For this case, the creep behavior exhibits the following characteristics: (i) both the apparent stress exponent and the apparent activation energy are high and increase with decreasing applied stress, (ii) the plots of $\dot{\gamma}^{1/n}$ vs. τ on a double linear scale lead to a straight line fit only when n = 4.5 (or 5); for n = 3, n =7, n = 8 or n = 10, the data of the plot would exhibit curvature, and (iii) when experimental data obtained are corrected for the effect of the threshold stress and its dependence on temperature, a stress exponent of about 5 and an activation energy which is close to that for self diffusion are obtained. This type of behavior is observed when the upper and lower bounds of creep rates correspond to points *a* and *b*, respectively in Figure 3, i.e., the true creep behavior is controlled by dislocation climb (lattice diffusion) [1, 3, 17, 27].

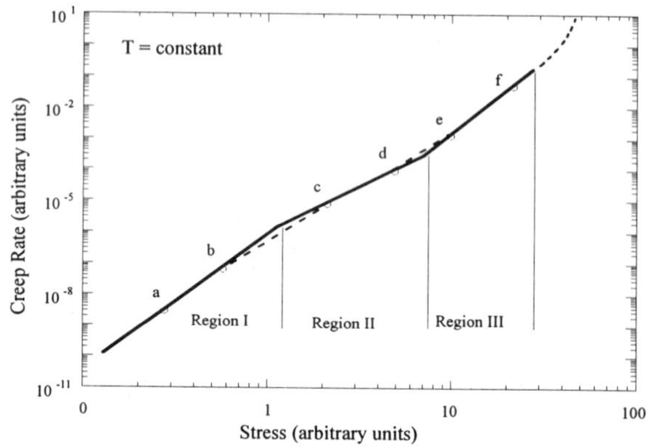

Figure 3. *Schematic presentation of creep behavior in Al-based solid solution alloys.*

Case II. This case is associated with the same trends mentioned in case I with the exception that the plots of $\dot{\gamma}^{1/n}$ vs. τ on a double linear scale lead to a straight line fit only when n = 3; for n = 5 and n = 8, the data of the plot would exhibit curvature. Also, when experimental data obtained are corrected for the effect of the threshold stress and its dependence on temperature, a stress exponent of 3 and an activation energy which is close to that for solute diffusion are obtained. These characteristics are noted under the condition that the upper and lower bounds of creep rates are points *c* and *d* (Figure 3), respectively, i.e., the true creep behavior is controlled by viscous glide [1, 2, 17, 28].

Case III. This case leads to characteristics identical with those associated with case I with the exception that correcting the experimental data for the effect of the threshold stress and its dependence on temperature may result in a true activation energy which is higher than that for self diffusion. According to Figure 3, these characteristics are noted when the upper and lower bounds of creep rates are points *e* and *f*, respectively (the true creep behavior is related to region III).

Case IV. While the creep characteristics associated with this case are similar in trend to those related to the above cases with regard to the variation in the apparent stress exponent and the apparent activation energy with applied stress, a major difference exists. Such a difference would be reflected in the plots of $\dot{\gamma}^{1/n}$ vs. τ on a double linear scale: for all selected values of the true stress exponent used (n = 3, 5, 7, 8 or 10), the data of the plot would show curvature. The characteristics of this case are observed under the condition that two or more creep processes contribute to the overall creep rate. This situation is encountered, for example, when: (a) the upper and lower bounds of creep rates are points *b* and *c*, respectively (the true creep behavior arises from the operation of both viscous glide and climb) or (b) the upper and lower bounds of

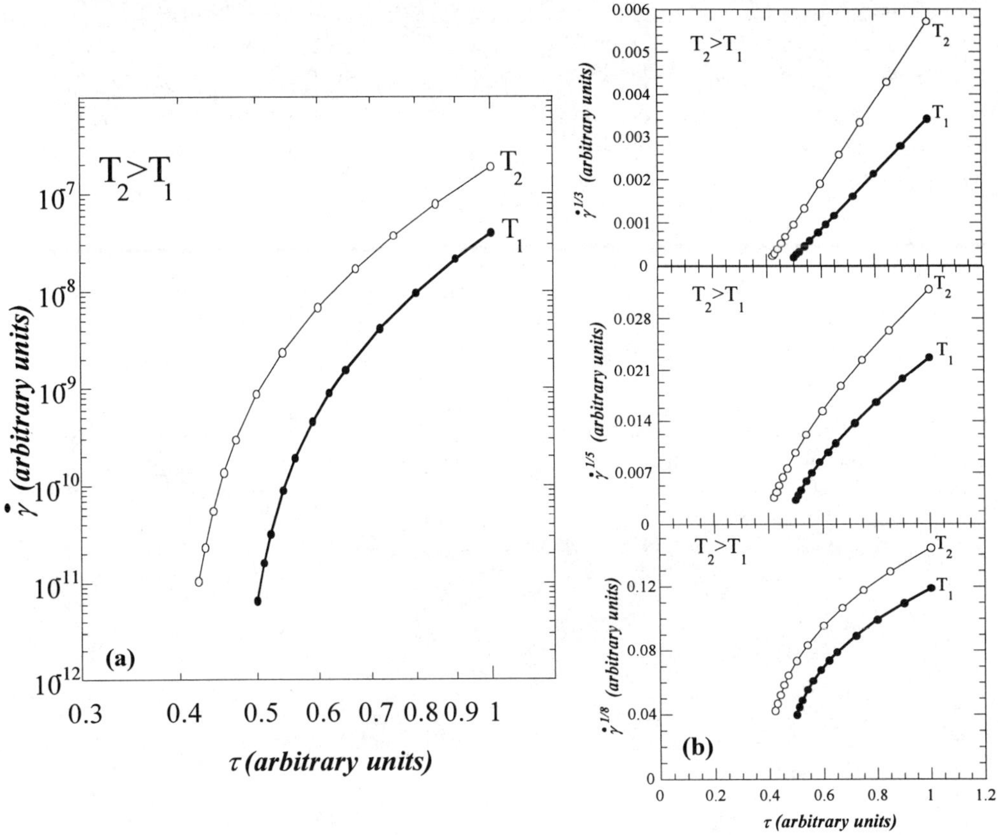

Figure 4. *Predictions of creep characteristics of PM Al alloys under the condition that the creep behavior is controlled by a single deformation mechanism (in this case n = 3). (a) shear stress vs. shear strain on a logarithmic scale, (b) double linear scale: (n = 3; n = 5; and n = 8).*

creep rates are points *d* and *e*, respectively (the creep behavior reflects the presence of two deformation regions II and III). It is clear that for both situations, analyses based on mechanical data are not sufficient to unambiguously identify the dominant creep processes. In this case, systematic data on creep substructure over the experimental ranges of the investigation is needed to provide guiding information that shed additional light on the origin of deformation mechanisms. For example, a transition in creep microstructure from subgrain formation to random distribution of dislocation with increasing stress indicates [25, 29] that the creep behavior of the materials is controlled by the sequential processes of viscous glide and dislocation climb.

Figure 4 serves as an illustration for cases I, II, and III, respectively (one of the plots of $\dot{\gamma}^{1/n}$ vs. τ on a double linear scale leads to a straight line fit for a particular value of *n*) while Figure 5 depicts an example for case IV (the plots of $\dot{\gamma}^{1/n}$ vs. τ on a double linear scale exhibit curvature for all vales of *n*). The results shown in these two figures (and other figures that are not reported for the sake of brevity) were obtained by applying the following considerations and conditions:

(a) The values of A_I (region I; 2.8×10^7), A_{II} (region II; 8), A_{III} (region III; 8×10^7), Q_I (142 kJ/mol), Q_{II} (142 kJ/mol), and Q_{III} (160 kJ/mol) are in the range of those reported for various Al-based solid solution alloys [18-23, 25, 26].

(b) The shear modulus was calculated from the equation available for Al [3].

(c) The temperature dependence of the threshold stress is given by Equation 4. Q_o is 20 kJ/mol, a value which falls in the range of Q_o measured for various alloys [16, 24], and B_o was judicially selected to govern the lower bound of the strain rate range.

(d) The total creep rate of an Al alloy $\dot{\gamma}_t$ in regions I (viscous glide controlled) and II (climb-controlled) is represented by the sequential summation [30, 31] of Equations 1 and 3 given in Table 1:

$$\dot{\gamma}_t = \frac{\dot{\gamma}_I \dot{\gamma}_{II}}{\dot{\gamma}_I + \dot{\gamma}_{II}} \tag{5}$$

(e) In the absence of more definitive information regarding the details of the process controlling region III, it is assumed that such a process and the process of viscous glide (controlling region II) operate independently.

$$\dot{\gamma}_t = \dot{\gamma}_{II} + \dot{\gamma}_{III} \tag{6}$$

(f) Under the operation of a modified creep process that incorporates a threshold stress, the relation between the apparent stress exponent, n_a, the true stress exponent, n, applied stress, τ, and the threshold stress, τ_o, is given by [12]:

$$n_a = \frac{n}{(1 - \tau_o/\tau)} \tag{7}$$

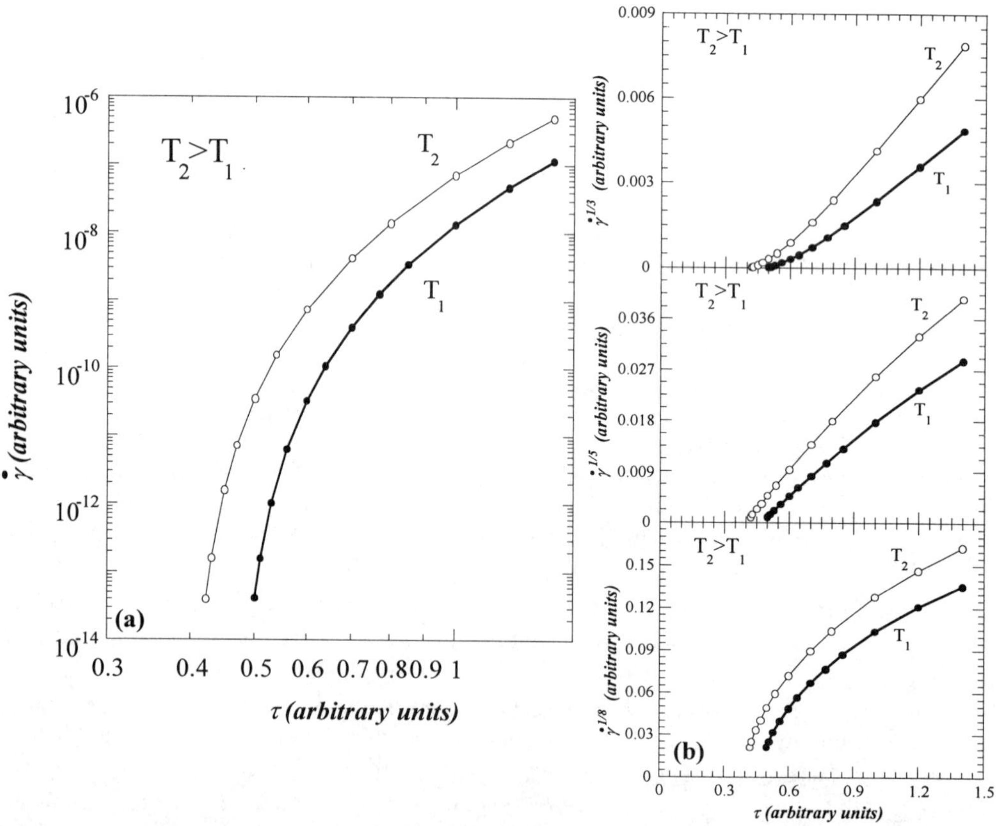

Figure 5. *Predictions of creep characteristics of PM Al alloys under the condition that the creep behavior is controlled by two deformation mechanisms (in this case, sequential interaction between n = 3 and n = 5 occurs). (a) shear stress vs. shear strain on a logarithmic scale, (b) double linear scale: (n = 3; n = 5; and n = 8).*

Also, when the temperature dependence of the threshold stress is represented by Equation 8, the apparent activation energy, Q_a, can be expressed in terms of Q_c, n, τ, τ_o and Q_o according to [14-16]:

$$Q_a = Q_c - \frac{nRT^2}{G}\frac{\partial G}{\partial T}\left(1+\frac{1}{\tau/\tau_o}\right)+\frac{nQ_o}{\tau/\tau_o - 1} \qquad (8)$$

Experimental trends

In this paper, only the experimental data of PM 6061 Al [15] are considered. These data are replotted in Figure 6(a) as shear strain rate, $\dot{\gamma}$, against effective stress, $\tau - \tau_o$, on a logarithmic scale; for 646 K and 678 K, τ_o = 8.21 MPa and 5 MPa, respectively [15]. Consideration of such a plot in the light of available information on the creep of solid-solution alloys (see Tables 1 and 2), suggests that the creep behavior of PM 6061 Al is consistent with that of a solid solution alloy which exhibits a transition from region II (viscous glide creep; n = 3) to region III (n >3). Two approaches are adopted below to further check this consistency.

An approximate estimate of the activation energy for creep from the separation of the low-stress data (n = 3) yields 147 kJ/mol. This value is in good agreement with those reported for Al-Mg alloys in the viscous glide region [18, 25, 29]. On the other hand, the activation energy inferred from the high-stress data (n > 3) is about 180 kJ/mol, a value which fall in the range of energies reported for Al-based solid-solution alloys in region III [20-22] (160-200 kJ/mol).

For solid-solution alloys, the transition in creep behavior from region II (viscous glide; n = 3) to region III (Table 2) has been accounted for by the breakaway of dislocations from their solute atom atmospheres [19-22, 26, 32]. The change in stress exponent (from 3 to 5) in PM 6061 Al is examined in the light of this explanation by applying Equation (6) (Table 2) under the following conditions: (a) the applied stress is replaced by an effective stress, τ_e, ($\tau_e = \tau - \tau_o$), (b) the transition effective stresses, τ_e, are approximately [Figure 6(a)] 3.1 MPa and 3.6 MPa for 678 K and 648 K, respectively; (b) the major solute element in PM 6061 Al is Mg and accordingly, e (the atom misfit parameter) = 0.12 [17]; and (c) the shear modulus is calculated from the equation available for Al [3]. Using the above conditions and values in Equation (6) gives c = 0.5 at. %. The value of c is less than the Mg content of the alloy (1.0 -1.2 at. %) but is in agreement with the inference that not all of Mg is in solid-solution with Al; available information [24] indicates the presence in 6061 Al of intermetallic spheroids that contain Mg.

Having demonstrated that the trend of experimental data on PM 6061 Al is consistent with a transition from a region controlled by viscous glide to a high-stress region related to the

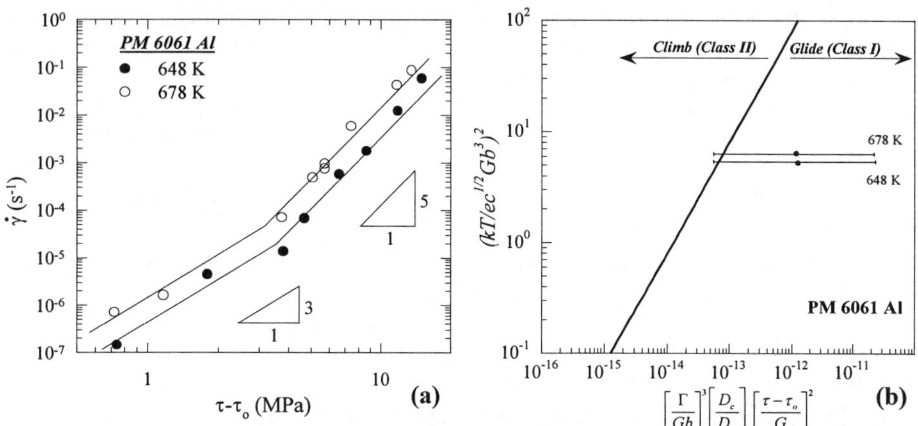

Figure 6. (a) Creep data on PM 6061 Al plotted as shear strain rate against effective stress on a logarithmic scale. (b) The modified glide climb criterion in terms of an effective stress. Solid circles represent the transition to the high stress region (region III).

breakaway of dislocations from solute atom atmospheres, attention is now focused on examining the correlation between the data and the prediction of the viscous glide/climb criterion [17, 23] for deformation in solid-solution alloys. In Figure 6(b), the modified form of Equation 6 (Table 2) is depicted graphically by plotting $(kT/ec^{1/2}Gb^3)^2$ vs. $(\Gamma/Gb)^3 D_c/D_g (\tau-\tau_o/G)^2$ on a logarithmic scale. On this form of plot, the equation is represented by a solid line whose slope is 1; the position of the line was determined from creep on Al-Mg alloys [17, 20-23, 33]. The effective stress range covering the experimental data of the alloy along with the transition point from n =3 to n > 3 was superimposed for each test temperature on Figure 6(b) using the following values: e = 0.12, c = 0.005, Γ = 190 mJ/m^2 [17], D_c / D_g =1 [17]. It is clear that the experimental data fall in the viscous glide region; data obtained at effective stresses higher than the transition effective stress between region II (n = 3) and region III (n > 3) (identified as a solid circle in the figure) are in the high-stress region whose advent is related to the breakaway of dislocations from solute atom atmospheres.

The preceding discussion along with consideration of Figure 6(b) raises the possibility of the occurrence of a transition in the creep behavior of PM Al alloys from that of Class I alloys (alloy Class, viscous glide control) to that of Class II alloys (metal Class, climb control). Such a transition was observed in Al-based solid-solution alloys [17-23, 25] when the applied stress was decreased below a critical value. In this case, it has been found that the transition in creep behavior is determined by a favorable combination of experimental conditions such as temperature and material parameters such as the atom misfit parameter and the stacking fault energy. However, in case of PM Al alloys, an additional parameter is expected to play an important role in deciding whether the transition from Class I behavior to Class II behavior would occur during the creep of these materials. Such a parameter is related to oxide particles that result from preparing PM Al alloys by powder metallurgy. The above prediction is equally valid in case of the creep of PM discontinuous SiC (or other ceramic reinforcement) Al composites partly because the microstructure of these composites, like that of PM Al alloys, contain incoherent oxide particles as a result of processing by powder metallurgy and partly because, according to experimental evidence [14, 34-36], deformation in the matrix Al alloy controls the creep of the composite.

SUMMARY

(i) It is suggested that the creep behavior of powder metallurgy Al alloys represents a modified form of the creep behavior of Al-based solid solution alloys, and that such a modification in creep behavior arises from the presence of a threshold stress for creep, τ_o, in the former alloys; the origin of the threshold stress is related to the interaction between moving dislocations and oxides particles which are present in Al alloys as a result of processing the alloys by powder metallurgy. The aforementioned suggestion leads to two implications. First, the rate-controlling equations proposed to describe the behavior of solid-solution alloys can be used to account for the creep behavior of powder metallurgy Al alloys provided that the applied stress, τ, in these equations is replaced by an effective stress, τ_e ($\tau_e = \tau - \tau_o$). Second, the creep characteristics of PM Al alloys can be predicted from information and analysis documented for the creep behavior of solid solution alloys.

(ii) It is predicted that a transition in the creep behavior of PM Al alloys from that of Class I (alloy Class) to that of Class II (metal Class) may occur when the effective stress is reduced below a critical value. This critical value for the effective stress is in part determined by the value of the threshold stress for creep which in turn is controlled by the volume fraction of oxide particles introduced in PM alloys.

(iii) Present analysis should also be applicable to the description of the creep behavior of PM discontinuous SiC (or other ceramic reinforcement) Al composites for two reasons. First, the microstructure of these composites, like that of PM Al alloys, contain incoherent oxide particles as a result of processing by powder metallurgy. Second, according to experimental evidence, deformation in the matrix Al alloy controls the creep of the composite.

ACKNOWLEDGMENTS

This work was supported in part by the National Science Foundation under Grants No. DMR-9500316 and in part by the Army Office under Grant No. DAAH04-95-1-0424. Thanks are extended to my graduate students, Ahmadali Yousefiani and Zhigang Lin, for their assistance.

REFERENCES

1. O.D. Sherby and P.M. Burke, Prog. Mater. Sci. 13 (1968) 325.
2. A.K. Mukherjee, J.E. Bird and J.E. Dorn, Trans. Am. Soc. Metals, 62 (1969) 155.
3. J.E. Bird, A.K. Mukherjee and J.E. Dorn, in Quantitative Relation Between Properties and Microstructure (D.G. Brandon and A. Rosen, eds., Israel Univ. Press, Jerusalem 1969), 255.
4. R. A. Wilcox and A. H. Clauer, Acta Metall., 20 (1971) 743.
5. R. W. Lund and W. D. Nix, Acta Metall., 24 (1976) 469
6. O. D. Sherby and J. Wadsworth: Progress Mater. Sci., 1989, vol. 33, pp. 169-221.
7. E Arzt and D. S. Wilkinson, Acta Metall., 34 (1986) 1893.
8. E Arzt and J. Rosler, Acta Metall., 36 (1988) 1053.
9. E. Orowan, Dislocations in Metals (M. Cohen, ed., AIME, New York, 1954), 1954.
10. E. Arzt and M. F. Ashby, Scripta Metall., 16 (1983)582.
11. F. A. Mohamed, J. Mater. Sci.18 (1980) 1061.
12. P. K. Chaudhury and F .A. Mohamed, Acta Metall. 36 (1988) 1099.
13. P. K. Chaudhury, K. T. Park, and F .A. Mohamed, Metall. Trans. 25A (1994) 2391.
14. F. A. Mohamed, K. T. Park and E. J. Lavernia, Mater. Sci. Eng., A150 (1992) 21.
15. K.T. Park, E.J. Lavernia and F.A. Mohamed, Acta Metall.Mater., 42 (1994) 677.
16. Y. Li, S.R. Nutt and F.A. Mohamed, Acta Metall.Mater., 45 (1997) 2607.
17. F.A. Mohamed and T.G. Langdon, Acta Metall., 22 (1974) 779.
18. K.L. Murty, F.A. Mohamed and J.E. Dorn, Acta Metall., 20 (1972) 1009.
19. K.L. Murty, Scr. Metall.,7 (1973) 899.
20. M.S. Soliman and F.A. Mohamed, Metall. Trans. A, 15 (1984) 1893.
21. P. K. Chaudhury and F.A. Mohamed, Metal. Trans. A, 18 (1987) 2105.
22. P. K. Chaudhury and F.A. Mohamed, Mater. Sci. Eng., 101 (1988) 13.
23. F.A. Mohamed, Mater. Sci. Eng., 61 (1983) 165.
24. F. A. Mohamed, Mater. Sci. Eng(in press).
25. P. Yavari, F.A. Mohamed and T.G. Langdon, Acta. Metall., 29 (1981) 1495.
26. P. Yavari and T.G. Langdon, Acta Metall., 30 (1982) 2181. 27.
27. J. Weertman, J. Appl. Phys., 28 (1955) 1213
28. J. Weertman, J. Appl. Phys., 28 (1957) 1185.
29. M.S. Soliman and F.A. Mohamed, Mater. Sci. Eng.55 (1982) 111.
30. F. A. Mohamed, J. Mater.Sci., 16 (1981) 2890.
31. T.G. Langdon and F.A. Mohamed, J. Aust. Inst. Met. 22 (1977) 189.
32. F.A. Mohamed, Mater. Sci. Eng., 38 (1979) 73.
33. F.A. Mohamed, Metall. Trans. A, 9 (1978) 1013.
34. Y. Li and F. A. Mohamed, Acta Mater. (in press).
35. K. T. Park and F. A. Mohamed, Metal. Mater. Trans. A, 26 (1995) 3119.
36. K. T. Park, E. J. Lavernia and F. A. Mohamed, Acta Metall., 38 (1990) 2149.

A MICROSTRUCTURAL MODEL FOR PRIMARY AND STEADY–STATE CREEP OF PARTICLE STRENGTHENED SUPERALLOYS

M. Heilmaier

Institute of Solid State and Materials Research Dresden, Institute of Metallic Materials, P. O. Box 27 00 16, D–01171 Dresden; Germany

Abstract

The Haasen–Alexander–Ilschner (HAI–) theory of internal and effective stress is combined with the modified threshold stress (σ_p–) concept in order to describe steady–state as well as transient creep of particle–strengthened nickel–base superalloys under constant stress (σ) conditions. In single–phase material the classical Taylor approach of a homogeneous dislocation distribution enables the prediction of the measured creep behavior via the long–range back stress σ_ρ. Thus, the dislocation density ρ and its microstructural evolution during creep plays the major role in the model. Second–phase particles influence creep in a 2–fold manner via reducing σ_{eff}, namely directly by the stress for particle overcoming, and indirectly by increasing ρ. The proposed concept is exemplified with IN 738 LC and MA 754, which represent typical nickel–base superalloys strengthened by coherent γ'–precipitates and incoherent oxide dispersoids, respectively. The model accounts for the observed pronounced normal primary creep as well as it correctly reflects the steady–state relationship of $\dot{\varepsilon}$ and σ at large plastic strains.

Modeling the Mechanical Response of
Structural Materials
Edited by E.M. Taleff and R.K. Mahidhara
The Minerals, Metals & Materials Society, 1997

Introduction

A comprehensive description of the mechanical response of structural materials requires the knowledge of appropriate constitutive laws establishing the relationship between the basic deformation parameters, materials constants and microstructure [1, 2]. Under high temperature creep conditions metallic materials approach a dynamic equilibrium of strain hardening and recovery, called steady–state. The resulting transient, however, by which steady–state will be achieved depends on the kinetic and evolution of internal back stresses involved. In the present work a simple model, which has been applied on a single–phase nickel–base alloy [3], is extended to account for the influence of particle–strengthening in superalloys on creep.

The HAI–Model

Basic features

The derivation of a kinetic law of plastic deformation starts from the fundamental Orowan equation:

$$\dot{\varepsilon} = \frac{b}{M}\, \rho_m\, v\,(\sigma_{\text{eff}}) \tag{1}$$

where ρ_m is the mean density of mobile dislocations and v their average dislocation velocity; b Burgers vector and M Taylor factor are material constants. According to **H**aasen and **A**lexander [4] and **I**lschner [5] the effective stress σ_{eff}, which controls the thermally–activated motion of dislocations, can be simply expressed as the difference between the applied stress σ and the athermal internal stress σ_{i}. In the original formulation of the effective stress model σ_{i} is identified with the long–range stress field σ_ρ of single dislocations. However, under many deformation conditions subgrains evolve during plastic deformation. Consequently, a back stress $\sigma_{\text{b}}^{\text{sub}}$ due to the evolution of that heterogeneous dislocation structure consisting of hard subgrains and soft subgrain interior should be taken into account [2]. In order to extend the model to particle–strengthened materials we further modify the effective stress model by introducing a threshold stress for particles σ_{p} to yield

$$\sigma_{\text{eff}} = \sigma - \sigma_{\text{i}} = \sigma - \sigma_\rho - \sigma_{\text{b}}^{\text{sub}} - \sigma_{\text{p}}\,. \tag{2}$$

Back stresses due to dislocations

The athermal stress component σ_ρ of free dislocations is defined by

$$\sigma_\rho = \alpha G\, b\, M\, \sqrt{\rho}\ , \tag{3}$$

with α elastic interaction constant, G shear modulus, b Burgers vector, M Taylor factor, ρ the mean total dislocation density. The amount of back stress $\sigma_{\text{b}}^{\text{sub}}$ exerted from subgrain walls into the subgrain interior can be calculated to

$$\sigma_{\text{b}}^{\text{sub}} = \frac{f_{\text{h}}}{1 - f_{\text{h}}}\, \sigma\,(k_{\text{h}} - 1), \tag{4}$$

with $f_{\text{h}} = f_{\text{sub}} \cdot 2a/w \approx 0.05$ as a reasonable value for a fully developed substructure; w is the subgrain size, a is the width of subgrain boundaries and f_{sub} is the volume fraction of subgrain–containing regions, respectively.

Particle hardening contribution

The classical threshold stress concept [6] assumes that no significant influence of the particles occurs on the rate of work hardening or more specifically on the kinetics of dislocation slip. In other words simply a constant extra stress is necessary to account for the hardening contribution of the respective particle species present in the material. However, compared to single–phase material the following microstructural evidences clearly indicate that this simplified approach has to be modified: (a) the $\dot{\varepsilon}$–decrease during primary creep is extremely high, usually several orders of magnitude, see the solid curves in figs. 1a and 4a; (b) the steady–state dislocation density ρ_{ss} is increased at equivalent strain rates $\dot{\varepsilon}$; and (c) in constant strain rate tests the yield stress increment at zero plastic strain ΔR_{p0} is significantly lower than the stress increment at steady–state σ_p [7].
Hence we follow Ashby [8], who introduced the idea of additional "geometrically necessary dislocations" ρ_p during plastically non–homogeneous deformation in metals containing hard particles in order to avoid hole formation around the particles. We suggest the effect of particle–strengthening on creep to act twofold: firstly directly via a constant σ_p^*–value representing ΔR_{p0} as the contribution for the mere particle overcoming process (explaining observation c) and secondly due to a back stress $\sigma_{p,\rho}$–term reflecting the evolution of additional dislocations during plastic deformation (satisfying observation b):

$$\sigma_p = \sigma_p^* + \sigma_{p,\rho}\,. \tag{5}$$

However, closer examination in the case studies discussed below will reveal that the evolution of $\sigma_{p,\rho}$ depends sensitively on the lattice type of the particles compared to the matrix.

Dislocation velocity

The dependence of dislocation velocity v on stress has been treated differently in the literature. In contrast to the Kocks–Mecking–Estrin school [1, 9] which favors a potential law, we choose:

$$v \;=\; v_0 \;\; \sinh(\beta\,\sigma_{\text{eff}}) \tag{6}$$

where the parameter $v_0 = \frac{20 \cdot D}{b}$ is dependent on temperature via the diffusion coefficient D [10] and $\beta = \frac{b \cdot \Delta a}{M\,k\,T}$ is related linearly to the activation area for creep Δa. Both parameters can be determined from the materials creep response immediately after an immediate change of stress provided that it has been deformed in a steady–state condition before. In this case we may assume a constant dislocation structure ($\rho_1 = \rho_2, \sigma_b^{\text{sub}} = const.$). Heilmaier et al. [3] have analysed the transient creep after stress changes in the single–phase matrix material Nimonic 75 at 850 °C using equs. 1 and 6. A satisfactory description of the creep response yields $\alpha = 0.2$ and $\beta = 0.08$ MPa^{-1} for given materials constants, $M = 3, G = 55.8$ GPa and $b = 0.254$ nm. The prefactor v_0 can be calculated analogously to be $v_0 = 2 \cdot 10^{-9}$ m/s. Unless otherwise stated these parameters have been applied for the modeling of the corresponding particle–strengthened materials in the following sections assuming equivalent dislocation kinetics to operate in the matrix between the particles.

Dislocation density

The steady–state dislocation spacings $\rho_{ss}^{-0.5}$ (i.e. inverse square root of the dislocation density ρ) have been determined by TEM in previous works [3, 11]. They commonly follow an inverse stress–proportionality [2]:

$$\rho_{ss}^{-0.5} \;=\; k_\rho\, b\, G\,/\sigma\;, \tag{7}$$

with $k_\rho = 1 \ldots 2$ for nickel–base alloys [3, 11]. Using this experimentally established relation the calculation of the Taylor back stress σ_ρ according to equ. 3 is straightforward.

Creep deformation law

Inserting equs. 3 to 5 into equ. 2 and, finally, combining it with equs. 6 and 1 we obtain the deformation law for particle–strengthened materials:

$$\dot{\varepsilon}_{\mathrm{HAI}} = \frac{\rho\, b\, v_0}{M} \sinh(\beta\,(\sigma - \alpha\, G\, b\, M\, \sqrt{\rho} - \frac{f_{\mathrm{h}}}{1 - f_{\mathrm{h}}}\, \sigma\,(k_{\mathrm{h}} - 1) - \sigma_{\mathrm{p}}^{*} - \sigma_{\mathrm{p},\rho}))\,. \tag{8}$$

Law of structural evolution

To complete the model for transient deformation one has to choose an appropriate law for structural evolution. As for many other diffusion–controlled processes at high temperatures we use a Johnson–Mehl–Avrami kinetic:

$$x = x_{ss} - (x_{ss} - x_0)\, \exp(-\, \varepsilon_{pl}\, /\, k)^q \tag{9}$$

where x_0 and x_{ss} are the initial and the steady–state values, respectively, representing $\rho^{-0.5}, \rho_{\mathrm{p}}^{-0.5}, f_{\mathrm{h}}, a$. k, q are rate constants which, however, must be evaluated by simultaneously fitting to the measured creep curves.

Case study 1: stable oxide dispersoids in MA 754

ODS MA 754 is essentially a nickel–chromium solid solution additionally strengthened by incoherent Y_2O_3 dispersoids. The MA process enables an extremely fine distribution of the thermally stable particles resulting in mean values of diameter d = 14 nm, volume fraction $f = 1.0\,\%$, and interparticle spacing L_p = 83 nm [12]. The final recrystallisation heat treatment yields on the one hand coarse elongated grains with typical dimensions of 1.5 mm in longitudinal direction and 0.12 mm perpendicular to that. On the other hand it leads to the formation of a strong {110}<100> cube on edge texture. Thus, $M = 2.45$ is applied for modeling.

Fig. 1a displays two measured tensile creep curves of MA 754 (solid lines) at 850 ^{o}C for constant true stresses of σ = 219 and 192 MPa yielding σ_{p} = 205 and 184 MPa, respectively. Post mortem TEM observations have revealed that the dispersoids cause slip dispersal in the matrix during creep. In other words nearly no subgrain formation was observed even after extensive compressive deformation up to 30 % of plastic creep strain, see fig. 2. Consequently, we set $\sigma_{\mathrm{b}}^{\mathrm{sub}} = 0$ in eqn. 8.

The corresponding model curves according to equ. 8 and the stress contributions $\sigma_\rho, \sigma_{\mathrm{p},\rho}, \sigma_{\mathrm{eff}}$ are plotted as dashed and dotted lines in figs. 1a and b, respectively. According to the yield stress increment measurements in [7] a constant fraction of $\sigma_{\mathrm{p}}^{*} = 0.6 \cdot \sigma_{\mathrm{p}}$ has been set. In fig. 1b $\sigma_{\mathrm{p},\rho}$ approaches its steady–state value of $0.4 \cdot \sigma_{\mathrm{p}}$ analogously to σ_ρ ($\equiv \alpha G b M \sqrt{\rho}$), but starts from zero using a different Avrami kinetic for the additional dislocation density ρ_{p}. Note the effect of the higher applied creep stress σ on the stress contributions of the HAI–model: besides the increased value for the particle overcoming (σ_{p}^{*},not plotted here), both dislocation terms ($\sigma_\rho, \sigma_{\mathrm{p},\rho}$) are also higher, compare the respective dashed and dotted curves in fig. 1b. Consequently, the effective stress σ_{eff} for dislocation movement is lower in the case of lower applied creep stress σ. In turn, the shape of the creep curves in fig. 1a follow the course of σ_{eff}: obviously the dashed and dotted curves describe the pronounced

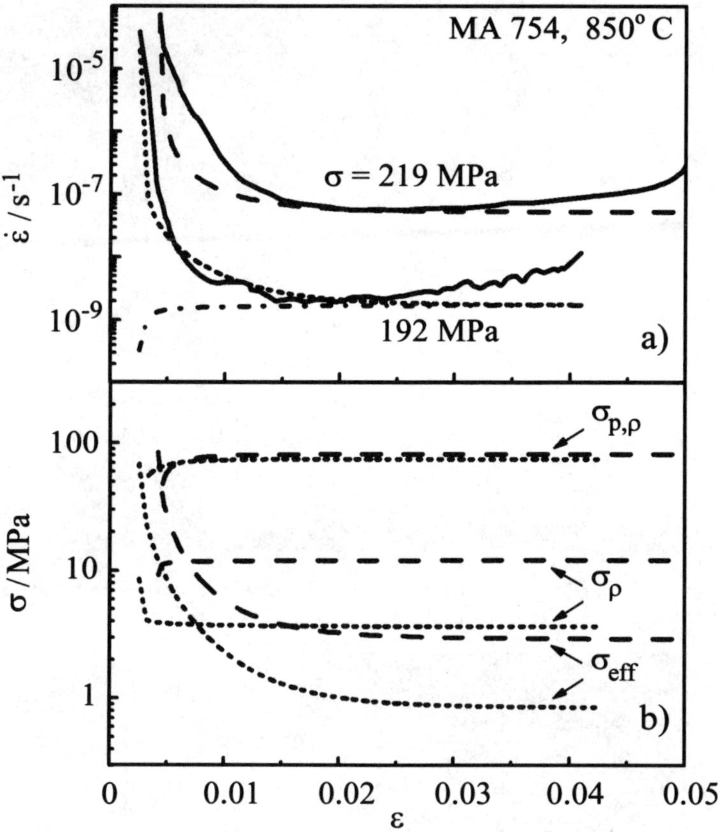

Figure 1: a) Differentiated creep curves of MA 754. Continous lines: measured; dashed and dotted lines: calculated according to the particle–modified HAI–model, equ. 8; b) stress contributions.

normal transients as well as the steady–state creep level (solid lines) adequately. In contrast, applying the original threshold stress concept without splitting σ_p into σ_p^* and $\sigma_{p,\rho}$ [6] the dash–dotted model curve for $\sigma = 192\,\mathrm{MPa}$ reflects inverse creep which has been observed so far only for single–phase Nimonic 75 at low stresses [3].

Case study 2: unstable γ'–precipitates in IN 738 LC

IN 738 LC is a conventionally cast nickel–base superalloy typically used for turbine blades in land–based gas turbines. The enhanced Al– und Ti–content leads to the precipitation of a large volume fraction ($f = 42\,\%$) of the ordered intermetallic phase $L1_2$–$Ni_3(Al,Ti)$. These γ'–particles are coherent and exhibit an average particle diameter $d_{\gamma'}$ in the fully heat–treated condition of $\approx 0.1\,\mu m$; the average grain size of IN 738 LC amounts to about 0.5 mm. During creep dislocation motion is mainly concentrated in the narrow matrix channels. In fact, subgrain formation is suppressed in IN 738 LC as the channel width (or interparticle spacing) L_p is always about one order of magnitude lower than the subgrain size w [11]. Thus, $\sigma_b^{sub} = 0$ in equ. 8. However, TEM observations indicate a particular variant of the particle–induced back stress $\sigma_{p,\rho}$ introduced in the previous section due to the coherent γ'–particles, fig. 3a. The shell model [13] accounts for the built–up

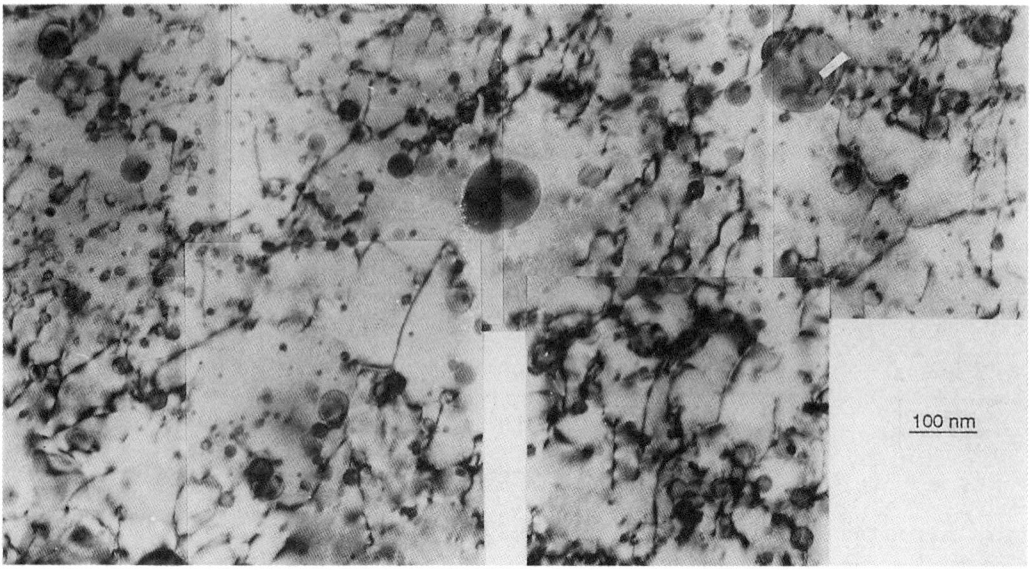

Figure 2: Bright field TEM micrograph of particle and dislocation microstructure in MA 754 evolved after compressive deformation of $\varepsilon = 0.3$ at $\dot{\varepsilon} = 0.01/s$.

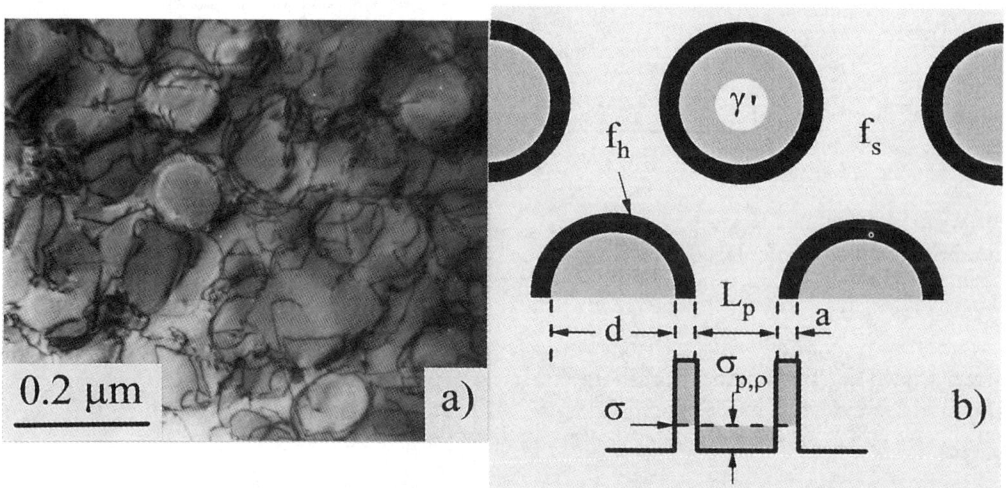

Figure 3: Particle and dislocation microstructure in IN 738 LC: a) TEM micrograph of creep deformed IN 738 LC, $\sigma = 100\,\mathrm{MPa}$, $\varepsilon = 0.02$; b) scheme of the shell model accounting for the evolution of an long–range back stress in the matrix channels due to hard dislocation networks around particles.

of a dislocation network of thickness a at the particle–matrix interface during proceeding plastic deformation arising from coherency/misfit dislocations, geometrically necessary dislocations [8] and glide dislocations, see fig. 3b. Thus, the dislocation arrangement is considered to be heterogeneous consisting of hard shells of volume fraction $f_{\mathrm{h},\gamma'}$ and soft channels of volume fraction f_{s}. Consequently, we evaluate $\sigma_{\mathrm{p},\rho}$ analogously to $\sigma_{\mathrm{b}}^{\mathrm{sub}}$ in equ.

4 simply replacing f_h by $f_{h,\gamma'}$. $f_{h,\gamma'}$ can be calculated by a third order polynomial of $\frac{a}{d_{\gamma'}}$ assuming a spherical shape of the γ'–particles [13]. For a equ. 9 applies.

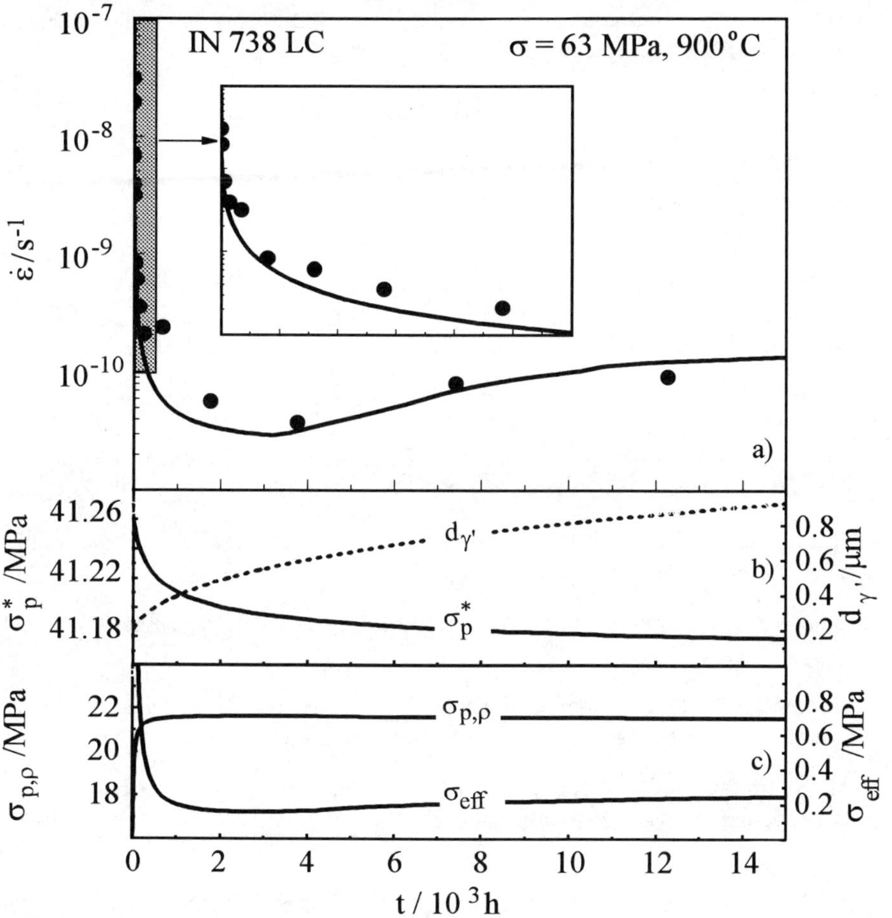

Figure 4: a) Differentiated creep curve of IN 738 LC in $\log \dot{\varepsilon} - t$–representation. Continous lines: measured; dashed lines: calculated according to the HAI–model, eq. 8, using $G = 54\,\text{GPa}$, $v_0 = 9 \cdot 10^{-9}\,\text{m/s}$, $\beta = 0.077\,\text{MPa}^{-1}$; b) and c): particle diameter and stress components.

However, in contrast to the case of thermally stable oxide dispersoids in MA 754 the situation gets more complicated as the γ'–particles show pronounced Ostwald ripening such that $d^3_{\gamma'} \propto t$. Thus, the observed microstructural changes act twice on σ_{eff} in eqn. 8: firstly the particle hardening contribution is reduced assuming an Orowan–type dependence $\sigma^*_{\text{p}} \propto 1/d_{\gamma'}(t)$ and secondly the dislocation network evolution around the particles accounts for $\sigma_{\text{p},\rho} \propto f(\frac{a}{d_{\gamma'}(t)})$. This complex situation is visualized for a long–term creep test at typical service conditions of 900 ^oC and $\sigma = 63\,\text{MPa}$ in fig. 4. Due to the mainly time dependent changes of the γ'–particle microstructure discussed above the creep curve is displayed in the $\log \dot{\varepsilon} - t$–representation (fig. 4a). Its shape results from the superimposed effects of particle coarsening and dislocation evolution via σ^*_{p} (Ostwald ripening, fig. 4b), $\sigma_{\text{p},\rho}$ (shell formation, fig. 4c) and σ_ρ (not plotted here) on σ_{eff} (fig. 4c). In fact, the course of σ_{eff} clearly reflects the features of the creep curve, namely: a pronounced decrease in the primary creep stage due to the shell formation followed by a smooth increase in the steady–state regime due to γ'–particle coarsening and, as a transition state a minimum at $\approx$ 4000 h in between, compare the corresponding curves of $\dot{\varepsilon}$ and σ_{eff} in figs. 4a and c, respectively. Damage accumulation like cavity growth only controls the latest creep stage [13] which has not been plotted here.

References

[1] H. Mecking and U.F. Kocks, "Kinetics of Flow and Strain–Hardening", Acta met., 29 (1981), 1865–1875.

[2] W. Blum, "High Temperature Deformation and Creep of Crystalline Solids", Materials Science and Technology, ed. R.W. Cahn et al., Vol. 6: Plastic Deformation and Fracture of Materials, ed. H. Mughrabi (Weinheim, Germany: VCH–Verlagsgesellschaft, 1992), 359–405.

[3] M. Heilmaier, K. Wetzel and B. Reppich, "Micromechanical Modelling of the Creep Behaviour of Superalloys Part I: Single Phase Material", Proc. 10th Int. Conf. on the Strength of Materials, ed. H. Oikawa et al. (Sendai, Japan: The Japan Institute of Metals, 1994), 563–566.

[4] H. Alexander, P. Haasen, "Dislocations and Plastic Flow in the Diamond Structure", Solid State Physics, 22 (1968), 27–158.

[5] B. Ilschner, Hochtemperaturplastizität (Berlin, Germany: Springer–Verlag, 1973).

[6] L.M. Brown, R.K. Ham, "Dislocation–Particle Interactions", Strengthening Methods in Crystals, ed. A. Kelly and R.B. Nicholson (London, UK: Applied Science Publishers, 1971), 9–135.

[7] M. Heilmaier et al., "Application of the HAI–model to the creep behaviour of particle hardened nickel–base superalloys", Computational Mater. Sci., 7 (1996), 159-166.

[8] M.F. Ashby, "The Deformation of Plastically Non–homogeneous Alloys", ibid. [6], 137–192.

[9] Y. Estrin and H. Mecking, "A Unified Phenomenological Description of Work Hardening and Creep based on One–Parameter Models", Acta met., 32 (1984), 57–70.

[10] C.R. Barrett and W.D. Nix, "A Model for Steady–State Creep based on the Motion of Jogged Screw Dislocations", Acta met., 25 (1977), 1247–1258.

[11] M. Heilmaier and B. Reppich, "On the Microstructural Origin of Primary Creep in Nickel–Base Superalloys", Mater. Sci. Eng., A234–236 (1997), 501–504.

[12] M. Heilmaier, B. Reppich, "Creep Lifetime Prediction of Oxide Dispersion Strengthened Nickel–Base Superalloys: A Micromechanically Based Approach", Met. and Mater. Trans., 27A (1996), 3861-3870.

[13] K.-D. Stein, "Zeitstandverhalten und Mikrostruktur langzeitbeanspruchter Superlegierungen", Fortschritt–Berichte VDI, (Düsseldorf, Germany: VDI–Verlag, 1996), Reihe 5 Nr. 428.

DEFORMATION CHARACTERISTICS OF Sn5%Sb SOLDER ALLOY

K. Linga Murty, M. D. Mathew, Y. Wang and Fahmy M. Haggag*

North Carolina State University, Raleigh, North Carolina 27695-7909
*Advanced Technology Corporation, 661 Emory Valley Road, Oak Ridge, TN 37830

Abstract

Deformation behavior of Sn5%Sb alloy has been investigated at various temperatures from ambient to 423 K from constant load creep, and a recently developed automated ball indentation (ABI) tests. A power law relationship between minimum creep rate and applied stress was obtained from the creep tests with a stress exponent n=5 and an activation energy of 12.6±1.1 kCal/mole. A transition from n=5 to n=3 was observed in the low stress region at 473 K. ABI tests showed a power law relationship between strain rate and ultimate tensile stress with a value of stress exponent n=5.4 and an activation energy of 13.0±1.8 kCal/mole. Based on the values of the stress exponent and activation energy, a new mechanism of creep deformation is proposed. It is suggested that the rate controlling deformation mechanism is viscous glide of dislocations assisted by dislocation core diffusion. Excellent agreement between creep and ABI data indicates that ABI can be successfully applied for studying the deformation behavior of solder joints in actual components in the electronic packaging industry.

Modeling the Mechanical Response of
Structural Materials
Edited by E.M. Taleff and R.K. Mahidhara
The Minerals, Metals & Materials Society, 1997

Introduction

Lead-tin alloys are commonly used as solder materials in microelectronic packaging and these alloys were deemed to have the appropriate mechanical, thermal and physical properties.[1] Because of the lead-poisoning-related environmental concerns, intense research on lead-free solder materials has been underway at electronic industries, national laboratories and academic institutions.[2] Despite extensive search, no real substitute could so far be found as *drop-in* replacement for Pb-Sn alloys. Tin-antimony alloys, including Sn5%Sb, are emerging as alternate materials for soldering applications in the electronic packaging industry.[3] During normal service conditions of the electronic components, these low melting solder alloys deform due to induced stresses from differential thermal expansion occurring between the substrate and the solder joints. This leads to eventual cracking of the joint and can become life limiting. Although thermal fatigue is the major cause of concern for electronic solders, creep behavior also constitutes an important consideration since ambient operating temperature and local heating during operation correspond to high homologous temperatures for these alloys.[4]

Tin alloyed with 5% antimony has a near peritectic composition and a relatively high melting point of 518 K. Compared to lead bearing alloys, tin-antimony alloys have better creep resistance and thermal fatigue characteristics.[5,6] The room temperature ductility of the alloy is relatively high.[7] Although the contact angle for the alloy is higher (43±4°) compared to that of lead-tin eutectic (17±4°), it is considered adequate from wettability considerations.[6] Recently, we have investigated the tensile and creep characteristics of Sn5%Sb alloy.[7,8] In addition, we reported the room-temperature mechanical characteristics determined using the stress-strain microprobe which employs the automated ball indentation (ABI) method, an essentially non-destructive technique.[9] The ABI tests are now extended to higher temperatures, and we report here the strain-rate variation of the flow stress at various temperatures from ambient to 473 K. The ABI results are correlated with tensile creep data from which the underlying deformation mechanism(s) could be evaluated.

Experimental

Sn-5%Sb alloy was obtained in the form of 1 mm thick sheet from Alpha Metals Inc. and was tested in the as-received condition.

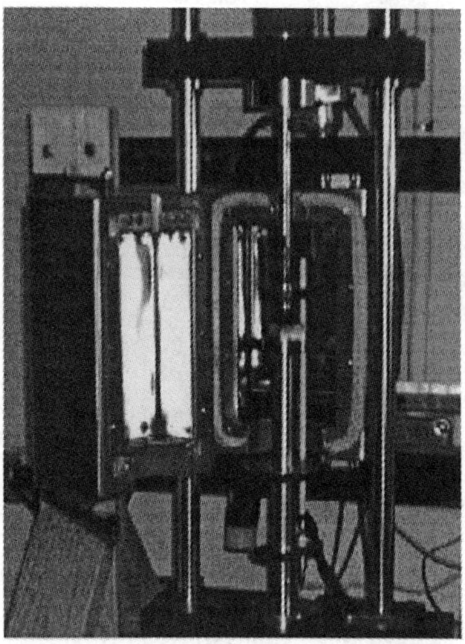

Figure 1. Stress-Strain Microprobe

Tensile creep specimens with a gauge length of 12.4 mm were prepared from the sheet parallel to the rolling direction. Constant load creep tests were performed at 298, 373, 423 and 473 K at stresses varying from 1.25 to 30 MPa, and the creep strains were measured using LVDT.

Automated ball indentation tests were carried out at 298, 323, 373 and 423 K. For ABI tests, no specific specimen preparation was required except that the specimen should be flat and the surfaces must be parallel.[10,11] A light mechanical polish (600 grade polish) was given in order to ensure that surface contamination did not influence the ABI test results and also to achieve flat parallel surfaces. ABI tests involve pushing a spherical indenter into a specimen while monitoring the load versus depth of penetration. Tests were carried out using a table top Stress-Strain Microprobe (SSM) System, Model PortaFlow-P1® procured from Advanced Technology Corporation, Oak Ridge, TN. A tungsten carbide ball indenter of 1.575 mm diameter was used and the maximum depth of penetration of the indenter was about 0.1 mm; the corresponding diameter of the indentation was about 0.375 mm. Each indentation covered a large number of grains since the grain size of the material was small (10 μm). Figure 1 is a photograph of the SSM system with environmental chamber for low and high temperature testing.

Results and Discussion

Creep Results

The creep curves were generally characterized by normal primary creep, short steady-state regime and relatively long tertiary region as shown in Fig. 2. This was typical of all the test conditions. The minimum creep rates were evaluated from the creep curves, and Figure 3 shows the variation of minimum creep rate with stress at different temperatures. A power law relationship of the form,

$$\dot{\varepsilon} = A\,\sigma^n\,, \quad (1)$$

was found to be obeyed over the entire stress range at 298, 373 and 423 K with a value of $n \approx 5$. At 473 K, a transition occurred from a value of $n=5$ in the high stress region to a value of $n=3$ in the low stress region (below 2 MPa). These results are similar to the tensile test data which also indicated the transition at 473 K.[12]

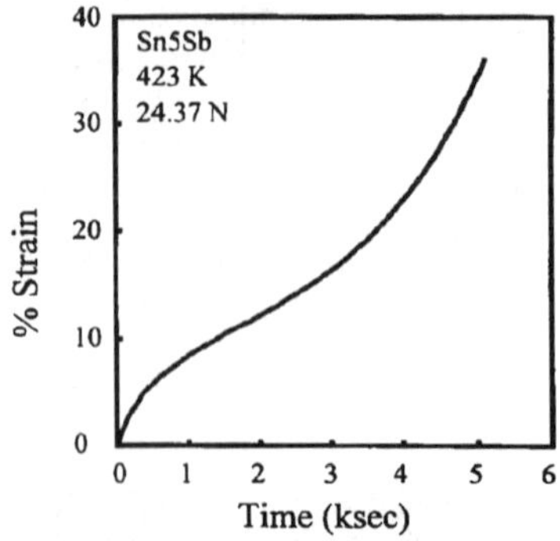

Figure 2
Typical Creep Curve

The activation energy for creep was determined in the high stress (n=5) region, from an Arrhenius plot of ln(strain-rate) against reciprocal of absolute temperature (Figure 4). Least squares analysis of the data yielded a value of Q=12.6±1.1 kCal/mole. As reported earlier, the tensile test data yielded much lower values (~6.4 kCal/mole).[12] Since data for the n=3 region was available only at one temperature, activation energy corresponding to this region could not be evaluated.

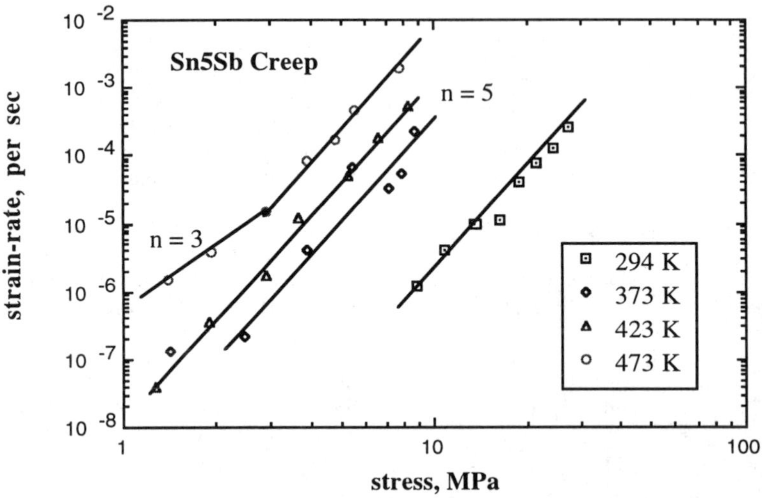

Figure 3
Minimum creep rate vs stress at various temperatures

Figure 4
Activation Energy Plot

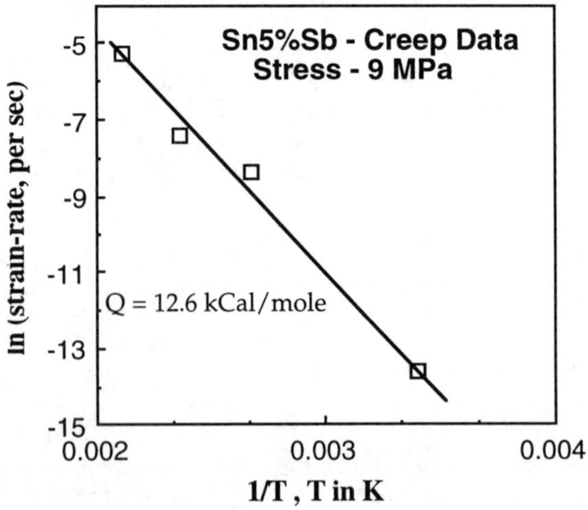

ABI Results

From previous studies, it has been well established that ABI can be employed for studying the stress-strain behavior of materials and that the results agree very well with conventional tensile results.[10] ABI is a relatively simple, rapid and non-destructive technique and requires small amounts of material with minimum material preparation. The principle of the test technique and the calculation procedures have been reported previously.[9-12] In the present investigation, ABI tests were carried out at 294, 323, 373 and 423 K over a range of indenter velocities (strain-rates). The primary data generated in an ABI test is the indentation load versus depth of penetration curve, and all the relevant mechanical properties are evaluated by further analysis of this data. Indentation load versus depth of penetration curves for Sn5%Sb obtained at two different indenter speeds are shown in Fig. 5. Single cycle method is used in lieu of the standard partial unloading technique because of the relatively fast rates employed in characterizing the strain-rate variation of flow behavior. The corresponding true stress and true strain values are calculated by the system software.

The true plastic strain is determined from the equation,

$$\varepsilon_p = 0.2\frac{d_p}{D}, \quad (2)$$

where d_p and D are the plastic indentation diameter and diameter of indenter respectively. The corresponding flow stress is calculated from the following equation,

$$\sigma = \frac{4P}{\pi d_p^2 \delta}, \quad (3)$$

where σ is the true indentation stress and δ is a parameter which depends among other things also on the system compliance. With ABI technique, yield stress, tensile strength, and the work hardening parameters of the material are evaluated directly from the flow curve. The flow curve is usually represented by the equation,

$$\sigma = K\varepsilon^n, \quad (4)$$

where K and n are the strength coefficient and strain hardening exponent respectively. From the fact that the strain hardening parameter n is equal to the true uniform strain (ε_u) at the ultimate tensile stress of the material, the true ultimate tensile stress can be evaluated as,

$$\sigma_{TS} = K\,n^n. \quad (5)$$

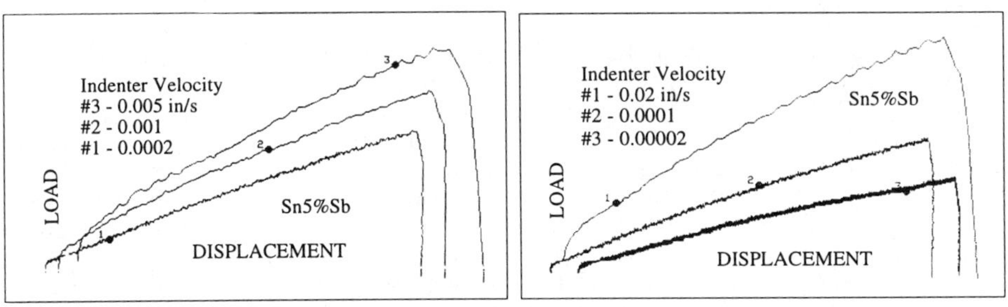

Figure 5. Typical indentation load versus depth curves at various indenter speeds.

The ultimate tensile strength (UTS, the engineering value) is then obtained from the equation,

$$UTS = K\left\{\frac{n}{e}\right\}^n. \quad (6)$$

The indenter velocity (v_i) is converted into tensile strain rate using the relationship,[9]

$$\dot{\varepsilon} = \frac{2\ v_i}{5\ d_p}. \quad (7)$$

Figure 6 shows the variation of strain rate with true ultimate tensile stress at these temperatures. A power law relationship was found to be obeyed with a value of n=5.4 which agrees closely with that determined from creep experiments at high stresses. From the Arrhenius plot (Fig. 7), the activation energy for deformation was determined to be 13.1±1.8 kCal/mole corresponding to a stress level of 60 MPa which is also in good agreement with that for creep. The values of n and Q obtained from both ABI and creep point out to the same deformation mechanism which we propose to be viscous glide as described below.

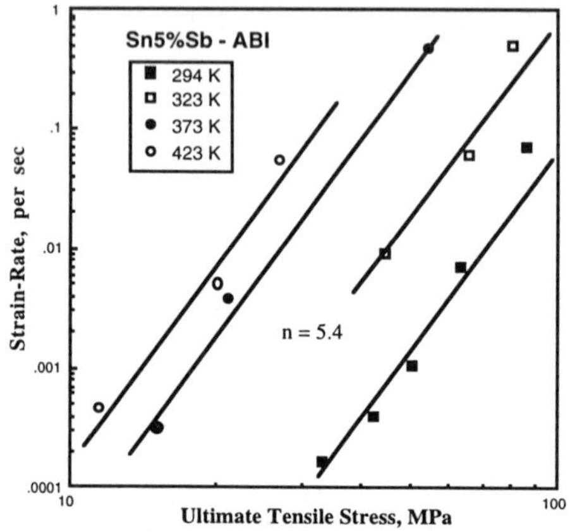

Figure 6
Variation of strain rate with UTS from ABI tests

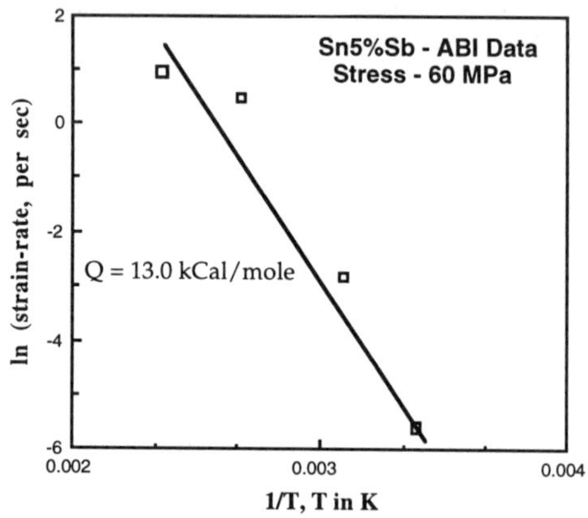

Figure 7
Arrhenius plot - ABI Data

SSM is a versatile tool in the sense that apart from stress-strain properties, it is capable of evaluating stress relaxation behavior of the material. Stress relaxation tests offer the possibility of studying the deformation behavior over a wider range of strain rates compared to conventional creep tests. It is clear from the study that by decreasing the ball diameters to micron-size, SSM can be adopted for in-service or in-situ application for characterizing the *solder-condition* leading to the predictability of their remaining life.

Deformation Mechanisms

Creep deformation mechanisms are identified mainly by the value of the stress exponent n and the activation energy Q.[4,13] Dislocation creep occurs as a result of glide and climb of edge dislocations.[14] These are sequential processes so that the slower of the two acts as the rate controlling step. Dislocation creep in pure metals is characterized by a value of n=5, and climb of edge dislocations being slower, becomes the rate controlling mechanism. Solid solution alloys are broadly categorized as class-M alloys if the value of n is equal to 5, and class-A alloys if the value of n is equal to 3.[15] In class-M alloys, the rate controlling mechanism is the climb of edge dislocations just as in the case of pure metals. In class-A alloys, the rate controlling mechanism is solute-diffusion controlled dislocation glide since the dislocations are locked by solute atmospheres such that the rate of glide becomes slower than that of climb. Such a mechanism which is also known as microcreep,[16] predicts an activation energy for creep which is equal to that for diffusion of solute elements. In either case, recovery takes place by the climb of dislocations. On the basis of the n value in the low stress region at 473 K, creep deformation is perceived to be controlled by viscous glide of dislocations.[9] The activation energy for deformation would be equal to that for solute diffusion via vacancy mechanism. Since the low stress behavior has been exhibited only at one temperature, the activation energy for the mechanism could not be evaluated.

Assuming that the low stress region is in fact controlled by viscous glide of dislocations, one would expect the dislocations to break away from the solute atmospheres as high stresses are applied.[17] Hence, at higher stresses, climb would be the rate controlling mechanism with a value of $n \approx 5$. It may hence be concluded that in the present study the rate controlling mechanism is dislocation climb at all the test conditions except at very low stresses at 473 K (Fig. 3). As Murty[17] has shown in the case of an Al-Mg alloy, the critical stress for the transition from viscous glide to climb should correspond to that for the breakaway of the dislocations from the solute atmospheres,

$$\tau_c = \frac{W_m^2 \, c_o}{2\beta \, kT \, b^3} , \quad (8)$$

where W_m is the binding energy between the solute atom and the dislocation, c_o is the solute concentration and β is a factor dependent on the nature of the impurity cloud with a value between 2 and 4. Due to the uncertainty in the values for the various parameters in this equation, an exact transition stress (τ_c) cannot be determined. However, the range for this critical stress corresponds to $2\times10^{-4}E$ to $1\times10^{-3}E$; E is the elastic modulus.[17,18] Such transitions were reported in many class-A alloys.[19] A review on the transition mechanisms in class-A alloys has been made recently by Murty.[20] The transitional stress in the present investigation does fall in this regime, and thus we believe that the high stress region is due to the climb of edge dislocations. However, the activation energy of 13 kCal/mole determined from both creep and ABI experiments is nearly half that for self diffusion (Q_{SD}) in this alloy system. This value of Q agrees with the activation energy for low temperature dislocation climb assisted by vacancy diffusion through dislocation core. However, in that case, the stress exponent will be higher by a value of 2 or equal to 7.[9,15]

It is possible that viscous glide is the rate controlling mechanism in both the regimes. At low temperatures and/or high stresses, solute diffusion through dislocation pipes could dominate so that the strain-rate would be proportional to the dislocation density,

$$\dot{\varepsilon} = A_1 \, \rho \, e^{-Q_{pipe}} \sigma^3 . \quad (9a)$$

Since $\rho \propto \sigma^2$,

$$\dot{\varepsilon} = A' \, e^{-Q_{pipe}} \sigma^5 . \quad (9b)$$

This implies that at low stresses and high temperatures, viscous glide operates with the steady-state creep-rate varying as σ^3 with an activation energy close to that for solute diffusion through the lattice. Our data is not sufficient to determine the activation energy for creep while cube-rate dependence *is* observed. At high stresses, however, the stress exponent of 5 along with the activation energy for creep of about one-half of that for lattice diffusion are in good agreement with *low-temperature viscous glide*. Further work is needed to unequivocally prove this contention.

Conclusions

Deformation of tin-5% antimony alloy in the temperature range of ambient to 473 K takes place by dislocation creep. Stress exponent, n, of 5 and activation energy Q of 13 kCal/mole have been obtained from creep and ABI tests. Based on these values of n and Q, a new deformation mechanism has been proposed involving viscous glide of dislocations controlled by dislocation pipe diffusion. ABI results agreed very well with the creep data. ABI technology can thus be adopted for characterizing solder reliability in real situations by decreasing the indenter size.

Acknowledgments

The experimental work on Sn5%Sb was initiated in collaboration with Dr. Rao K. Mahidhara using tensile and creep tests and we acknowledge many of his contributions and discussions.

References

1. Solder Joint Reliability : Theory and Applications, J. H. Lau (Ed.), Van Nostrand Reinhold, 1991.
2. S. Jin, J. Metals, Vol. 45 (1993) 13.
3. K. L. Murty, H. Yang, P. Deane and P. Magill, *Deformation Mechanisms in Sn-based Solder Materials,* EEP-Vol. 19-1, Advanced Electronic Packaging - 1997, ASME Vol. 1 (1997) 1221.
4. K. L. Murty and I. Turlik, *Deformation Mechanisms in PbSn Alloys : Application to Solder Reliability in Electronic Packaging,* Proc. Joint ASME/JSME, Vol. 1 (1992) 309.
5. D. Hanson and E. J. Sanford, J. Inst. Metals, 62 (1938) 215.*Reliability in Electronic Packaging,* Proc. Joint ASME/JSME, Vol. 1 (1992) 309.
6. J. S. Hwang, Solder Paste in Electronic Packaging, van Nostrand Reinhold, NY (1989).
7. R. K. Mahidhara, S. M. L. Sastry and K. L. Murty, Scripta Met. Mat., Vol. 31 (1994) 1145.
8. R. K. Mahidhara, P. C. Markle, K. L. Murty and I. Turlik, *Tensile and Creep Behavior of Sn5%Sb Solder,* in *Microstructures and Mechanical Properties of Aging Materials,* P. K. Liaw, et al (Eds.), TMS (1996) 111.
9. K. L. Murty, F. M. Haggag and R. K. Mahidhara, J. Elec. Mat., Vol. 26 (1997) 839.
10. F. M. Haggag, in *Small Specimen Test Technique Applied to Nuclear Reactor Vessel Thermal Annealing and Plant Life Extension,* ASTM STP 1204 (1993) 27.
11. F.M. Haggag and K. L. Murty, *A Novel Stress-Strain Microprobe for Nondestructive Evaluation of Mechanical Properties of Materials,* in Proc. Of the Non-destructive Evaluation and Materials Properties III, P. K. Liaw et. al., (Eds.) TMS (1996) 101.
12. R. K. Mahidhara, K. L. Murty and F. M. Haggag, *Creep and Mechanical Properties of Sn5%Sb Solder,* in *Design and Reliability of Solders and solder Interconnections,* R. K. Mahidhara et. al., TMS (1997) 75.
13. J. E. Bird, A. K. Mukherjee and J. E. Dorn, in *Quantitative Relation Between Properties and Microstructure,* Israel University Press, Jerusalem (1969) 255.
14. K. L. Murty and O. Kanert, J. Appl. Phys., Vol. 67 (1990) 2866.
15. O. D. Sherby and P.M. Burke, Progress in Mater. Sci., Vol. 13 (1967) 325.
16. J. Weertman, J. Appl. Phys., Vol. 18 (1957) 1185.
17. K. L. Murty, Scripta Met., Vol. 7 (1973) 899.
18. P. Yavari and T. G. Langdon, Acta Met., Vol. 30 (1982) 2181.
19. T. G. Langdon, in *Dislocations and Properties,* The Institute of Metals, London (1985) 221.
20. K. L. Murty, in Proc. Seventh International Conference on *Creep and Fracture of Engineering Materials and Structures,* J. C. Earthman and F. A. Mohamed (Eds.) TMS (1997) 69.

ANISOTROPIC BIAXIAL CREEP OF TEXTURED Cp-TITANIUM TUBING : EFFECT OF COLD-WORK

K. Linga Murty, B. V. Tanikella,* and Ratnaji R. Kola+

North Carolina State University, Raleigh, North Carolina 27695-7909
*now at St. Gobain / Norton Company, Northboro MA 01532
+now at Lucent Technologies, Murray Hill NJ

Abstract

Biaxial creep behavior of thin-walled tubing of recrystallized (Rx) cp-titanium alloy was investigated using internal pressurization superimposed with axial load at 723 K from which the creep locus was constructed at a constant energy dissipation. Similar study was made at 573 K on cold-worked stress-relieved (CWSR) tubing. The experimental loci were fit to the modified Hill's equation with anisotropy parameters, R and P, which are also the contractile strain(-rate) ratios. The normalized loci for both the CWSR and Rx clearly deviated from that for isotropy. The crystallographic texture of the tubing was characterized through inverse and direct pole figures from which the crystallite orientation distribution functions (CODF) were derived. The CODF was combined with plasticity model based on power-law stress dependence of the strain-rate and dominance of basal, prism and pyramidal slip systems were considered. Experimental results deviated from the prism-model predictions albeit the deformed microstructures revealed only prism dislocations. These differences are attributed to the relatively high O-equivalent content while grain-shape anisotropy due to cold-work seems to have a negligible influence.

Modeling the Mechanical Response of
Structural Materials
Edited by E.M. Taleff and R.K. Mahidhara
The Minerals, Metals & Materials Society, 1997

Introduction

Hexagonal close packed (hcp) metals such as titanium alloys are commonly used for structural applications in aerospace and energy technologies due to their relatively high specific strengths and their resistance to corrosion. Because of the limited number of slip systems and the inherent anisotropy of the hcp single crystal, these materials become highly textured during thermomechanical processes which result in anisotropic deformation and creep. Recrystallization annealing is known to modify the texture resulting in creep anisotropy different from that of stress-relieved. Extensive studies on biaxial creep of zirconium alloys by Murty[1] and Murty et al[2] clearly indicated distinct deviations from isotropic behavior, and the creep anisotropy of the recrystallized material correlated well with the crystal plasticity model based on dominance of prismatic slip as expected for low c/a-ratio hcp metals. On the contrary, the anisotropic creep characteristics of the cold-worked stress-relieved (CWSR) material deviated from predictions based on prism slip dominance even though the texture modifications following cold-working are taken into account.

Recent work by Murty and coworkers[3] revealed that grain shape anisotropy had distinct superimposed effect on crystallographic texture resulting in the observed differences between the experimental and predicted (based on texture) creep loci. The present study was undertaken to examine these effects in a commercial pure titanium (cp-Ti). Biaxial creep characteristics of cp-Ti tubing are investigated using internally pressurized thin-walled tubing with axial load from which the creep loci were constructed at constant energy dissipation. The experimental loci were fit to the modified Hill's equation with anisotropy parameters, R and P, which are also the contractile strain(-rate) ratios.[1] The crystallographic texture of the material is studied using both inverse and direct pole figures from which the crystallite orientation distribution functions (CODF) are determined. The derived CODFs are combined with power-law hardening crystal-plastic models based on slip dominant on prismatic, basal and pyramidal planes to predict the mechanical and creep anisotropy. Although excellent agreement between the experimental and model predictions was reported for recrystallized Zircaloy tubing,[1,3] distinct deviations are noted in cp-Ti even after complete recrystallization. These deviations are believed to arise from the relatively high O-equivalent content resulting in reduced CRSS for basal slip which leads to non-negligible contribution of basal slip to overall deformation.

Experimental Details

Materials

CWSR Cp-Ti was received in the form of thin walled tubing of 9.5 mm outer diameter and 1.829 mm thickness. The composition of the material is shown in Table 1. The tubes were cut to a length of 165 mm from the as received tubing, vacuum sealed (10^{-5} torr) in quartz tubing and annealed at 923 K (650 °C) for 4 hours and cooled in the furnace to ambient temperature. This was done to recrystallize the material. Optical metallography revealed elongated grains in CWSR along rolling (axial) direction, which became equiaxed following recrystallization anneal.

Table 1
Chemical Composition (wt %) of Cp-Ti

O	N	H	Fe	Ti
0.12	0.01	0.002	0.05	Balance

Crystallographic Texture

The tendency of grains in a polycrystalline aggregate to cluster about a particular orientation or orientations is called texture. Crystallographic textures are characterized by pole-figures, which describe the preferred orientations. Both inverse and direct pole figures were determined, and Figure 1 depicts the basal (00.2), prismatic (10.0), and pyramidal (10.2) pole figures for CWSR and Rx materials respectively in terms of times-random units which illustrate the bimodal distribution of basal poles typical of low c/a ratio hcp metals.[1] The effect of recrystallization treatment is seen to result in the 30° rotation of the hexagonal crystal around the c-axis as noted in the prism (10.0) pole figures. A more complete description of texture is possible through CODF analysis where a set of three angles, θ, ψ, and ϕ (Euler angles) that are related to the crystal coordinate axes and the sample coordinate axes (which are the principal working directions of the material). $\omega(\theta, \psi, \phi)$, named as crystallite orientation distribution function, expresses the probability that a crystallite has orientation (θ, ψ, ϕ) with respect to the specimen axes. In the methods of Roe and Bunge, the CODFs are obtained from a set of conventional pole figure data; Roe's notation is used here.[4] This method generates the CODF as a series of generalized spherical harmonics,

$$\omega(\theta,\psi,\phi) = \sum_{l=0}^{\infty} \sum_{m=-l}^{l} \sum_{n=-l}^{l} W_{lmn} Z_{lmn}(\cos\theta)\, e^{-im\psi} e^{-in\phi}, \qquad (1)$$

where W_{lmn} are the appropriate series coefficients, Z_{lmn} are the augmented Jacobi polynomials and (θ,ψ,ϕ) are Euler angles connecting coordinate axes embedded in the crystallite with axes coincident with the specimen and the principal working directions of the material. The data from the three pole figures (basal, prismatic and pyramidal) made it possible to evaluate the series coefficients W_{lmn} up to l=16. The accuracy of the generated CODFs was checked by reconstructing the basal pole figure from the derived CODF.[5] CODFs are generally represented graphically in the form of Euler plots at fixed intervals of the angle, ϕ, of rotation around the c-axis up to $\phi = 55°$; beyond this, repetition occurs due to hexagonal symmetry. Euler plots are compiled in Figure 2 for CWSR and Rx cp-Ti respectively, where iso-intensity contours are plotted in 'times-random' units at constant intervals of 1 with the base contour also at 1. The three orthogonal tube directions, $\hat{r},\hat{\theta},\hat{z}$ correspond to (θ,ψ) angles of (90°,0°), (0°,0°) and (0°,90°) respectively. The constant ϕ sections of the Euler plots can also be considered in terms of the ideal orientations (hkil)[uvtw], where (hkil) refers to the crystallographic plane lying in the surface of the sheet and [uvtw] is a crystallographic direction in this plane and parallel to $\hat{z}$. While relatively more spread of intensities is noted following rerystallization, the effect of recrystallization is noted to have a rather minor effect on the peak orientations which seems to be different from that noted in Zircaloys.[1,4]

One of the main advantages of the formulation of the crystallographic texture in terms of CODF $[\omega(\theta,\psi,\phi)]$ is in the quantitative predictability of the properties of the textured bulk polycrystalline materials from the appropriate information on the single crystals. To the first order, the bulk property is related to the microscopic property, $\rho(\theta,\psi,\phi)$, by statistical averaging using the CODF so that the average property of the polycrystalline aggregate, $<\rho(\theta,\psi,\phi)>$, is given by

$$<\rho(\theta,\psi,\phi)> = \int_{0}^{2\pi}\int_{0}^{2\pi}\int_{-1}^{1} \rho(\theta,\psi,\phi)\,\omega(\theta,\psi,\phi)\,d(\cos\theta)\,d\psi\,d\phi. \quad (2)$$

As will be described later, we use the CODFs to predict the strain-rate ratio (creep anisotropy) and thus the property ρ of interest is the strain-rate of the textured polycrystal evaluated from that of the single crystal due to the pertinent underlying slip systems that are expected to be dominant in these crystal systems.

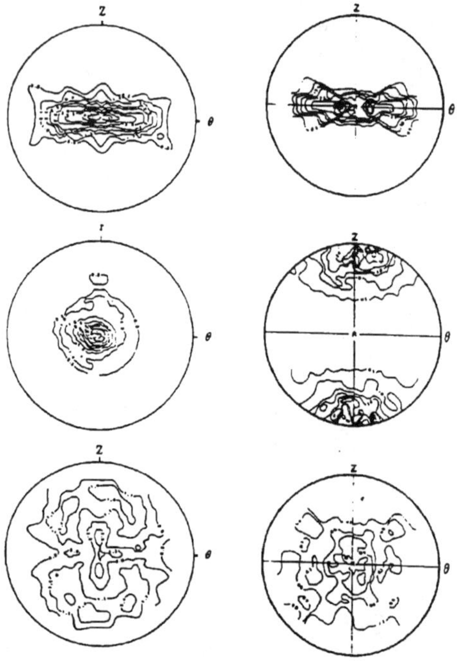

CWSR Figure 1 Recrystallized
Direct Pole Figures of cp-Titanium Tubing
(from top - Basal, Prism and Pyramidal)

<u>Biaxial Creep</u>

Biaxial creep tests were performed on the tubing samples at hoop to axial stress-ratios ($\alpha = \sigma_\theta/\sigma_z$) of 0, 0.5,1.0 and 2.0 on an ATS (Applied Test System@) lever arm machine,[6] in the stress range from 48 MPa to 210 MPa. A schematic of the test system is shown in figure 3. The tubing samples were fitted with Swagelok@ union connectors on both ends for biaxial testing and aluminum mandrels were inserted in the tubes to minimize gas volume to reduce the risk of damage to the experimental setup in the event of failure of the specimen. The specimen was loaded axially by a dead weight pan attached to the lever arm. The hoop stress was achieved by internally pressurizing the specimen with inert Argon (to avoid any reaction with the material) using a sustained pressurizing system. Tests could be performed to pressures $\leq$ 34.5 MPa (5000 psi). The specimen temperature was maintained at the test temperature in an ATS 2961 series three zone furnace such that the temperature variation of the sample was less than $\pm$1 K. The gage length of the specimen was restricted to 50.8 mm (2").

Instrumentation for the lever arm creep machine included an axial extensometer, ATS Model No. 4100 with a linear variable differential transducer (Daytronic LVDT) for measuring axial displacement and a Lasermike Extensometer (Techmet Corp.) to monitor the diametral change during creep tests. Signals from the LVDT and Lasermike extensometers were fed into an IBM PC through an analog to digital I/O System (DT 2801 Series) for on-line data analysis. Appropriate algorithm was developed to monitor *in-situ* the time dependence of the hoop and axial creep strain as well as their rates thereby making it possible to note the onset of the steady-state creep stage.

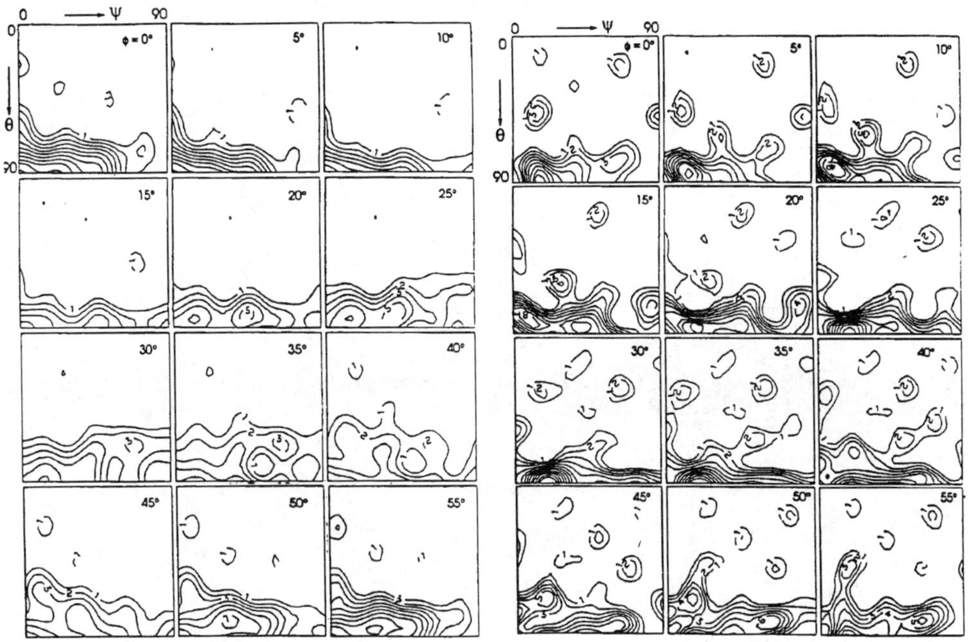

CWSR Figure 2 Recrystallized

Euler Plots for CWSR and Rx Cp-Ti tubing at constant ϕ intervals

For a tubing of inner and outer diameters, a and b, the hoop and axial stresses are given by,

$$\sigma_\theta = \frac{a^2+b^2}{b^2-a^2} P \text{ and } \sigma_z = \frac{\sigma_\theta}{2} + \frac{4W_a}{\pi(b^2-a^2)} \,. \quad (3)$$

In equation 3, P is the internal pressure and W_a is the axial load. For a thin-walled tubing such as that used here, σ_r is taken to be 0 and thus the stress ratio, $\alpha = \sigma_\theta/\sigma_z$ which is varied between 0 (uniaxial) and 2 (closed-end internal pressurization). True stresses and strains were evaluated on the basis of constant volume during plastic deformation; however, the corrections were noted to be relatively minor. The steady-state creep-rates along hoop and axial directions for loading at various stress-ratios and stress levels enable evaluation of creep locus similar to an yield locus. However, since creep can occur under any stress level, constant or invariant dissipative work criterion is used. The dissipative work function, W is defined as,

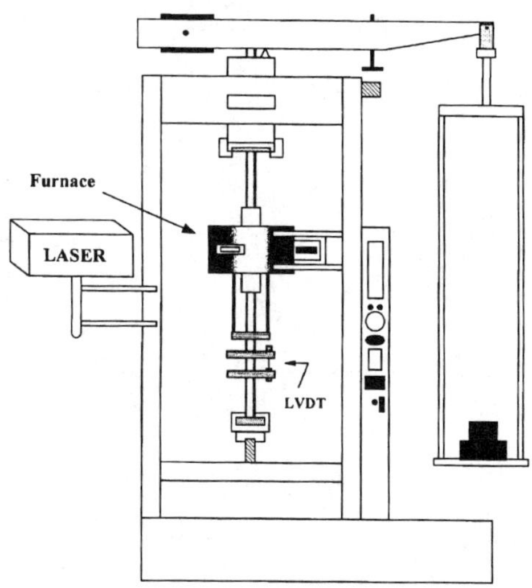

Figure 3
Schematic of Biaxial Creep Machine

$$W = \sigma_g \dot{\varepsilon}_g = \sigma_\theta \dot{\varepsilon}_\theta + \sigma_z \dot{\varepsilon}_z + \sigma_r \dot{\varepsilon}_r, \quad (4)$$

where $\dot{\varepsilon}_g$ is the generalized creep rate equivalent to the uniaxial creep rate determined in a cylindrical specimen loaded axially to a stress equivalent in magnitude to σ_g. The generalized or equivalent stress is defined using Hill's formulation,[7]

$$\sigma_g^2 = \frac{R(\sigma_r-\sigma_\theta)^2 + RP(\sigma_\theta-\sigma_z)^2 + P(\sigma_z-\sigma_r)^2}{P(R+1)}, \quad (5)$$

where σ_g stands for the resultant or the generalized stress with R and P being the anisotropy parameters. Using the Prandtl - Reuss[8] energy balance and the above equation, the strain increments or rates along the radial, hoop and axial directions are related to the respective stresses according to the following:

$$\begin{bmatrix} \dot{\varepsilon}_r \\ \dot{\varepsilon}_\theta \\ \dot{\varepsilon}_z \end{bmatrix} = \frac{\dot{\varepsilon}_g}{P(R+1)} \begin{bmatrix} R+P & -R & -P \\ -R & R(1+P) & -RP \\ -P & -RP & (R+1)P \end{bmatrix} \begin{bmatrix} \sigma_r \\ \sigma_\theta \\ \sigma_z \end{bmatrix}. \quad (6)$$

The above matrix representation forms the basis for creep locus. It is noted from this equation that the anisotropy coefficients R and P are the contractile strain ratios in the uniaxial axial and hoop tests, respectively,

$$R = \frac{\dot{\varepsilon}_\theta}{\dot{\varepsilon}_r}\Big|_{\varepsilon_z}, \text{ at } \sigma_\theta = \sigma_r = 0; \quad \text{and } P = \frac{\dot{\varepsilon}_z}{\dot{\varepsilon}_r}\Big|_{\varepsilon_\theta}, \text{ at } \sigma_z = \sigma_r = 0. \quad (7)$$

For thin walled tubing ($\sigma_r = 0$), the equation reduces to,

$$\beta = \frac{\dot{\varepsilon}_\theta}{\dot{\varepsilon}_z} = \frac{\frac{P+1}{P}\alpha - 1}{\frac{R+1}{R} - \alpha}. \quad (8)$$

Thus for an isotropic material with R = P =1, the creep-rate ratio is unity with identical rates along hoop and axial directions.

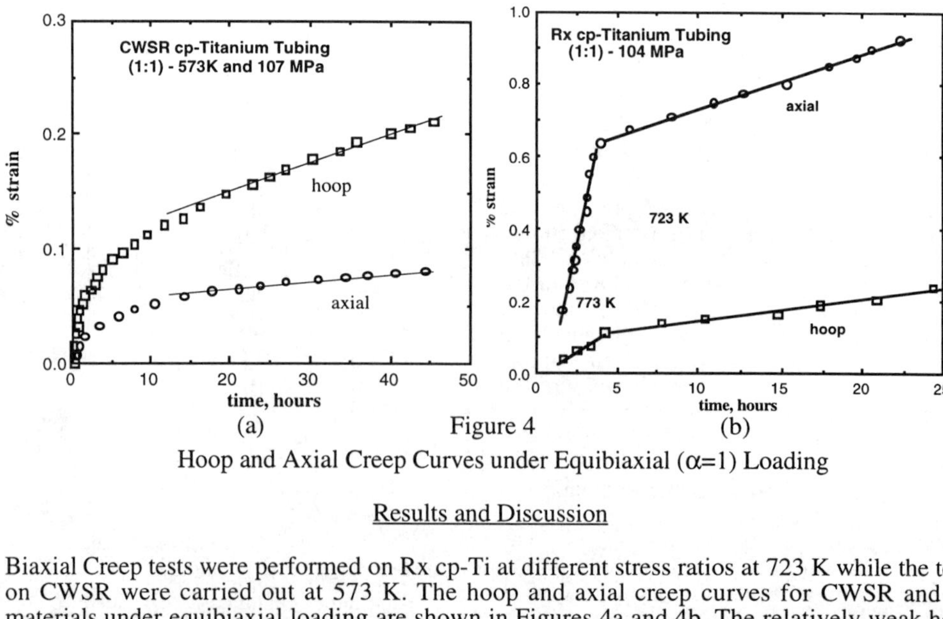

(a) Figure 4 (b)

Hoop and Axial Creep Curves under Equibiaxial (α=1) Loading

Results and Discussion

Biaxial Creep tests were performed on Rx cp-Ti at different stress ratios at 723 K while the tests on CWSR were carried out at 573 K. The hoop and axial creep curves for CWSR and Rx materials under equibiaxial loading are shown in Figures 4a and 4b. The relatively weak hoop became stronger following recrystallization anneal. The anisotropic creep behavior is not influenced by temperature within the range of temperatures studied here. Similar curves were obtained for other stress ratios from which the steady-state creep-rates along hoop and axial directions were evaluated. Limited amount of tests were made at two other temperatures on Rx material under equibiaxial (α = 1) loading to evaluate the activation energy for creep which yielded a value of 46.15±0.61 kCal/mole in agreement with that for self-diffusion.

The hoop and axial strain-rates followed power-law stress-dependence, and Figures 5a and 5b are double-log plots of W versus axial stress for CWSR and Rx materials respectively. As noted in these figures, the data at different stress-ratios followed straight lines with essentially identical slopes but displaced along the stress axis. As pointed out earlier, a creep locus corresponds to a set of hoop and axial stresses for a specified W value, and Figures 6a and 6b include such loci for the CWSR and Rx materials. The loci for both CWSR and Rx materials deviated from isotropy.

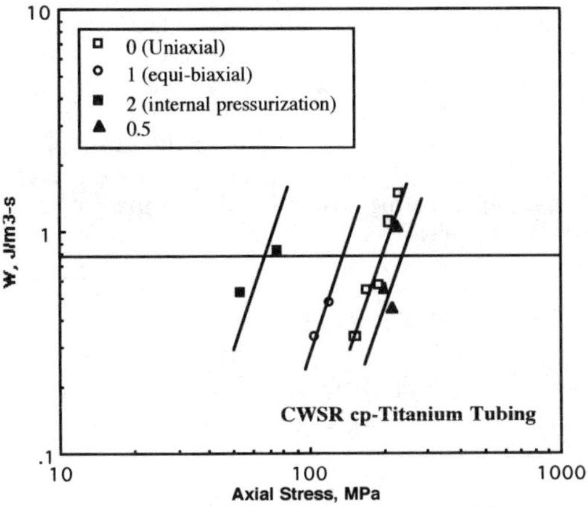

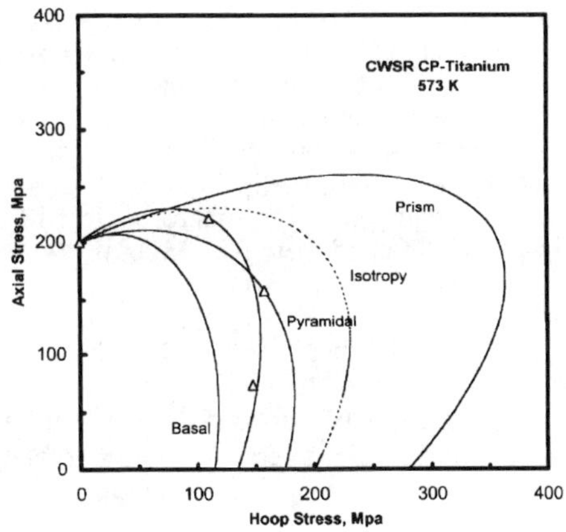

Figure 5a *CWSR cp-Ti Tubing* Figure 6a

Double-Log Plots of W versus σ_z Experimental Loci and Model Predictions

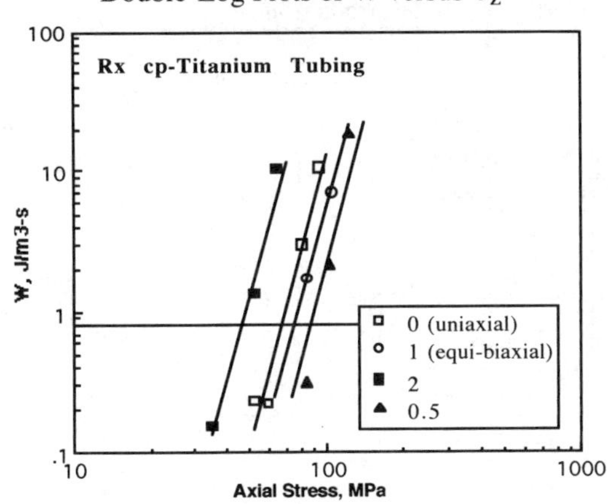

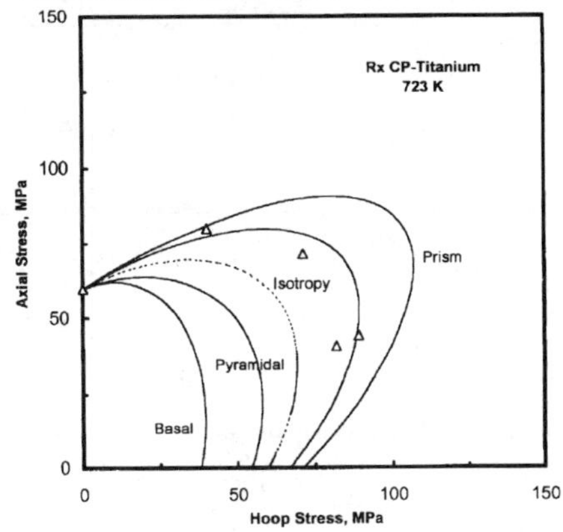

Figure 5b *Rx cp-Ti Tubing* Figure 6b

Double-Log Plots of W versus σ_z Experimental Loci and Model Predictions

<u>CODF-Creep Model Predictions</u>

The shear strain-rate on the sth slip system is related to the resolved shear stress through a power-law with exponent, n, so that,

$$\dot{\gamma}^{s} = \dot{\gamma}_{o}^{s}\left(\frac{\tau^{s}}{\tau_{o}^{s}}\right)^{n}, \qquad (9)$$

where τ_o^s is the reference shear stress on the sth slip system and it specifies the relative ease of plasticity on a given slip system. In the above equation, $\dot{\gamma}_o^s$ is the reference strain-rate which depends on the test temperature generally through an Arrhenius function. The total strain-rate in the crystal is expressed as the sum of the contributions from each of the active slip systems,

$$\dot{\varepsilon}_{ij}^{c} = \sum_{kl} \mu_{ij}^{kl}\dot{\gamma}_{ij}^{kl} = M_{ijkl}\sigma_{kl}, \qquad (10)$$

where μ_{ij} is the geometric tensor connecting slip coordinates with crystal coordinates. The bulk strain-rate is then obtained by averaging the strain-rate of the individual grains weighted by CODF using Eq. 3. In the present analyses, basal ({00.2}<11.0>), prismatic ({10.0}<11.0>) and pyramidal ({11.2}<11.3>) slip modes are considered. Values of the reference stresses (τ_o) are chosen to emphasize the contribution of individual or particular combinations of the slip modes to the polycrystalline behavior. It should be pointed out that, due to the paucity of information on the strain-rate-sensitivity for specific slip systems, n is taken to be the same ($\sim$ 8) for all of the operating slip systems. In the lower bounding model, the stress state in each crystallite is equated to the macroscopic stress state, so that the lower bound estimate of the strain-rate is obtained from the CODF using the averaging process (Eq. 3) :

$$\dot{\varepsilon}_{ij} = <\dot{\varepsilon}_{ij}> = \int_0^{2\pi}\int_0^{2\pi}\int_{-1}^{1} \omega(\theta,\psi,\phi)\, \dot{\varepsilon}_{ij}^{c}(\theta,\psi,\phi)\, d(\cos\theta)\, d\psi\, d\phi, \qquad (11)$$

where $\omega(\theta,\psi,\phi)$ is the fraction of the crystallites with orientation (θ,ψ,ϕ). It is to be emphasized that the grain-to-grain interactions are not taken into account in this simple model which was noted to yield good correlation in recrystallized Zircaloys.[1]

Table 2
Anisotropy Parameters for CWSR and Rx cp-Ti Tubing - Experimental Results Compared with Model Predictions

	Experimental			CODF-Creep Model Predictions			
Material	R	P	$\frac{R}{P}$	Slip System	R	P	$\frac{R}{P}$
CWSR	2.91±0.7	0.51±0.05	2.48	basal	**0.80**	**0.17**	**4.71**
				prism	0.85	8.60	0.10
				pyramidal	**0.55**	**0.37**	**1.49**
Rx	1.45±0.3	2.79±0.6	0.52	basal	0.70	0.20	3.50
				prism	**1.70**	**7.75**	**0.22**
				pyramidal	0.60	0.45	1.33

$\frac{R}{P} = \{\frac{\dot{\varepsilon}_\theta}{\dot{\varepsilon}_z}\}_{1:1} \Leftrightarrow$ CWSR : $\dot{\varepsilon}_\theta > \dot{\varepsilon}_z$ and R_X : $\dot{\varepsilon}_\theta < \dot{\varepsilon}_z$

The CODF-Creep model predictions are included in figure 6 for comparison with the experimental loci. None of the predictions based on the dominance of any of the slip systems agree with the experimental results. This is in contradiction to the behavior reported for recrystallized Zircaloy tubing where an excellent agreement is noted[1] between the experimental locus and that derived from the model based on prism slip dominance as expected in low c/a-ratio hcp metals. The experimental creep loci for CWSR and Rx cp-Ti tubing shown in Figure 6 were fit to the Hill's equation (Eq. 5) to derive the anisotropy parameters, and they are compared with those obtained from model predictions based on the dominance of various slip systems (Table 2). From equations 6 and 7, we find that a plot of the strain-rate ratio (r) along the axial

($\dot{\varepsilon}_z$) and radial ($\dot{\varepsilon}_r$) directions versus α would yield R and P at intersections of $\alpha = 0$ and at α when $r = 0$:[9]

$$r = \frac{-\beta}{1+\beta}; \quad R = \frac{\dot{\varepsilon}_z}{\dot{\varepsilon}_r} = r_{\alpha=0} \quad \text{and } P = \frac{\alpha_{r=0}}{1-\alpha_{r=0}} \quad . \quad (12)$$

The R and P values evaluated for Rx tubing using such an approach are in agreement with those derived from creep loci (Table 2). However, the CWSR tubing yielded values of 0.62 and 0.25 for R and P respectively.

As the summary in Table 2 illustrates, the experimental anisotropy parameters for both Rx and CWSR materials are far removed from any of the model predictions. However, we note that the trend (P > R) is indeed predicted for Rx material by CODF-Creep model based on the dominance of prism slip albeit the quantitative agreement is relatively poor; whereas, prism slip model for CWSR material predicts wrongly when compared to the experimental data. Although both basal and pyramidal model predictions do indicate reasonable agreement with the experimental results, no experimental evidence exists for their operation and/or dominance in low c/a ratio materials such as titanium alloys. Indeed, extensive electron microscopy observations in crept Ti3Al2.5V tubing revealed dislocations on prism planes only with Burgers vectors along <11.0> in both Rx and CWSR materials.[10] Similar findings were reported in Zr-based alloys also.[11] However, the relatively high concentration of O-equivalent in the present material might induce basal slip along with prism. We have not performed any model predictions with combinations of slip systems since no quantitative information on relative ease of the various slip systems is available to-date.

Recent study by Murty and coworkers[3] indicated that the grain shape anisotropy in CWSR material could lead to stress enhancements in some planes along some directions due to grain boundary sliding occurring on these planes. Incorporation of such stress enhancement anisotropy in CWSR Zircaloy along with texture helped to explain the observed experimental results with prism slip still being the dominant slip mode. However, preliminary calculations along these lines did not explain the observed discrepancy in cp-Ti.[12]

Acknowledgments

This work was supported by the National Science Foundation grant #DMR-9504818, and acknowledgments are due to Dr. Bruce MacDonald for support and encouragement.

References

1. Murty, K.L., "Technological Applications of Crystallographic Textures of Zirconium Alloys," Materials Forum, Vol. 15 (1991) 217.
2. Murty, K.L., R.R. Kola, and S.T. Mahmood, Textures and Microstructures, Vol. 7 (1987) 211.
3. Murty et al, Acta Met., Vol. 42 (1994) 3653.
4. H.J. Bunge: Three Dimensional Texture Analysis, Intl. Mat. Rev., Vol. 32, (1987) 265.
5. S.T. Mahmood and K. L. Murty: J. Mat. Eng., Vol. 11 (1989) 315.
6. V. Venkatesan, S.T. Mahmood and K.L. Murty, Met. Trans., Vol. 21A (1990) 3001.
7. W.A. Backofen, W.F. Hosford and J.J. Burke, Trans. ASM, Vol. 55 (1948) 64.
8. A.T. Winter, Phil. Mag., Vol. 30 (1974) 719.
9. K. L. Murty, Trans. IIM, Vol. 39 (1986) 357.
10. S. Nangalia, K. L. Murty and A. Paradkar, "Creep Characteristics and Dislocation Microstructures in Textured Ti3Al2.5V Tubing Under Biaxial Loading," in Proceedings of Harold Margolin Symposium on *Microstructure-Property Relationships of Titanium Alloys,* S. Ankem and J.A. Hall (eds.) TMS, 1994, pp. 99-106.
11. D. S. Sarma, Khalid M. Al-Otaibi and K. L. Murty, Met. Mat. Processes, Vol. 4 (1992) 217.
12. K. Linga Murty and Ratnaji R. Kola, "Anisotropic Biaxial Creep of Textured Cp-Ti Tubing," paper presented at TMS Symposium on *High Temperature Titanium Alloys,* Orlando, Feb. 1997.

Superplasticity and Microstructural Evolution

MICROSTRUCTURAL EVOLUTION IN PURE ALUMINUM DURING EQUAL-CHANNEL ANGULAR PRESSING

Minoru Furukawa,[1] Zenji Horita,[2] Minoru Nemoto[2]
and Terence G. Langdon[3]

[1]Department of Technology, Fukuoka University of Education
Munakata, Fukuoka 811-41, Japan
[2]Department of Materials Science and Engineering, Faculty of Engineering
Kyushu University, Fukuoka 812-81, Japan
[3]Departments of Materials Science and Mechanical Engineering
University of Southern California, Los Angeles, CA 90089-1453, U.S.A.

Abstract

Samples of high purity aluminum were deformed to different levels of strain using the procedure of equal-channel angular (ECA) pressing in which an intense plastic strain is introduced by pressing a sample through a special die. The results demonstrate the potential for using the ECA pressing technique to reduce the grain size from ~ 1 mm initially to a final equiaxed array of grains having a size close to ~ 1 μm.

Modeling the Mechanical Response of
Structural Materials
Edited by E.M. Taleff and R.K. Mahidhara
The Minerals, Metals & Materials Society, 1997

Introduction

Equal-channel angular (ECA) pressing is a processing procedure developed several years ago by Segal and co-workers [1]. The principle of this procedure is based on the observation that metal working by simple shear provides close to an ideal method for structure and texture formation [2]. An important advantage of the process is that the intense plastic straining introduced using ECA pressing is capable of producing an ultra-fine grain size at the submicrometer or even nanometer level [3-5].

Although there have been several reports to date of the properties developed in materials after ECA pressing, there are relatively few reports dealing with the microstructural development during the pressing process [6-8]. Therefore, this report describes some of the results obtained in a detailed investigation designed to examine the nature of the formation and development of arrays of ultra-fine grains during the ECA pressing of samples of high purity aluminum.

Experimental material and procedures

The experiments were conducted using aluminum of 99.99% purity. The material was received in the form of an ingot which was subjected to cold rolling and subsequent annealing for 1 hour at 773 K. Metallographic inspection revealed a large initial grain size of ~1.0 mm.

The ECA pressing was performed using a die of the type described earlier [9]. Briefly, two blocks of tool steel were bolted together such that an internal channel, of square cross-section, was continuous through the block and formed an L-shape configuration subtending an angle of 90° between the two channels. There was also an angle of 20° defining the outer arc of curvature at the point where the two channels intersected. A relationship was derived earlier [10] giving the magnitude of the strain introduced into a sample when it is pressed through the die: the validity of this relationship has been confirmed in a series of model experiments using plasticine layers contained within a plexiglass pressing facility [11]. In the present investigation, a single pressing through the die gave a strain of the order of ~1.

The ECA pressing was conducted using Al samples having dimensions of 10 × 10 mm^2 and lengths of the order of ~75 mm. Specimens were polished on the longitudinal faces prior to pressing at room temperature using an MoS_2 lubricant and specimens were pressed up to several times through the die to give a high cumulative strain. In practice, it is possible to define three distinct processing routes, as illustrated in Fig. 1: route A refers to the situation where the repetitive pressings are conducted without any rotation of the sample, route B denotes pressings where there is a rotation of the sample by 90° in the same direction between each separate pressing and route C denotes repetitive pressings with a rotation between pressings of 180°. Samples were pressed using routes A, B or C for up to 4 passes through the die.

After pressing to a selected total strain, samples were thinned for inspection by transmission electron microscopy (TEM). Observations were made by TEM using samples oriented in three mutually perpendicular planes as defined in Fig. 2, where x, y and z are the planes perpendicular to the longitudinal axis of the sample, parallel to the side face at the point of exit from the die or parallel to the top face at the point of exit, respectively. The procedure for preparation of the TEM samples was described earlier [8]. Representative microstructures were recorded by TEM and selected area electron diffraction (SAED) patterns were taken from regions having diameters of 12.3 μm. Subgrain and grain sizes were measured directly from the TEM photomicrographs using the standard linear intercept method.

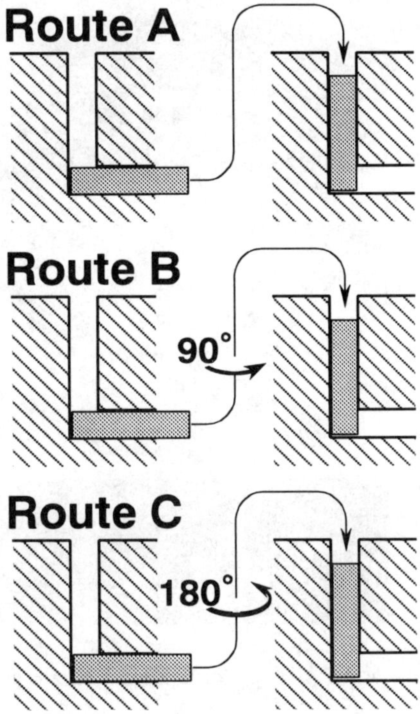

Figure 1 - Processing routes A, B and C in ECA pressing.

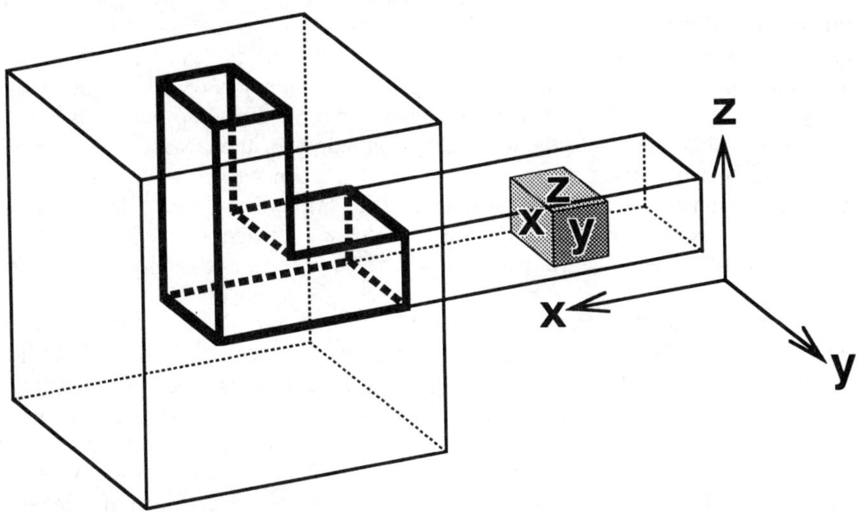

Figure 2 - Definitions of planes *x*, *y* and *z*.

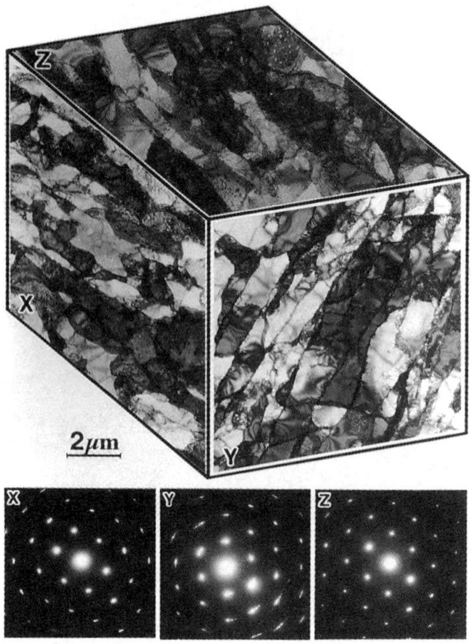

Figure 3 - Microstructure and SAED patterns after a single passage through the die.

Experimental results and discussion

Figure 3 shows the microstructures in the *x*, *y* and *z* planes after a single pressing, together with the corresponding SAED patterns for each plane. Inspection shows that a single passage through the die, giving a total strain of ~1, leads to the formation of bands of subgrains where, based on the SAED patterns, the boundaries have low angles of misorientation. These subgrain bands lie essentially at 45° to the top and bottom edges of the *y* plane and therefore the bands are oriented in the direction of shearing as the sample proceeds through the die. These bands are also parallel to the top and bottom edges of the *x* plane and perpendicular to the pressing direction in the *z* plane. The marker for 2 μm serves to illustrate the very small subgrain size which is achieved on a single passage through the die. Thus, one pass and a strain of ~1 gives an array of subgrains having an average size of <1 μm. Detailed microstructural measurements are given elsewhere [8] for routes A and C up to 10 pressings.

Figures 4 - 6 illustrate the effect of 4 pressings through the die to a total strain of ~4 for samples processed using routes A, B and C, respectively. A comparison of these three sets of photomicrographs shows that microstructural evolution is most rapid when using route B whereby the sample is rotated in the same direction by 90° between each separate pressing. In Fig. 5, illustrating route B, the subgrain bands are no longer visible after 4 pressings and instead there are arrays of grains in the x, *y* and *z* planes which appear to be reasonably equiaxed. The SAED patterns for each plane in Fig. 5 show diffracted beams scattered around rings, thereby demonstrating that the boundaries are now in high angle misorientations. By contrast, there is very clear evidence for the continued presence of subgrain bands after 4 pressings when using route A, as illustrated in Fig. 4, and some subgrain bands are also visible when using route C, as illustrated in Fig. 6.

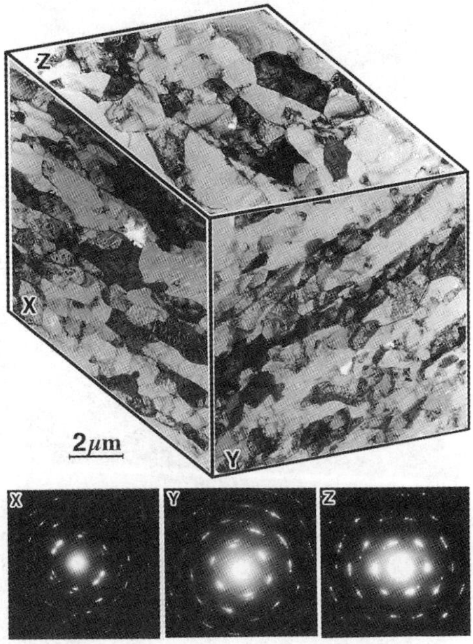

Figure 4 - Microstructure and SAED patterns after 4 pressings through the die using route A.

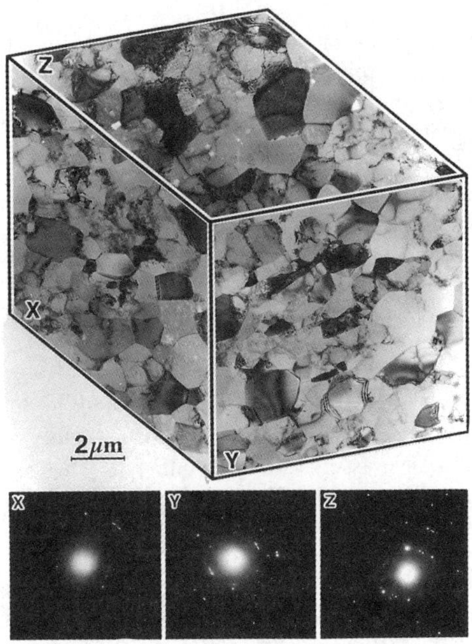

Figure 5 - Microstructure and SAED patterns after 4 pressings through the die using route B.

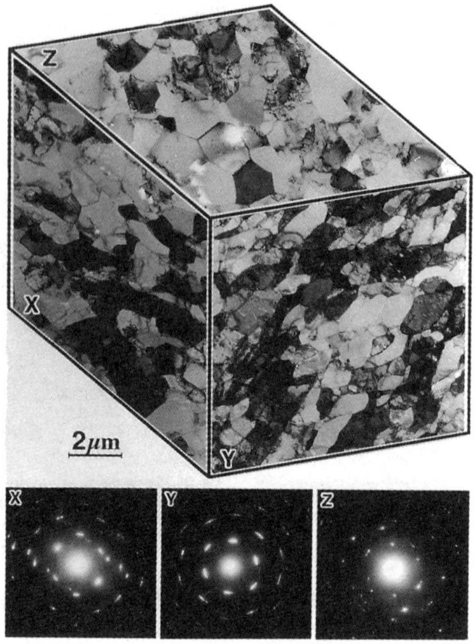

Figure 6 - Microstructure and SAED patterns after 4 pressings through the die using route C.

It is apparent from these observations that ECA pressing of high purity aluminum leads rapidly to the formation of arrays of subgrain bands within each grain and these subgrain boundaries evolve with further pressings into high angle grain boundaries. Measurements were taken of the subgrain sizes and grain sizes after 1, 2, 3 and 4 pressings and the results are shown in Fig. 7 for ECA pressing conducted using (a) route A, (b) route B and (c) route C, respectively. Separate points are shown in Fig. 7 for the x, y and z planes, and the open, half-closed and closed symbols are used to differentiate between microstructures consisting only of subgrains, of a mixture of subgrains and grains and of fully developed grains separated by high angle grain boundaries, respectively: where the subgrains were elongated, as in the early stages of ECA pressing, the sizes reported in Fig. 7 are the average lengths of the short axis, equivalent to the widths of the subgrain bands visible in Fig. 4.

Inspection of Fig. 7 confirms that an array of grains with high angle boundaries is achieved most rapidly when using processing route B and least rapidly with route A. With route B, there is an evolution after 4 pressings to a reasonably equiaxed structure with a grain size of ~1.3 μm in each plane of sectioning. Therefore, the grain size has been reduced from ~1.0 mm to ~1.3 μm in only 4 pressings through the die but with a rotation of 90° between each pressing. The results suggest that an evolution to a similar final grain size will occur also when using routes A and C but more pressings are then required to attain an array of high angle grain boundaries. These results are therefore consistent with those described earlier when using routes A and C [8]. Close inspection of Fig. 7 suggests also a difference between the three different planes of observation. In general, the smallest subgrains tend to be observed in the y plane where the subgrain bands are most clearly delineated, and these subgrains remain essentially unchanged in size at ~0.6 μm when using routes A and C but they increase in size with route B to the final equiaxed grain size of ~1.3 μm.

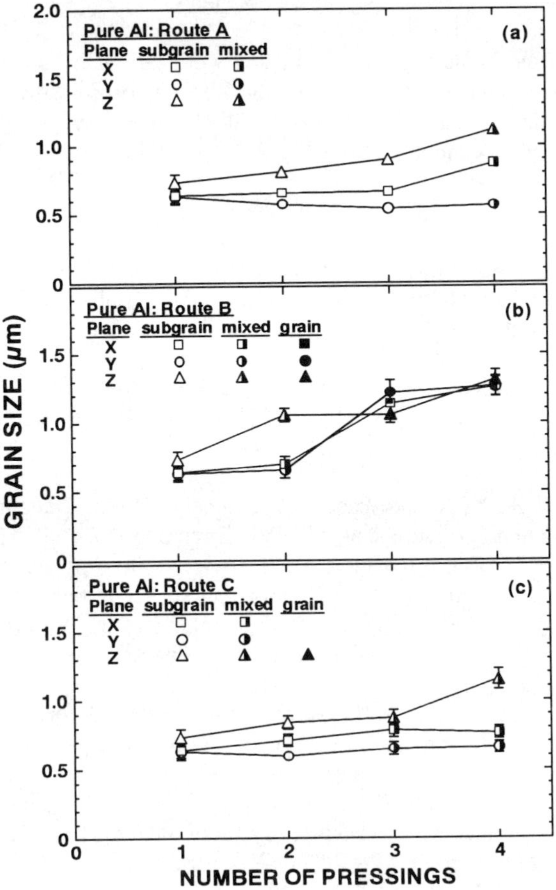

Figure 7 - Grain size measurements for routes A, B and C.

The results obtained in this investigation after small strains have similarities with earlier reports of microstructural evolution in pure Al after cold rolling, where it has been shown that reductions of ~15% - 30% lead to the formation of bands of elongated subgrains with sizes of, typically, ~1 - 2 μm [12-14]. However, a more detailed analysis of ECA pressing is required, including inspection by optical microscopy, in order to develop a complete understanding of the nature of grain refinement in the ECA pressing procedure.

Conclusions

Samples of high purity aluminum were subjected to equal-channel angular (ECA) pressing up to total strains of ~4 and using three different processing procedures: route A without rotation of the sample between consecutive pressings and routes B and C with rotations of 90° and 180° between each pressing. Microstructural observations show that subgrain bands are formed after a single pressing through the die, and these bands evolve after 4 pressings into an array of equiaxed grains with an average size of ~1.3 μm when using processing route B.

Acknowledgements

This work was supported in part by the Light Metals Educational Foundation of Japan, in part by a Grant-in-Aid for Scientific Research from the Ministry of Education, Science, Sports and Culture of Japan, in part by the Japan Society for the Promotion of Science and in part by the National Science Foundation under Grants No. DMR-9625969 and INT-9602919.

References

1. V.M. Segal, V.I. Reznikov, A.E. Drobyshevskiy and V.I. Kopylov, "Plastic Working of Metals by Simple Shear," Russian Metallurgy (Metally), 1 (1981) 99-105.
2. V.M. Segal, "Materials Processing by Simple Shear," Mater. Sci. Engng, A197 (1995) 157-164.
3. R.Z. Valiev and N.K. Tsenev, "Structure and Superplasticity of Al-based Submicron Grained Alloys," in Hot Deformation of Aluminum Alloys, ed. T.G. Langdon, H.D. Merchant, J.G. Morris and M.A. Zaidi (Warrendale, PA: The Minerals, Metals and Materials Society, 1991), 319-329.
4. R.Z. Valiev, N.A. Krasilnikov and N.K. Tsenev, "Plastic Deformation of Alloys with Submicron-grained Structure," Mater. Sci. Engng, A137 (1991) 35-40.
5. V.M. Segal, R.E. Goforth and K.T. Hartwig, "The Application of Equal Channel Angular Extrusion to Produce Extraordinary Properties in Advanced Metallic Materials," in First International Conference on Processing Materials for Properties, ed. H. Henein and T. Oki (Warrendale, PA: The Minerals, Metals and Materials Society, 1993), 971-974.
6. S. Ferrasse, V.M. Segal, K.T. Hartwig and R.E. Goforth, "Microstructure and Properties of Copper and Aluminum Alloy 3003 Heavily Worked by Equal Channel Angular Extrusion," Metall. Mater. Trans., 28A (1997) 1047-1057.
7. S. Ferrasse, V.M. Segal, K.T. Hartwig and R.E. Goforth, "Development of a Submicrometer-grained Microstructure in Aluminum 6061 Using Equal Channel Angular Extrusion," J. Mater. Res., 12 (1997) 1253-1261.
8. Y. Iwahashi, Z. Horita, M. Nemoto and T.G. Langdon, "An Investigation of Microstructural Evolution During Equal-Channel Angular Pressing," Acta Mater., in press.
9. J. Wang, Y. Iwahashi, Z. Horita, M. Furukawa, M. Nemoto, R.Z. Valiev and T.G. Langdon, "An Investigation of Microstructural Stability in an Al-Mg Alloy with Submicrometer Grain Size," Acta Mater., 44 (1996) 2973-2982.
10. Y. Iwahashi, J. Wang, Z. Horita, M. Nemoto and T.G. Langdon, "Principle of Equal-Channel Angular Pressing for the Processing of Ultra-fine Grained Materials," Scripta Mater., 35 (1996) 143-146.
11. Y. Wu and I. Baker, "An Experimental Study of Equal Channel Angular Extrusion," Scripta Mater., 37 (1997) 437-441.
12. B. Bay and N. Hansen, "Microstructures in Cold-Rolled Polycrystalline Aluminium," in Deformation of Polycrystals: Mechanisms and Microstructures, ed. N. Hansen, A. Horsewell, T. Leffers and H. Lilholt (Roskilde, Denmark: Risø National Laboratory, 1981), 137-144.
13. B. Bay and N. Hansen, "Deformed and Recovered Microstructures in Pure Aluminium," in Annealing Processes - Recovery, Recrystallization and Growth, ed. N. Hansen, D. Juul Jensen, T. Leffers and B. Ralph (Roskilde, Denmark: Risø National Laboratory, 1986), 215-220.
14. B. Bay, N. Hansen, D.A. Hughes and D. Kuhlmann-Wilsdorf, "Evolution of f.c.c. Deformation Structures in Polyslip," Acta Metall. Mater., 40 (1992) 205-219.

APPLICATION OF EQUAL-CHANNEL ANGULAR PRESSING FOR PRODUCING SUPERPLASTIC ALUMINUM ALLOYS

P. B. Berbon[1], M. Furukawa[2], Z. Horita[3], M. Nemoto[3],
N. K. Tsenev[4], R. Z. Valiev[5] and T. G. Langdon[1]

[1]Departments of Materials Science and Mechanical Engineering
University of Southern California, Los Angeles, CA 90089-1453, U.S.A.

[2]Department of Technology, Fukuoka University of Education
Munakata, Fukuoka 811-41, Japan

[3]Department of Materials Science and Engineering, Faculty of Engineering 36
Kyushu University, Fukuoka 812-81, Japan

[4]Institute of Chemical Technology
Ufa State Petroleum Technical University, 450062 Ufa, Russia

[5]Institute of Physics of Advanced Materials
Ufa State Aviation Technical University, 450000 Ufa, Russia

Abstract

The equal-channel angular (ECA) pressing technique was used to process two different Al alloys. Different routes were performed to observe the effect of the ECA procedure parameters. Observations by transmission electron microscopy (TEM) showed that a fine-grained structure was introduced in all processed alloys when there was a sufficient amount of strain. Mechanical tests performed at low temperatures and relatively fast strain rates demonstrated that very large ductility may be achieved. This paper describes results obtained on Al-5%Zn and Al-5.5%Mg-2.2%Li-0.12%Zr alloys. The first material showed the appearance of subgrains after a small strain was introduced, followed by an evolution of the structure into an array of high angle grain boundaries. The second material showed the effect of the strain introduced in this complex alloy. In the most successful case, an elongation of more than 1180% was obtained in the Al-Mg-Li-Zr alloy.

Modeling the Mechanical Response of
Structural Materials
Edited by E.M. Taleff and R.K. Mahidhara
The Minerals, Metals & Materials Society, 1997

Introduction

High ductilities have been observed in some coarse-grained materials [1] but by far the most common phenomenon for very high ductilities is superplasticity. Among the most stringent requirement for superplasticity is the necessity to have a very small grain size [2]. In industry, this translates into additional preparation steps and additional cost and it has given rise to an extensive search for ways to reduce the grain size of a material. Several powder metallugy techniques have emerged but their optimization is a long task and the shortcomings of residual porosity, very small sample size and the introduction of oxide particles remain [3]. Other techniques are based on the mechanical alloying method but again the shortcomings are numerous [4]. Thus, the most promising path may be the introduction of severe plastic deformation inside the material [5]. The strain introduced often gives rise to the appearance of small subgrains or grains. The two leading techniques for introducing a high strain are torsion straining (TS) and equal-channel angular (ECA) pressing [6]. TS uses a combination of compression and rotation whereas ECA pressing introduces strain through simple shear by forcing the material through an angled die [7]. In this study, several Al alloys were processed following different procedures in order to observe the effect of the various ECA parameters. The first goal was the production of a fine grained structure and the second goal was the introduction of superplastic behavior.

Experimental materials and procedures

Experiments were conducted using two different Al-based alloys: an Al-5%Zn solid solution alloy and a commercial Russian aluminum alloy known as Al-1420 and containing 5.5%Mg, 2.2%Li and 0.12%Zr [8]. All ECA pressing was performed using dies in which the two channels intersected at an angle of 90°. In this condition, each pass introduced a strain of ~ 1 inside the deformed piece [9] and therefore the total strain is equal to the number of passes.

The various ECA pressing procedures used for the Al-5%Zn are summarized in Table I. The number of passes through the die was varied from 4 to 8 and the specimen was rotated by 90° in the same direction between each pass in a procedure designated route B. The material was prepared for observation by transmission electron microscopy (TEM) by polishing to 300 μm thickness, trimming to 3 mm diameter, polishing to 150 μm, and electropolishing at -15 °C in a mixture of 20% nitric acid and 80% methyl alcohol. All TEM specimens were prepared from within the grip sections. The observations were performed using a Hitachi H8100 operating at 200 kV.

The various ECA pressing procedures used for the Al-5.5%Mg-2.2%Li-0.12%Zr alloy are summarized in Table II. For this material, the number of passes through the die and the pressing temperature were varied and there was no rotation of the piece between passes in the procedure designated route A. The initial grain size of this material was ~ 400 μm. The material was prepared for TEM observations following the same technique as for the Al-5%Zn alloy with an electrolyte of 10% perchloric acid, 20% glycerin and 70% ethyl alcohol at ~ 5 °C. Rectangular tensile specimens with a gauge length of 4 mm, a gauge section of 2×3 mm^2, and a total length of 24 mm were machined from the pressed samples and tested in an Instron machine operating at a constant cross-head displacement. The testing temperatures were from 298 to 673 K controlled to within ± 2 K and the testing strain rates were from 10^{-4} to 1 s^{-1}.

Table I: The three procedures used for Al-5%Zn

Procedure	Route	Processing procedure (strain and temperature)
A1	B	4 @ 298 K
A2	B	6 @ 298 K
A3	B	8 @ 298 K

Table II: The three procedures used for the Al-1420 alloy

Procedure	Route	Processing procedure (strain and temperature)
B1	A	4 @ 673 K
B2	A	8 @ 673 K + 4 @ 473 K
B3	A	8 @ 573 K

Results and discussion

Al-5%Zn

The evolution of the microstructure with the number of passes (*i.e.* with the amount of strain introduced) is apparent from the photomicrographs in Fig. 1 after 4, 6 and 8 passes through the die. The photomicrographs are very similar in these three cases, with the grain size decreasing slightly from Fig. 1(a) to (c). However, important differences can be observed in the selected area electron diffraction (SAED) patterns. In fig. 1(a), the SAED pattern consists of a slightly elongated net pattern, showing the importance of subgrains. Misorientation angles lower than 10° dominate the structure. In Fig. 1(b), a mixture of grains and subgrains is observed. The pattern is not close to a net pattern but it is also not random. In Fig. 1(c), the structure consists of large angle grain boundaries as is evident from the random pattern.

This result gives an interesting insight into the mechanism leading to grain refinement via ECA pressing of Al alloys, and it is in agreement with the conclusions of Ferrasse et al. [10]. It appears that subgrains form first, followed by evolution into large-angled grain boundaries. It is likely that this sequence of events can be generalized to all cell-forming materials [11,12]. Therefore, ECA pressing represents a serious candidate procedure for the processing of all cell-forming alloys and for the introduction of small grains separated by large angle boundaries.

On the other hand, if the processed alloy contains appropriate particulates to hinder grain growth at high temperature it will be expected to exhibit a superplastic behavior [13]. No attempt has been made in this investigation to examine the precise structure of the grain boundaries in the Al-5%Zn alloy after ECA pressing. This is important to determine if they have a non-equilibrium character of the type described by Nazarov et al. [14].

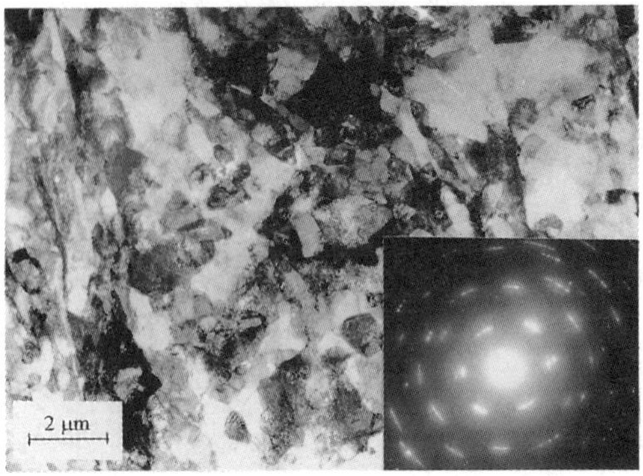

(a)

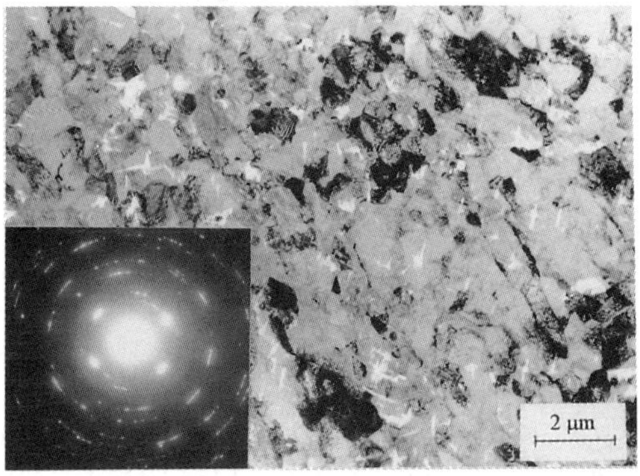

(b)

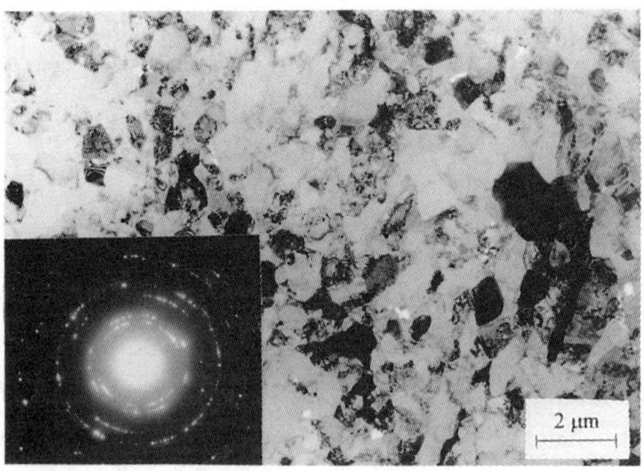

(c)

Figure 1: TEM photomicrographs and SAED patterns for Al-5%Zn after (a) 4, (b) 6 and (c) 8 passes, respectively

Al-1420

The Russian Al-1420 alloy appears as a good candidate for further study since it contains finely dispersed Al_3Li and Al_3Zr particulates. It has been observed that the ordered Al_3Zr particulates are particularly effective in pinning grain boundaries at high temperatures [15].

The microstructure in the as-pressed condition via procedure B1 showed a grain size of ~ 1.2 μm and a mixture of grains and subgrains [16]. Figure 2 shows the material processed via procedure B2 which has a similar grain size but all areas consisted of large-angled grain boundaries. This is in agreement with the preceding result on Al-5%Zn. It should be noted that the numerous white spots apparent in Fig. 2 are due to the preferential electropolishing of the stable AlLi particulates. Very little grain growth was detected in the material on testing up to 623 K, where the grain size was ~ 2 μm as shown in Fig. 3. However, a temperature of 673 K was sufficient to introduce abnormal grain growth as shown in Fig. 4.

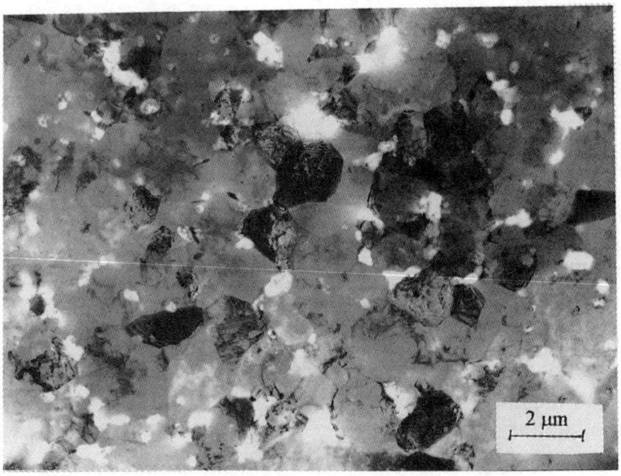

Figure 2: Al-1420 processed via procedure B2, as-pressed

Figure 3: Al-1420 processed via procedure B2, after testing at 623 K

Figure 4: Al-1420 processed via procedure B2, after testing at 673 K

The mechanical tests performed on the Al-1420 alloy are summarized in Table III where they are compared to the results for the initial unpressed material. It can be seen that the ductility of the material processed via procedure B1 is already greatly enhanced, and the elongation before fracture of 550% at 603 K demonstrates the possibility of limited superplasticity.

The properties of the material processed via procedure B2 show excellent elongations, reaching more than 1180% at 623K with a strain rate of 10^{-2} s^{-1}. The values at the strain rates of 10^{-1} and 1 s^{-1} demonstrate that this material can deform in the range of high strain rate superplasticity (HSR SP) [17]. These strain rates are more than two orders of magnitude higher than the conventional superplastic forming strain rates. It is of interest to note that, due to the low testing temperatures, the presence of a liquid phase is impossible in this material, therefore constituing evidence for the occurrence of HSR SP without an interboundary or triple junction liquid phase. Previous results on metal matrix composites and mechanically alloyed alloys showed that the presence of a thin liquid layer at triple junctions may enhance the superplastic characteristics through the phenomenon of stress relief inside the liquid [18, 19]. The present result is more likely to be evidence of normal superplasticity which takes place at a high strain rate. This is in agreement with the expected shift in optimum strain rate for superplasticity with decreasing grain size [20]. Tests at higher temperatures led to a decrease in superplastic behavior, in agreement with the advent of grain growth observed in the TEM analysis. This suggests a possible shortcoming of this result because higher temperatures are necessary in order to reduce the yield stresses. Table III shows that the yield stresses may be too high to use the material in conventional superplastic forming presses, where stresses lower than 10 MPa are preferable.

The preliminary results on the material processed via procedure B3 show that the superplastic state is also achieved after introducing a strain of 8. This is of interest for further development of the ECA procedure. It is a reasonable result since the TEM analysis of the material processed via route B1 (4 passes) already contained 60 to 70 % of regions with large angle boundaries. More tests need to be performed on this material to verify the evolution of the percentage high angle boundaries with strain.

Table III: Mechanical results on Al-1420

ECA preparation	T (K)	strain rate (s^{-1})	EF (%)	YS (MPa)
initial material	603	$3.3x10^{-4}$	210	21
		$3.3x10^{-3}$	180	62
		$3.3x10^{-2}$	140	95
procedure B1	603	$3.3x10^{-4}$	416	9
		$3.3x10^{-3}$	550	39
		$3.3x10^{-2}$	341	75
procedure B2	573	$1x10^{-2}$	1041	19
		$1x10^{-1}$	383	43
		1	197	113
	623	$1x10^{-2}$	> 1180	5
		$1x10^{-1}$	907	20
		1	339	61
	673	$1x10^{-1}$	209	69
		1	199	77
procedure B3	623	$1x10^{-2}$	1160	10

Conclusions

1. An investigation of equal-channel angular (ECA) pressing of Al-5%Zn gives insight into the mechanism responsible for grain refinement through the introduction of strain in the material: the formation of subgrains followed by the evolution of low angle sub-boundaries into large angle grain boundaries.
2. An investigation of Al-5.5%Mg-2.2%Li-0.12%Zr shows that the ECA pressing technique emerges as a viable method to produce superplastic Al alloys.
3. The introduction of a true strain of 8 in the Al-1420 alloy appears to be sufficient to produce a structure consisting entirely of high angle boundaries.
4. Superplastic behavior was achieved at a high strain rate, more than one order of magnitude higher than the usual strain rates for commercial Al alloys.
5. Superplasticity was achieved at a temperature of ~ 0.7 T_M, where T_M is the absolute melting temperature, therefore incipient melting cannot occur at triple junctions or grain boundaries. This is evidence for normal superplasticity occurring at a high strain rate.

Acknowledgements

This work was supported in part by the Light Metals Educational Foundation of Japan, in part by a Grant-in-Aid for Scientific Research from the Ministry of Education, Science, Sports and Culture of Japan, in part by the National Science Foundation of the United States under Grants No. DMR-9625969 and INT-9602919, and in part by the U.S. Army Research Office under Grants No. DAAH04-96-1-0332 and N68171-96-6-9006

References

1. E. M. Taleff, G. A. Henshall, D. R. Lesuer, T. G. Nieh and J. Wadsworth, "Enhanced tensile ductility of coarse-grain Al-Mg alloys", in Superplasticity and Superplastic

Forming, eds. A. K. Ghosh and T. R. Bieler, TMS, Warrendale, PA (1995), 3-10
2. T. G. Langdon, “The mechanical properties of superplastic materials”, Metall. Trans. A13 (1982), 689-701
3. P.G. Sanders, G.E. Fougere, L.J. Thompson, J.A. Eastman and J.R. Weertman, "Improvements in the synthesis and compaction of nanocrystalline materials," Nanostruct. Mater. 8 (1997), 243-252
4. S. K. Pabi and B. S. Murty, “Mechanism of mechanical alloying in Ni-Al and Cu-Zn systems”, Mater. Sci. Eng. A214 (1996), 146-152
5. R. Z. Valiev special editor, Ultra-fine grained materials produced by severe plastic deformation, Annales de Chimie, 21 (1996)
6. M. Furukawa, P. B. Berbon, Z. Horita, M. Nemoto, N. K. Tsenev, R. Z. Valiev and T. G. Langdon, “Production of Ultrafine-Grained Metallic Materials Using an Intense Plastic Straining Technique”, Mater. Sci. Forum, 233 (1997), 177-184
7. V. M. Segal, “Materials processing by simple shear”, Mater. Sci. Eng. A197 (1995), 157-164
8. I. N. Fridlyander, V. S. Sandler and T. I. Nikol’skaya, “Investigation of the ageing of aluminum-magnesium-lithium alloys”, Fiz. Metal. Metalloved, 32 (1971), 767-774
9. Y. Iwahashi, J. Wang, Z. Horita, M. Nemoto and T. G. Langdon, “Principle of equal-channel angular pressing for the processing of ultra-fine grained materials”, Scripta Mater. 35 (1996), 143-146
10. S. Ferrasse, V. M. Segal, K. T. Hartwig and R. E. Goforth, “Development of a submicrometer-grained microstructure in aluminum 6061 using equal channel angular extrusion”, J. Mater. Res. 12 (1997), 1253-1261
11. B. Bay, N. Hansen, D. A. Hughes and D. Kuhlmann-Wilsdorf, “Evolution of f.c.c. deformation structures in polyslip”, Acta Metall. Mater. 40 (1992), 205-219
12. F. R. N. Nabarro and D. Kuhlmann-Wilsdorf, “Nucleation of small-angle boundaries”, Scripta Mater. 35 (1996), 1331-1333
13. H. S. Yang, M. Shaarbaf and K. R. Brown, “On the fabrication aspect of commercial superplastic 5083 aluminum alloy sheets”, in Superplasticity and Superplastic Forming, eds. A. K. Ghosh and T. R. Bieler, TMS, Warrendale, PA (1995), 17-24
14. A. A. Nazarov, A. E. Romanov and R. Z. Valiev, “On the nature of high internal stresses in ultrafine grained materials”, Nanostruct. Mater. 4 (1994), 93-101
15. M. Furukawa, H. Wang and M. Nemoto, "Precipitation hardening of Al-0.5%Zr alloy", J. Japan Inst. Light Metals, 40 (1990), 20-26
16. P. B. Berbon, M. Furukawa, Z. Horita, M. Nemoto, N. K. Tsenev, R. Z. Valiev and T. G. Langdon, “An investigation of the properties of an Al-Mg-Li-Zr alloy after equal-channel angular pressing”, Mater. Sci. Forum, 217-222 (1996), 1013-1018
17. K. Higashi, “Positive exponent superplasticity in advanced aluminum alloys with nano or near-nano scale grained structures”, Mater. Sci. Eng. A166 (1993), 109-118
18. T. G. Nieh and J. Wadsworth, “High-strain-rate superplasticity in aluminum matrix composites”, Mater. Sci. Eng. A147 (1991), 129-142
19. K. Higashi, T. Okada, T. Mukai, S. Tanimura, T. G. Nieh and J. Wadsworth, “Superplastic behavior in a mechanically alloyed aluminum composite reinforced with SiC particulates”, Scripta Metall. Mater. 26 (1992), 185-190
20. K. Higashi, T. Okada, T. Mukai and S. Tanimura, “Positive exponent strain-rate superplasticity in mechanically alloyed aluminum IN 9021”, Scripta Metall. Mater. 25 (1991), 2053-2057

A Model for Texture-Related Grain Boundary Misorientations in a Superplastic Aluminum Alloy

M. T. Pérez-Prado*, M. E. McMahon** and T. R. McNelley**

*Department of Physical Metallurgy, Centro Nacional de Investigaciones Metalúrgical (CENIM), Avda. De Gregorio del Amo 8, 28040 Madrid, Spain

**Department of Mechanical Engineering, 700 Dyer Road, Naval Postgraduate School, Monterey, CA 93943-5146 USA

Abstract

Microtexture and grain boundary misorientation distributions in a commercial superplastic aluminum alloy, Supral 2004, have been studied by means of computer-aided electron backscatter diffraction (EBSD) analysis methods. The development of a superplastically enabled microstructure in this alloy has been shown to occur by a continuous reaction during elevated temperature annealing and superplastic straining of as-processed material. Bimodal grain boundary misorientations are frequently observed in association with such continuous reactions. Such misorientation distributions in as-processed and processed and annealed Supral 2004 are interpreted here by a model combining misorientations between symmetric variants of the three main texture components with misorientations obeying an empirical relationship reflecting the role of recovery on the deformation microstructure.

Modeling the Mechanical Response of
Structural Materials
Edited by E.M. Taleff and R.K. Mahidhara
The Minerals, Metals & Materials Society, 1997

Introduction

Two alternative microstructural transformation mechanisms that enable superplastic response in aluminum alloys have been proposed. The first is that associated with the Supral process and described by Grimes [1] and Watts, et al. [2] in terms of dynamic, continuous recrystallization. The process was developed for Al alloys containing nominally 6.0 wt. pct. Cu - 0.5 wt. pct. Zr and involves warm (≈300°C) and cold rolling to large strains. The presence of Zr in the form of fine Al_3Zr particles apparently retards high-angle boundary migration so that these alloys are resistant to recrystallization and recovery is the predominant softening mechanism during annealing. Upon heating and deforming at superplastic forming temperatures (typically 450°C) the rheological characteristics suggest that a superplastic microstructure only evolves during deformation at strain rates $\dot{\epsilon} \sim 10^{-2}\ s^{-1}$ and in a continuous manner. The other transformation mechanism is associated with the Rockwell process, developed later by Paton et al. [3,4] and initially used for processing of AA7475. Coarse particles are developed through aging to assist in recrystallization following deformation processing in a manner which may be described by the particle-stimulated nucleation theory of Humphreys [5].

Recrystallization, e.g. as defined by Doherty et al. [6,7], is generally agreed to involve both the formation and the migration of high-angle grain boundaries. Recently, it has been suggested that refined, cellular dislocation structures produced during deformation processing of metals may transform in either of two ways during subsequent annealing [8]. One involves an apparently unstable growth of isolated cells heterogeneously within the deformation microstructure; this would correspond to normal, or discontinuous, recrystallization and be described as a nucleation and growth process. The other consists of stable coarsening of the deformation microstructure in homogeneous fashion. Such stable coarsening would be favored if, on the average, the cellular structures are relatively highly misoriented and further characterized by a small spread in misorientation. This latter process has been described as continuous recrystallization due to the absence of obvious nucleation and growth stages but the details of the mechanisms have remained elusive. For instance, it has been proposed that high-angle boundaries (HABs, $\theta > 15°$) evolve from deformation-induced dislocation structures only during annealing and deformation by interactions among dislocations and dislocation networks. However, many low-angle boundaries (LABs, $\theta \leq 5°$) and moderately misoriented boundaries (MMBs, $5° < \theta \leq 15°$) have also been observed but little data has been reported on their evolution [9,10]. Recent investigation into the characteristics of grain boundaries in severely deformed pure metals has revealed that some highly misoriented cell boundaries represent interfaces between variants of the deformation texture. Other boundaries are apparently the result of random dislocation interaction [11].

The goal of the present study is to apply interactive electron backscatter diffraction (EBSD) analysis methods to examine the microtexture and grain boundary characteristics of as-processed as well as processed and annealed Supral 2004, a superplastic Al-Cu alloy. Discrete grain orientation measurements may provide insight into refinement processes that occur during processing and determine the grain boundary character as well as transformation behavior upon subsequent annealing. Future reports will examine transformations during deformation.

Experimental

Supral 2004 material was obtained as 2.0 mm sheet in the as-processed condition from Superform-USA, Inc. Chemical composition data are provided in Table I. Coupons of as-processed material and material given a 30 mins. annealing treatment at 450°C were prepared for EBSD examination at locations near the midplane of the rolled sheet. Details of EBSD examination have been

Table I. *Alloy composition (wt%) for the as-received SUPRAL 2004.*

	Cu	Zr	Fe	Si	Zn	Mn	Mg	Ti	Li	Al
Cast No. 2004F013	5.66	0.37	0.14	0.06	0.029	0.013	0.003	0.005	0.0001	bal

provided by Randle [12] and specific procedures employed here have been given previously [13-18]. The electron beam of the SEM, operating in the spot mode, is initially positioned and a diffraction pattern is captured, automatically indexed and stored. The data are in the form of the three Euler angles specifying the orientation of the crystal relative to axes associated with the final cold rolling stage of the prior TMP. The electron beam is then moved, pixel by pixel, until a new diffraction pattern is observed, collected, indexed and the corresponding data also stored. The distance between successive patterns is also recorded.

The resulting microtexture data may be displayed in the form of discrete pole figures or orientation distribution functions. The successive diffraction patterns correspond to the orientations of adjacent grains and so the disorientation of the intervening boundary may be calculated as the minimum rotation angle about an arbitrary axis to bring the neighboring crystals into coincidence. Here, the specific term disorientation is used to refer to the minimum rotation angle to bring *adjacent* crystals into coincidence; disorientations of adjacent grains were calculated and the data represented by histograms of relative number versus disorientation angle. The corresponding rotation axes were also determined and displayed within a unit triangle as a grain misorientation texture [12]. Finally, back scattered electron (BSE) micrographs were obtained of the regions examined.

Results

Microstructural and microtextural data for as-processed material are shown in Fig. 1. The BSE micrograph of the as-processed material, Fig. 1(a), reveals a dispersion of coarse θ (Al_2Cu) particles and a highly directional microstructure due to mechanical fibering introduced in the final cold rolling process. The original grains have been reduced to a thickness of ≈ 3.0 μm and elongated in the rolling direction. The texture data shown in the discrete Euler plot of Fig. 1 (b) comprise a deformation texture. The successive orientations were captured at ≈ 0.5 μm spacings, reflecting the presence of a fine, cellular structure in the as-processed material. While several components along the β fiber are present, the Brass (Br) component and nearby orientations related to Br by rotations about the sheet normal will be of particular interest here.

Corresponding microstructural and microtextural data for material annealed 30 mins. at 450°C are provided in Fig. 2. A refined substructure 0.5 to 1.0μm in size has become apparent as shown in Fig. 2(b) but the Euler plot in Fig. 2(b) reveals a deformation texture similar to that of the as-processed condition. Following annealing the texture has become somewhat sharper and the relative populations of the various orientations have shifted somewhat but no new texture components have become apparent.

The Relationship between Grain Boundaries and Microtexture

Examination of the microtexture data for these two conditions reveals that there are two components nearby Br but rotated relative to Br about the sheet normal. The locations of Br and these two nearby components, here designated Br(R_1) and Br(R_2), are provided in Table II. For

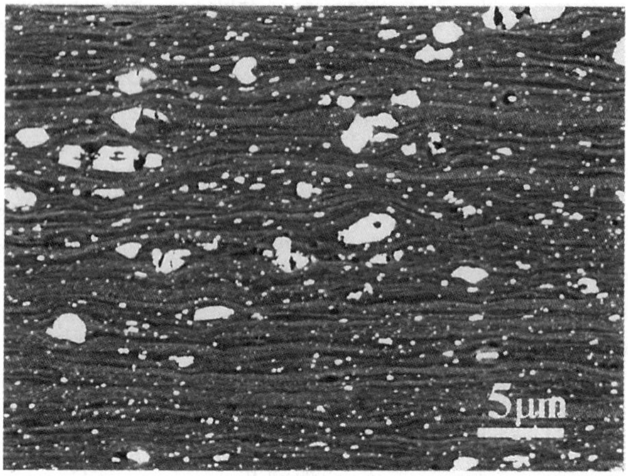

a)

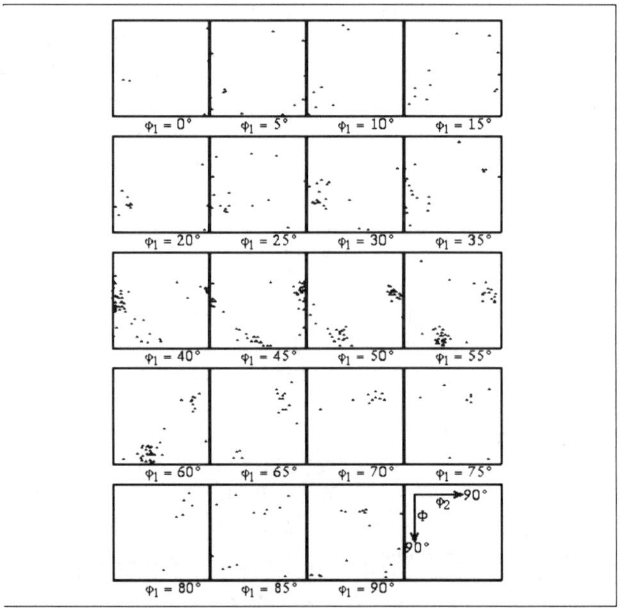

b)

Fig. 1 Microstructure of Supral 2004 as processed. a) BSE micrograph. The rolling direction is the horizontal. b) Discrete Euler plot showing the main texture components.

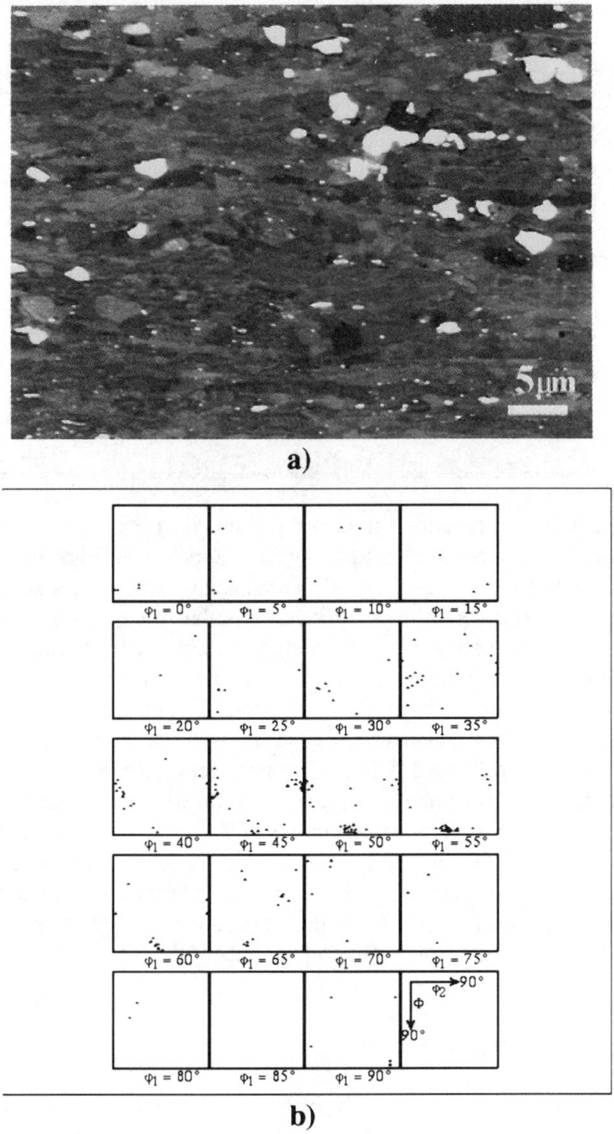

Fig. 2 Microstructure of Supral 2004 annealed for 30 min at 450C a) BSE micrograph. The rolling direction is the horizontal. b) Discrete Euler plot showing the main texture components.

each of these components there are two distinct variants related to one another by the orthotropic symmetry of the rolled sheet. The corresponding Euler angles are also given in Table II for each of the symmetric variants.

Table II. *Euler Angles for the Symmetric Variants of the Br and Nearby Texture Components.*

Component	Designation (Rolling Plane)[Rolling Direction]	ϕ_1	Φ	ϕ_2
Br1	(110) $[1\bar{1}2]$	55°	90°	45°
Br2	(011) $[2\bar{1}1]$	35°	45°	0°
Br(R_1)1	(110) $[3\bar{3}5]$	50°	90°	45°
Br(R_1)2	(011) $[5\bar{3}3]$	40°	45°	0°
Br(R_2)1	(110) $[3\bar{2}3]$	45°	90°	45°
Br(R_2)2	(011) $[3\bar{2}2]$	45°	45°	0°

Grain boundary data for these two conditions are presented in Figure 3. A bimodal distribution of disorientation angles in the as-processed material is apparent in Fig. 3(a). Again, successive orientation measurements were obtained at $\approx 0.5\mu m$ spacings and so these data reflect mainly the distribution of disorientations within a refined, cellular dislocation structure and not the disorientation distribution of the prior grain structure. The grain misorientation texture data for this processing condition are included in the inset at the right in Fig. 3(a) and show that the rotation axes for MMBs (data for boundaries of 5° - 15° are indicated by open symbols) are randomly distributed within the unit triangle. However, the rotation axes for HABs of 55° - 62.8° disorientation, shown by the filled symbols in the inset, generally lie along the arc from (111) to (110). The distribution of disorientation angles is more distinctly bimodal after annealing at 450°/C for 30 mins. (Fig. 3(b)); the population of MMBs appears to have increased while that of boundaries between 20° and 50° has gone down. The rotation axes for these MMBs are still randomly distributed in the unit triangle. However, the rotation axes for the 55° - 62.8° HABs now tend to be clustered nearer to (111). In the next section, it will be shown that these HABs may be modeled as interfaces between symmetric variants of the three texture components included in Table II.

A Model for the High-Angle Boundaries

Following an approach suggested in the work of Hughes and Hansen [11], the ideal orientations for both distinct variants of each of the three Br-related components are depicted in the {111} pole figures of Fig. 4. For the Br component, the two distinct variants, Br1 and Br2, are related to each other by a 60° rotation about a <111> axis; indeed, the particular rotation axis is that which lies parallel to the transverse direction (TD) of the rolled sheet. This 60°/<111> rotation about the TD is equivalent to the effect of a mirror plane normal to the TD (one of the elements of the orthotropic symmetry of the rolled sheet). Similarly, it is possible to show that the symmetric variants of Br(R_1), Br(R_1)1 and Br(R_1)2, are related by a 60.6° rotation about a <332> which, again, is parallel to TD. Also, the symmetric variants of the third component in Table II, Br(R_2)1 and Br(R_2)2, are related by a rotation of 62.7° about a <552>, which is parallel to TD. Additional possible disorientations are indicated in Fig. 4, such as that between Br1 and Br(R_2)2,

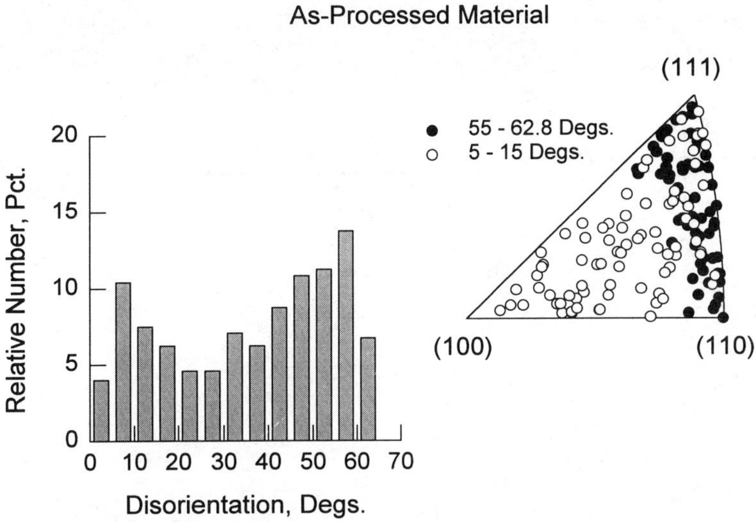

(a)

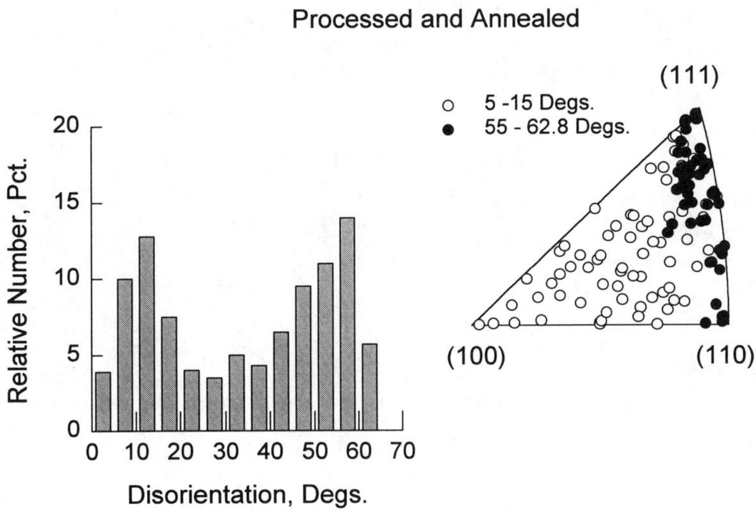

(b)

Fig. 3 Grain boundary misorientation and grain misorientation texture data for (a) as-processed and (b) processed and annealed Supral 2004 material.

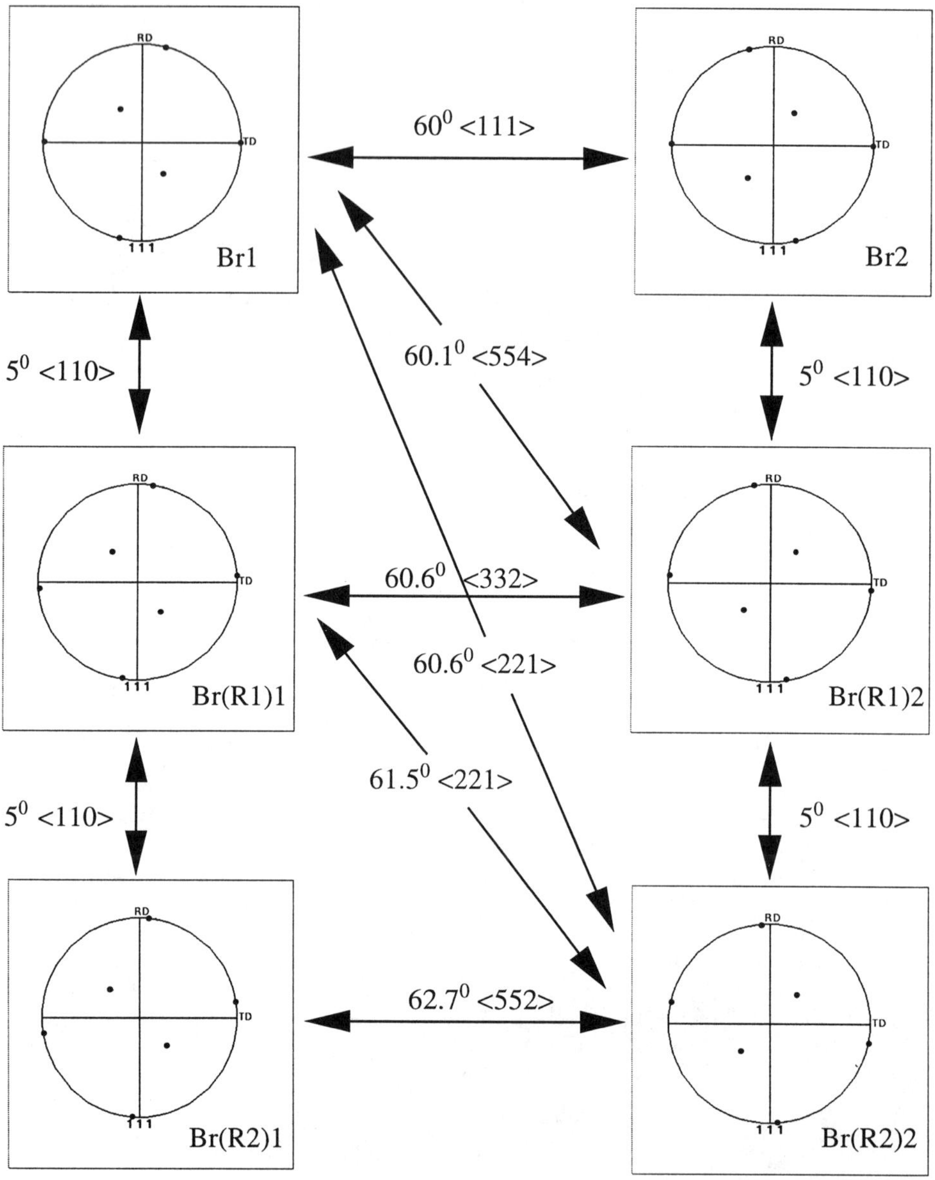

Fig.4 A schematic diagram illustrating the relative locations on {111} pole figures of the symmetric varients for the main texture components present in Supral 2004, and the corresponding axis/angle pairs of the boundaries separating grains oriented as symmetric varients..

60.6 about a <221>, and small rotations of either 5° or 10° about a <110>(specifically, the sheet normal).

Together, the data of Figs. 3 and 4 indicate that the HABs with disorientation angles of 55° - 62.8° may be modeled as interfaces between symmetric variants of these Br-type components of the texture. All of the rotation axes predicted in Fig. 4 for boundaries in this disorientation range lie along the arc connecting (111) and (110), as observed in the data of Fig. 3. It is noteworthy that there are no MMBs (disorientations of 5° - 15°) with rotation axes near (110). This suggests that individual grains, or blocks within grains, rotate toward stable end orientations near the Br orientation during the warm and cold rolling of the prior TMP and that adjacent grains or blocks rotate toward symmetric orientations. While these interfaces may not be well-defined at the conclusion of rolling, upon annealing they apparently evolve into grain boundaries and there are no low-angle boundaries separating regions within a given texture variant. After annealing there is an increased relative population of boundaries with rotation axes near those for the Br (<111>) and Br(R_1) (<554>), suggesting a preference for orientations near Br.

An idealized model for the distribution of boundary disorientation is depicted in Fig. 5. The model includes contributions for the boundaries of the deformation induced cellular dislocation structure produced by rolling as well as for HABs that are interfaces between symmetric variants of the Br related texture components here. The HAB distribution is indicated at the right in Fig. 5. If interfaces between symmetric variants are present in equal numbers, five of six ideal orientations have disorientation angles exceeding 60° and this model predicts a higher population of boundaries in the range > 60° than in the 55° - 60° range. However, this approach does not recognize that the observed rotation axes aren't exactly on the (111) - (110) arc. Also, this model does not consider other components along the β fiber, which may account for the spread in the population towards lower misorientation angles.

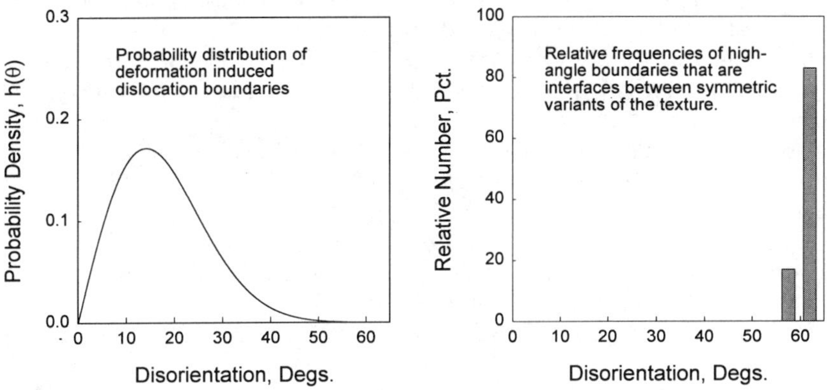

Fig. 5 The grain boundary misorientation distribution may be modeled as the sum of the probability distribution for dislocation boundaries and the relative number of boundaries representing interfaces between symmetric variants of the texture.

The LABs and MMBs in the Disorientation Distribution

An empirical relationship of the form

$$h(u) = u^{(n-1)} \exp(-u^{n})$$

where $u = \theta/\theta_{ave}$, θ is the disorientation angle and n is an exponent, has been shown to represent the probability distribution of the deformation-induced dislocation boundaries in several severely deformed pure metals. Such a distribution evidently reflects a buildup in the disorientations of cell walls due to dislocation interactions that occur during the straining of rolling and, in many cases, observed distributions of disorientation angles were well described by values of θ_{ave} in the range of 10° - 15° [11]. Here, this function is plotted in Fig. 5 using values of the parameters $n = 2$ and $\theta_{ave} = 15°$. When combined with a distribution related to the disorientation angles given by the interfaces between texture variants this relationship provides a reasonable initial approximation to the experimental disorientation distributions obtained by EBSD methods.

References

1. R. Grimes: *Advances and Future Directions in Superplastic Materials*, NATO-AGARD Lecture Series, 1988, No.168, pp. 8.1-8.16.
2. B.M. Watts, M.J. Stowell, B.L. Baike, D.G.E. Owen: *Metal Sci. J.*, 1976, vol. 10, No. 6, pp. 189-197.
3. N.E. Paton and C.H. Hamilton: US Patent 4,092,181, 1978.
4. J.A. Wert, N.E. Paton, C.H. Hamilton, and M.W. Mahoney: *Metall. Trans. A.*, 1981, vol. 12A, pp. 1267-1276.
5. F.J. Humphreys: *Acta Metall.*, 1977, vol. 25, pp. 1323-44.
6. R.D Doherty, G. Gottsttein, J. Hirsch, W.B. Hutchinson, K. Lucke, E. Nes, and P.J. Wilbrandt: *Report of Panel on Recrystallization Textures: Mechanisms and Experiments*, ICTOM 8, J.S. Kallend and G. Gottstein (eds.), TMS, 1988.
7. R.D. Doherty, D.A. Hughes, F.J.Humphreys, J.J. Jonas, D. Juul Jensen, M.E. Kassner, W.E. King, T.R. McNelley, H.J. McQueen and A.D. Rollett: *Mater. Sci. Engng*, 1997 (in press).
8. F.J. Humphreys: *Acta Mater.*,1997, vol. 45, pp.4231-4240.
9. R.H. Bricknell and J.W. Edington: *Acta Metall.*, 1979, vol. 27, pp. 1303-1311.
10. J.W. Edington: *Metall. Trans. A*, 1982, vol. 13A, pp.703-15.
11. D.A. Hughes and N. Hansen: *Acta Mater.*, 1997, vol. 45, pp.3871-3886.
12. V. Randle: *Microtexture Determination and Its Applications*, The Institute of Metals, 1992.
13. T.R. McNelley and M.E. McMahon: *J. of Metals*,1996, 48, 2.
14. T.R. McNelley and M.E. McMahon: *Metall. and Mater. Trans. A.*, 1996, vol. 27A, pp. 2252-2262.
15. M.E. McMahon, *Grain Boundary Development in Superplastic Aluminum Alloys*, Ph.D. Dissertation, Naval Postgraduate School, Monterey, CA, 1996
16. T.R. McNelley, M.E. McMahon and S.J. Hales: *Scri. Mater.*, 1997, vol. 36, pp. 369-375.
17. T.R. McNelley and M.E. McMahon: *Metall. and Mater. Trans. A.*, 1997 (in press)
18. M.T. Pérez-Prado, T.R. McNelley, O.A. Ruano and G. González-Doncel: *Metall. and Mater. Trans. A.*, 1997 (in review).

Behavior of Titanium Alloy under Uniaxial and Biaxial Superplastic Conditions [1]

Z. Chen, Z. Li, P. Dang and N. Chandra

Department of Mechanical Engineering, FAMU-FSU College of Engineering,
Florida A&M University, Florida State University
Tallahassee, FL 32310, U.S.A.

Abstract

The validity of extending constitutive relationships established under uniaxial condition to biaxial forming is an important issue from both basic and applied perspectives. The deformation of titanium alloy is studied under both uniaxial and biaxial states of stress to understand this relationship. Based on the mechanical constitutive equation obtained from uniaxial test, a model based on finite element method (FEM) was established to predict the superplastic deformation (SPD) of biaxial cones. The cones are designed to be subjected to balanced biaxial state of stress under a constant pressure loading. In the multi-cone test, four cones of different diameters are concurrently formed when they are subjected to different biaxial strain rate conditions. Sensitivity of material parameters obtained under uniaxial conditions are critically examined. Thickness profile and metallurgical state of the cones are also analyzed.

[1]Work supported by Army Research Office (Drs. Andrew Crowson and Wilbur Simmons).

INTRODUCTION

Superplastic forming (SPF) [1-4] is an advanced method of sheet metal forming resulting in tremendous cost and weight savings and has a variety of applications in aerospace, architecture, rail transportation, medicine, and even recreation industries. The determination of process parameters, e.g., pressure-time and thickness profile, is crucial not only in reducing the process time but also in forming complex components without failure. Finite element method is widely used as an analysis and design tool for application of SPF to industries. In the process modeling of complex components, a necessary input is the constitutive equation of superplastic materials, the geometry of the part, and the frictional conditions. Typically, the constitutive equations are formulated on the basis of uniaxial tests performed under isothermal conditions, in which the flow stress is related to the strain-rate in the form of a power law model. In other phenomenological models, the stress is additionally related to the strain level, grain size, and grain evolution. All the material parameters (e.g. K and m) are determined solely from the uniaxial tests.

Actual sheet forming operations involve a minimum of two dimensional state of loading, and the primary query addressed in this paper is whether the assumption that relationships established in the uniaxial state will be valid in biaxial state of stress or not. When a biaxial or triaxial state of stress needs to be converted to an equivalent uniaxial value, any energy based criterion (e.g. shear energy based von Mises criterion) is invoked. These criteria are strictly applicable to isotropic materials and do not consider the evolving state of material. The question then is how applicable is this theory to large deformation problems in general, and superplasticity in particular. We examine this problem by applying uniaxial material constants in modeling deformation under biaxial states of loading and compare the thickness distributions between the predicted values and the experimental results. We then modify the material parameters in order to better match the biaxial tests.

Another fundamental issue is the understanding of the grain growth kinetics under uniaxial and biaxial conditions. Grain growth in a material occurs under the influence of temperature and is usually enhanced by the applied strain level. Thus the rate of grain growth can be separated into that contributed primarily by: $\dot{d} = \dot{d}_t + \dot{d}_m$ where the $\dot{d}_t$ is the contribution under zero strain conditions (purely thermal), and $\dot{d}_m$ under specified strain level purely mechanical). We believe that though $\dot{d}_t$ may not be affected by uniaxial/biaxial states of loading, $\dot{d}_m$ will be influenced whether the loading is under uniaxial or biaxial conditions. It is to be realized that different materials show different levels of response to the loading conditions effect. It will be shown later that Ti-6Al-4V is affected marginally whereas Al 5083 shows marked difference.

In this paper, FEM is used as a design tool in the prediction of process parameters for superplastic sheet metal forming process [6-7]. Ti-6Al-4V alloy is selected as the model material. The constitutive relationship established under uniaxial condition was used to model to biaxial forming. The stress-strain rate relationship, optimal strain rate at specific temperature and strain-rate sensitivity were obtained from uniaxial test. The biaxial forming conditions and the resulting cone of the SPF were next examined.

PROCESS MODELING USING FEM

During the FEM modeling of SPD process, nonlinearities in the problem arise due to complex die geometry, materials, nonconservative pressure loading, contact and friction between sheet material and die. Each of these quantities introduces an additional stiffness term which in turn contributes to the increased complexity in formulation and solution procedures. The process modeling is carried out using non linear finite element method and implemented in a commercial code MARC using user subroutines. Friction between the ceramic die and the sheet metal is an input parameter which influences the final thickness profile. A friction value of f = 0.3 to 1.2 is assumed in the computations.

The forming pressure (kinetic quantity as input) required to maintain optimum strain rate needs to be determined, as a function of forming time. A user subroutine is developed in which the classical power-law constitutive equation is used. A four 60-degree-cone model was developed using MENTAT and MARC on a Silicon Graphics Origin 200 server. The diameter for each cone is 1.75, 2.0, 2.25 and 2.5", respectively. The center of each cone was located in the corner of a 4" by 4" square. Because of symmetry, a quarter section of each cone plus central region was considered in the model. Totally, 836 four-node isoparametric membrane elements were used. Figure 1 shows the finite element discretization of the domain. Initial thickness of Ti-6Al-4V sheet was 0.038". 400 increments are used for each solution, and approximately 40 min cpu time is required for each solution.

EXPERIMENTAL: UNIAXIAL AND BIAXIAL TESTING

Uniaxial Tensile Testing

A large volume of experimental data are available for uniaxial superplasticity of Ti-6Al-4V; hence, uniaxial test was conducted only at 920°C. Samples were cut from Ti-6Al-4V sheet with a thickness of 0.08 and average grain size of 10 μm. The dimension of sample is 0.6" x 0.2" (gauge length by width). Tensile testing was conducted in MTS machine under vacuum of 2 x 10^{-5} torr. Jump strain rate method [8, 9] was used to determine the relationship of true stress and strain rate. The sample was first prestrained to 20%. Each strain rate lasted four minutes before next strain rate was applied. Upon completion of the jump test, the sample had a total (engineering) strain of 250 per cent. Maximum load in each strain rate step was used to calculate true stress.

Biaxial Forming

Biaxial forming test was conducted in a press with a controlled gas pressure and die temperature. Sealing was obtained using superalloy die with a section area of 12" x 12". The four-cone workpiece was formed on fast-switchable ceramic die located inside the superalloy die. Argon gas from gas cylinder was exerted on the Ti-6Al-4V sheet and the pressure was controlled by a pressure controller. Pressure-time profile obtained from the FEM model was used to form the cones. Because of the big variation of interfacial friction between metal sheet and ceramic die, the effect of friction was studied in the model.

RESULTS AND DISCUSSION

The stress-strain rate relationship from unitensile test was shown in Figure 2. The true stress linearly increases with increasing strain rate in the $\log\sigma - \log\dot{\epsilon}$ plot. When a power law fitting ($\sigma = K\dot{\epsilon}^m$) was adopted, the obtained fitting equation was $3452.9\dot{\epsilon}^{0.72}$ MPa, with a correlative coefficient (R) equal to 0.9886 for our result. The results from Hamilton [5] for the grain size of 9 μm and 11.6 μm are also shown in Figure 2 for comparison purposes. The material parameters obtained in this work was next used in FEM to model biaxial forming process.

In FEM modeling, a target strain rate of 5x10^{-4} s^{-1} was set with cutoff strain rate of 10% of the target strain rate. When maximum strain rate over all the elements is larger than the cutoff strain rate, average strain rate over the elements with a strain rate bigger than the cutoff strain rate was used. Otherwise, an average over the entire FEM domain was used. Figure 3 shows simulation result from FEM after 270 increment, and Figure 4 shows the corresponding pressure-time profile from FEM computations. It is seen that pressure ripples with time. This phenomenon comes from the adjustment around the target strain rate. That is, if the calculated strain rate is above the target strain rate, pressure will go down from the previous value by a factor determined by the ratio of the target true stress to the calculated true stress.

Effect of m and f on Pressure-Time Profile

The higher the strain rate sensitivity (m), the more uniform superplastic deformation. Therefore, the free dome area has a larger thickness and hence a higher required pressure. On the other hand, increasing the level of coefficient of friction (f) resulted in decreasing pressure. For example, when f increases from 0.3 to 1.2, pressure decreased approximately 5 to 10 psi. Higher friction prevents slip of material and causes deformation to be less uniform. If metal sheet contacts die, further deformation is difficult. Therefore, metal sheet thins faster when the coefficient of friction is higher. Accordingly, the load required to maintain the target strain rate is reduced.

For cone forming, strain rate over FEM domain is not uniform for geometrical reasons. We selected the center node of each cone to represent the maximum strain rate in the corresponding cone (the center node is final point touching ceramic die). Figure 5 shows the variation of equivalent strain rate with time for each cone for f=1.2. The horizontal line in the figure shows the target strain rate. As mentioned before, increasing coefficient of friction would decrease pressure. Accordingly, equivalent strain rate would decrease with increasing coefficient of friction. For cones with different diameters, larger diameter cone has higher strain rate than smaller diameter cone under the same value of pressure. In our case, the biggest ratio of diameter is 2.5 /1.75 = 1.43. From simple estimation, the largest difference between strain rate will be 0.64 or 64 per cent, as shown in Table I.

Thickness prediction from FEM model is shown in Figure 6. As the forming proceeds into the cones, its thickness decreases. In our case, the smaller diameter cone thins faster than the larger diameter cone since the volume of material in smaller cone is smaller. Because it is impossible for the different diameter cones to reach the same depth under the same pressure, the thickness difference becomes less obvious below certain depth. Increasing coefficient of friction (f) or decreasing strain rate sensitivity (m) result in a quick thinning process because the material participating in the forming is limited to non-contacted parts of the sheet.

The pressure-time from FEM computation was not directly applied to the biaxial superplastic forming for practical reasons. In the real-time forming of Ti-6Al-4V sheet at 920°C with a target strain rate of $5x10^{-4}$ s^{-1}, pressure increased gradually to 110 psi, and this pressure was kept unchanged for about 70 min instead of 30 min from FEM modeling. After superplastic forming, the workpiece was kept in the furnace for extra 6 hours until temperature went down to 600°C. These adjustments were made based on the actual conditions in the forming press.

Figure 7 shows thickness profile for the formed workpiece compared with the results from FEM. Thickness of each cone decreases with the increasing depth and there is some evidence of uniform thinning. As shown in the figure, thickness decreases with increasing depth in both simulations and experiments. Our predicted values are very close to the experimental results, though actual thickness was found to be less than that predicted by the model. By decreasing strain rate sensitivity or increasing the coefficient of friction, the difference between the experimental results and FEM result can be reduced. In other words, cone is thinning at a lower rate in the model than in the experiment. As depth of the cone increases, the difference gets larger.

Finally, the microstructure of the superplastically formed cone for Ti-6Al-4V was examined and the results were given in Figure 8 (a) and (b). It is seen that the grain structure and grain size of the deformed cone are very similar to the undeformed material. This observation indicates that there is no significant effect of superplastic strain on the microstructure of Ti-6Al-4V. This is contrast to the Al-5083 material, in which grain growth occurs during the superplastic deformation, as shown in Figure 8 (c) and (d). The observation needs to be further studied to verify the grain growth in superplastic forming process.

SUMMARY

The deformation of Ti-6Al-4V was studied under both uniaxial and biaxial states of stress. The stress-strain rate relationship from unitensile test can be described by a power law fitting equation $3452.9\dot{\epsilon}^{0.72}$ MPa. A FEM model was also developed to predict SPD process of biaxial cones. When the material parameters obtained from uniaxial test were applied in the modeling of biaxial forming process, the predicted thickness profile is approximately equivalent to the experimental biaxial forming result. However, a better fit to experiments is seen when m is reduced from 0.72 to 0.55. This suggests that m may be a function of the state of stress. This assumption has wider ramifications and hence need to examined very carefully. The evolution of microstructure under uniaxial and biaxial conditions is seen to to different for different materials systems. For Ti-6Al-4V alloy (a highly superplastic materials) there appears to be no influence, whereas in Al5083 (limited superplasticity), there is a significant effect. This aspect also needs to be addressed carefully.

REFERENCES

1. J. Pilling, and N. Ridley, Superplasticity in Crystalline Solids, (London, UK: The Institute of Metals, 1989).

2. N. Chandra, and D. Kannan, "Superplastic Sheet Metal Forming of Generalized Cup, Part I: Uniform Thinning", Journal of Material Engineering and Performance, 1 (1993), 801-812.

3. N. Chandra, and D. Kannan, "Superplastic Sheet Metal Forming of Generalized Cup, Part II: Non-Uniform Thinning", Journal of Material Engineering and Performance, 1 (1993), 813-822.

4. S. C. Rama, and N. Chandra, "Development of a Pressure-Time Predicting Algorithm for a Superplastic Forming Process", International Journal of Non-linear Mechanics, 26 (1991), 711-725.

5. A. K. Ghosh, and C. H. Hamilton, "Mechanical Behavior and Hardening Characteristics of a Superplastic Ti-6Al-4V Alloy," Metallurgical Transactions A, 10A (1979), 699-706.

6. N. Chandra, and S. C. Rama, "Application of Finite Element Method to the Design of Superplastic Forming Process", ASME Journal for Engineering and Industry, 114 (1992), 452-459.

7. R. Sadeghi, and Z. Puesell, "Finite Element Modeling of Superplastic Forming Using Analytical Contact Surfaces," Materials Science Forum, 243-245 (1997), 719-728.

8. J. Hedworth, and M. J. Stowell, "The Measurement of Strain-Rate Sensitivity in Superplastic Alloys," Journal of Materials Science, 6 (1971), 1061-1069.

9. G. Rai, and N. J. Grant, "On the Measurement of Superplasticity in an Al-Cu Alloy," Metallurgical Transactions A, 6A (1975), 385-390.

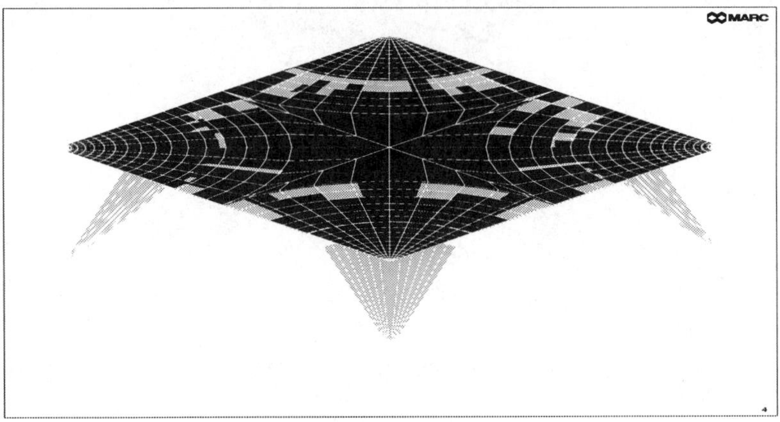

Figure 1 - FEM mesh.

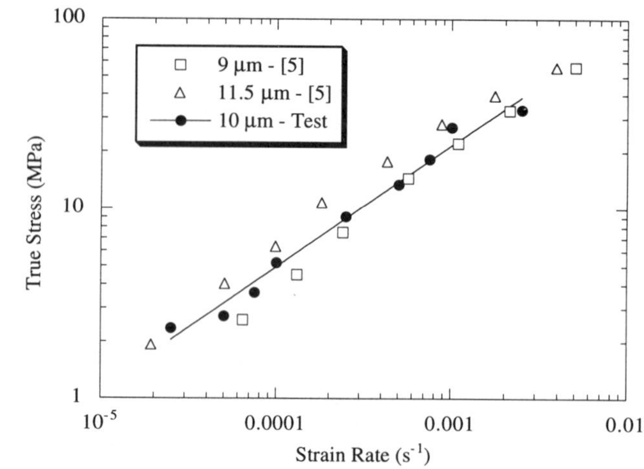

Figure 2 - Relationship between true stress and strain rate for Ti-6Al-4V.

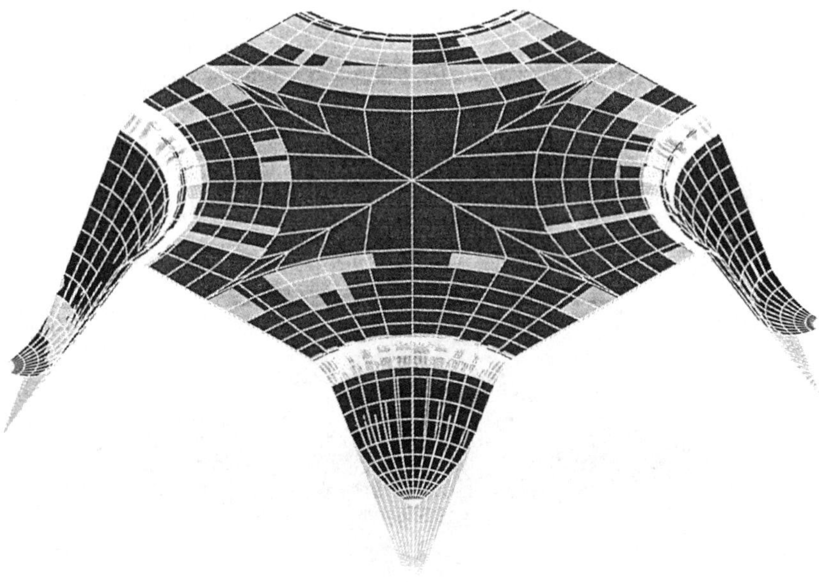

Figure 3 - Deformed Ti-6Al-4V alloy sheet in FEM simulation.

Table I Average equivalent strain rate and ratio to that of 1.75 in cone

	Coefficient of friction f = 0.3		Coefficient of friction f = 1.2	
Cone Diameter	Average Strain rate	ratio to 1.75 in	Average Strain rate	ratio to 1.75 in
1.75 in	0.000510	1.00	0.000447	1.00
2.00 in	0.000598	1.17	0.000526	1.18
2.25 in	0.000688	1.35	0.000615	1.38
2.50 in	0.000783	1.54	0.000710	1.59

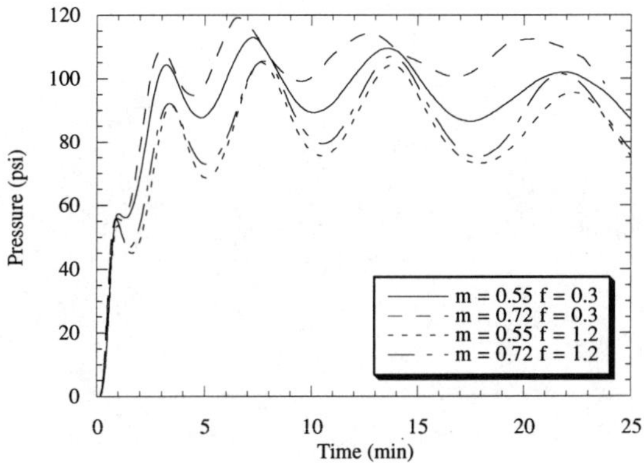

Figure 4 - Pressure-time profiles from FEM for different strain rate sensitivity and coefficient of friction.

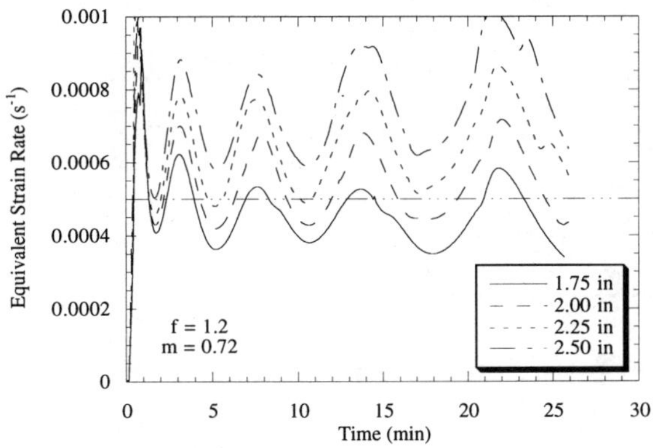

Figure 5 - Variation of equivalent strain rate with time for each cone at m = 0.72 and f = 1.2.

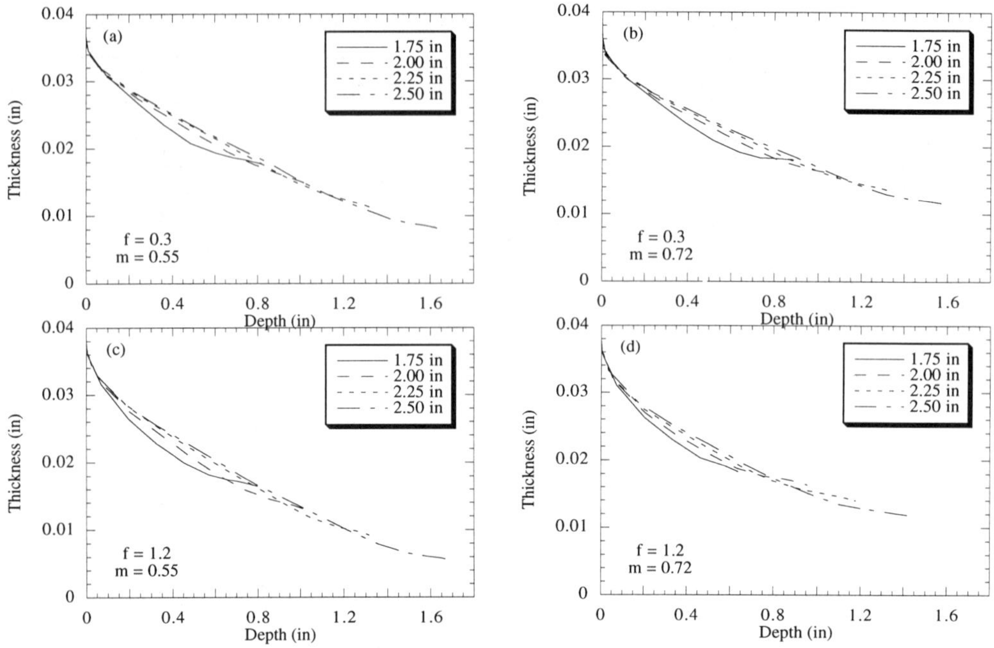

Figure 6 - Variation of thickness with depth of cone for m = 0.55 and 0.72 and f = 0.3 and 1.2.

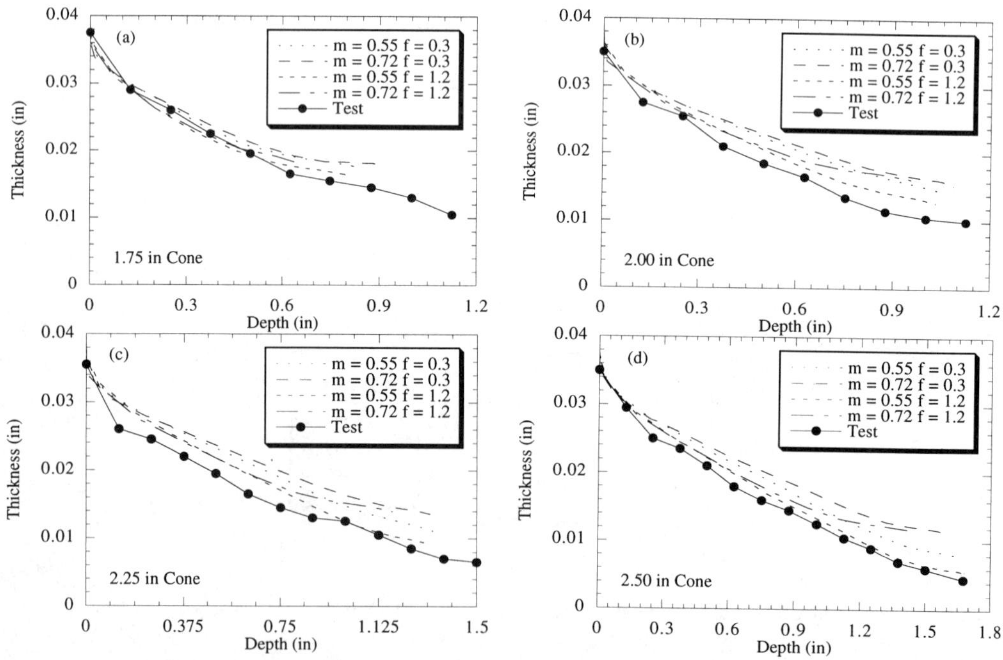

Figure 7 - Comparison between experimental thickness and FEM prediction and the effect of coefficient of friction.

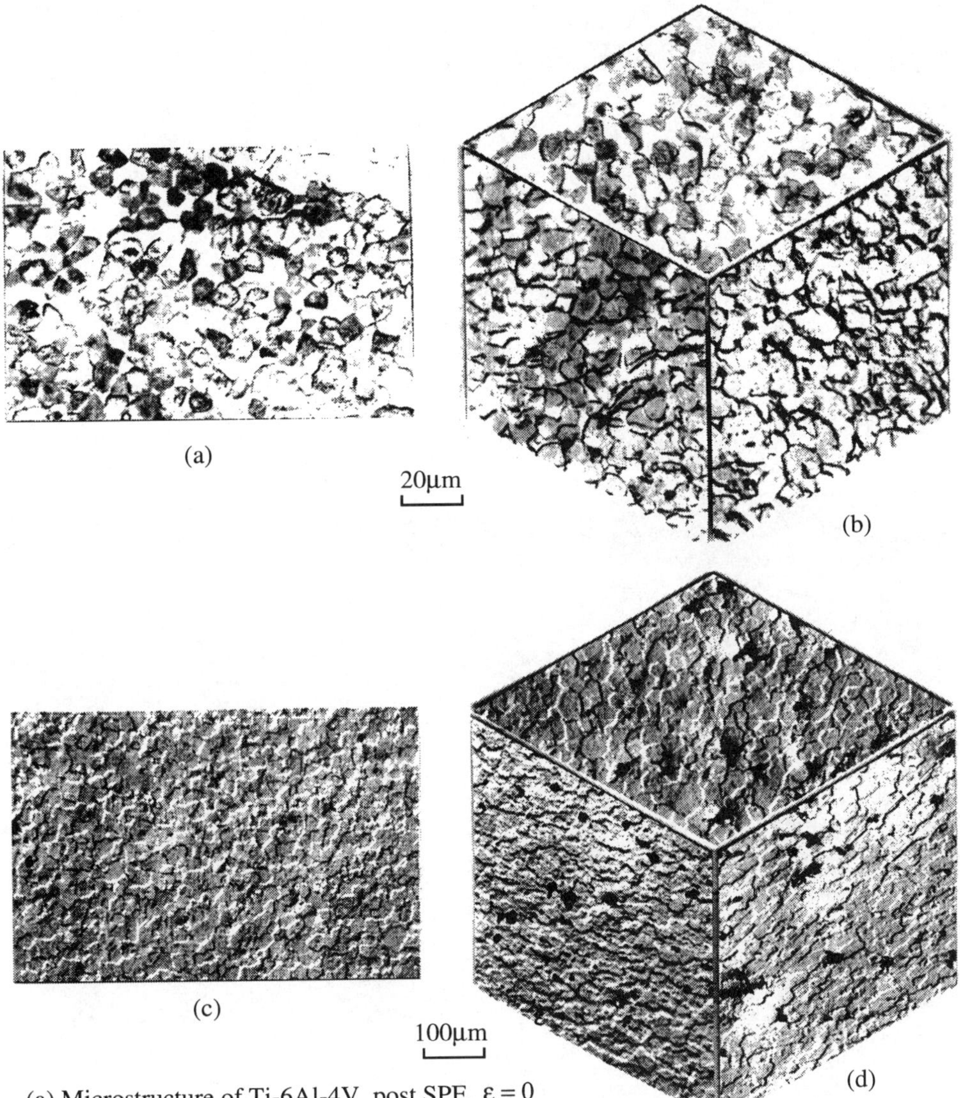

(a) Microstructure of Ti-6Al-4V, post SPF, $\varepsilon = 0$

(b) Microstructure of Ti-6Al-4V, post SPF $\varepsilon = 1.2$

(c) Al-5083, post SPF, $\varepsilon = 0$

(d) Al-5083, post SPF, $\varepsilon = 1.2$

Figure 8 - Microstructure of Ti-6Al-4V and Al-5083 after SPF.

CONSTITUTIVE EQUATIONS FOR SUPERPLASTIC FLOW IN ALUMINUM MATRIX COMPOSITES

Mamoru Mabuchi[#] and Kenji Higashi[+]

[#]National Industrial Research Institute of Nagoya, Hirate-cho, Kita-ku, Nagoya 462, Japan
[+]College of Engineering, Department of Mechanical Systems Engineering (present address : Department of Metallurgy and Materials Science), Osaka Prefecture University, Gakuen-cho, Sakai, Osaka 593, Japan

Abstract

A constitutive equation for superplasticity is analyzed for aluminum matrix composites with discontinuous reinforcements based on the results of the tensile tests, and the effects of the reinforcements on superplastic flow are investigated. The analysis suggests that the strongest strengthening process of the dislocation-pile up mechanism and the diffusional relaxation-limitation or dislocation by-pass mechanism affects the rate-controlling process.

Modeling the Mechanical Response of
Structural Materials
Edited by E.M. Taleff and R.K. Mahidhara
The Minerals, Metals & Materials Society, 1997

1. INTRODUCTION

It has been demonstrated that aluminum matrix composites with discontinuous reinforcements show superplastic behavior.(1-7) In particular, some composites showed superplasticity at high strain rates ($>10^{-2}$ s^{-1}). High strain rate superplasticity is very attractive for commercial applications because one of current drawbacks in superplastic forming technology is a slow forming rate which is typically $\sim 10^{-4}$ s^{-1}.

It has been reported that superplastic behavior is drastically changed by the presence of a liquid phase.(10) However, it is important to investigate superplastic properties in a solid state for understanding the mechanisms of superplasticity in metal matrix composites. Many data of the tensile tests are available in a temperature range below the melting temperature for superplastic aluminum matrix composites.(2-9) In the present investigation, a constitutive equation for superplasticity is analyzed for the aluminum matrix composites based on these experimental results. In addition, the effects of the reinforcements on superplastic flow are investigated.

2. ANALYSIS

The superplastic aluminum matrix composites used for analysis in the present study are listed in Table 1. The activation energy for superplastic flow in a temperature range below the partial

Table 1 The superplastic aluminum matrix composites used for analysis in the present study.

	Volume fraction of reinforcement/%	Morphology of reinforcement	Size of reinforcement/ μm	Strengthening coefficient	References
SiC*p*/Al-Zn-Cu-Mg	10	particulate	5	1.19	(2)
SiC*p*/Al-Zn-Cu-Mg	15	particulate	5	1.35	(2)
SiC*p*/Al-Zn-Cu-Mg	20	particulate	5	1.54	(2)
SiC*p*/Al-Cu-Mg	15	particulate	8	1.35	(3)
SiC*w*/Al-Zn-Cu-Mg	12	whisker	---	2.50	(4)****
Si3N4*p*/Al-Mg	20	particulate	0.2 **	1.54 ***	(5)
Si3N4*p*/Al-Mg-Si	20	particulate	0.2	1.54	(6)
Si3N4*p*/Al-Mg-Si	20	particulate	0.5	1.54	(6)
Si3N4*p*/Al-Mg-Si	20	particulate	1	1.54	(6)
Si3N4*p*/Al-Zn-Mg	20	particulate	1	1.54	(9)
Si3N4*w*/Al-Mg-Si	20	whisker	8/0.5*	4.22	(6)
Si3N4*w*/Al-Mg-Si	20	whisker	8/0.5*	4.22	(8)
Si3N4*w*/Al-Zn-Mg	20	whisker	8/0.5*	4.22	(7)

* length/diameter
** unknown data
*** The length and diameter are taken to be 8, 0.5μm, respectively.
**** In this paper, two composites (SiCp/Al-Cu-Mg and SiCp/Al-Zn-Mg-Cu) were examined, however, the testing temperature of SiCp/Al-Zn-Mg-Cu is higher than the solidus temperature of the matrix material. Therefore SiCp/Al-Zn-Mg-Cu is not analyzed in the present work.

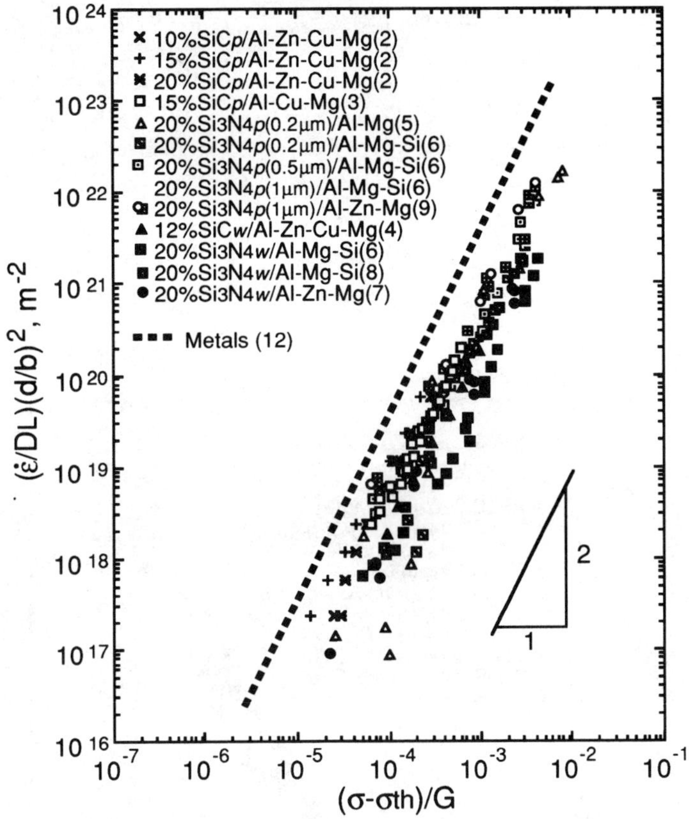

Fig. 1 The variation in $(\dot{\varepsilon}/D_L)(d/b)^2$ as a function of $(\sigma\text{-}\sigma_{th})/G$.

melting temperature for the aluminum matrix composites is close to that for lattice self-diffusion of aluminum.(11) Therefore, the $(\dot{\varepsilon}/D_L)(d/b)^2$ - $(\sigma\text{-}\sigma_{th})/G$ relation was analyzed from the data in a temperature range below the melting temperature, where $\dot{\varepsilon}$ is the strain rate, D_L is the lattice self-diffusion coefficient, d is the grain size, b is the Burgers vector, σ is the stress, σ_{th} is the threshold stress and G is the shear modulus. The plot is shown in Fig. 1. For reference, the relation for superplastic metals, which was developed by Sherby and Wadsworth(12), is plotted in the figure. It should be noted that as a whole, the strengthening for the particle-reinforced composites is lower than that for the whisker-reinforced composites. It is found that the superplastic behavior for the composites depends on the morphology of reinforcements.

The strain rate of composites may be given by(13,14)

$$\dot{\varepsilon} = \dot{\varepsilon}_o\left(\frac{\sigma}{a\sigma_o}\right)^n \qquad (1)$$

where a is the strengthening coefficient, σ_o is the reference stress and $\dot{\varepsilon}_o$ is the strain rate according to the reference stress. The strengthening coefficient is approximated from the calculation of elastic-perfectly plastic law by[13,14]

$$a \approx 1+2(2+c)f^{3/2} \qquad (2)$$

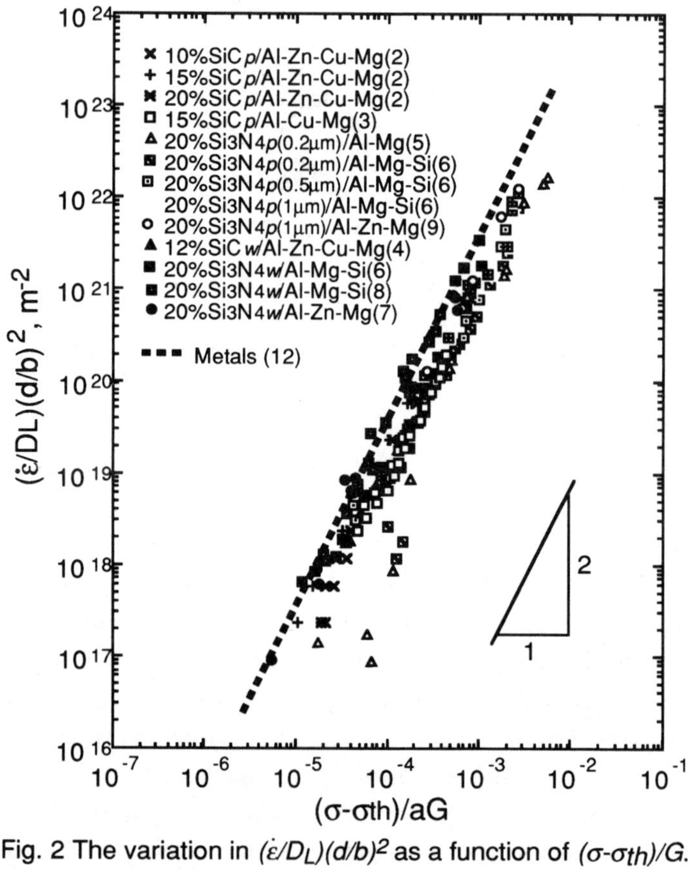

Fig. 2 The variation in $(\dot{\varepsilon}/D_L)(d/b)^2$ as a function of $(\sigma-\sigma_{th})/G$.

for fibers, where c is the aspect ratio of reinforcements and f is the volume fraction of reinforcements. A value of the strengthening coefficient for each composite is listed in Table 1. The strengthening coefficient depends on the morphology and volume fraction of reinforcements. Therefore the strain rate of composites may be given by

$$\dot{\varepsilon} = B\left(\frac{\sigma - \sigma_{th}}{aG}\right)^n \left(\frac{b}{d}\right)^p D \tag{3}$$

where B is a constant. The variation in $(\dot{\varepsilon}/D_L)(d/b)^2$ as a function of $(\sigma - \sigma_{th})/aG$ is shown in Fig. 2. It appears that the mechanical properties of the particulate-reinforced composites are roughly in agreement with those of the whisker-reinforced composites, compensated by the strengthening coefficient, however, there are still some deviations in the data shown in Fig. 2. The strengthening coefficient defined by eq. (2) does not depend on the reinforcement spacing and size, so effects of the reinforcement spacing and size are not taken into consideration in the above analyses. However, there is no reason why the reinforcement spacing and size have no effect on the deformation process. Pandy *et al.*(15) experimentally revealed that the reinforcement size affected the creep behavior of an aluminum matrix composite. Therefore, it is suggested that the deviations in Fig. 2 are attributed to effects of the reinforcement spacing and size on the deformation process.

Moving dislocations pile up due to the presence of reinforcements during superplastic deformation of metal matrix composites. This situation is similar to the case of dislocation pile up at grain boundaries at low temperatures, for example, at room temperature for metals. Hence the strengthening is likely increased with decreasing reinforcement spacing. On the other hand, Rösler *et al.*(14) analyzed limitation in relaxation of stress concentrations by diffusion and they revealed that the critical strain rate, below which the strengthening is eliminated and above which the strengthening is retained, increases with decreasing particle size. This suggests that the strengthening is retained more easily for larger particle size. Also Arzt *et al.*(16) theoretically analyzed by-pass of dislocations over reinforcements and they showed that the strain rate is inversely proportional to the reinforcement size. In these cases, the strengthening is expected to be increased with increasing reinforcement size.

The variation in $(\dot{\varepsilon}/D_L)(d/b)^2$ as a function of the reinforcement spacing, λ, is shown in Fig. 3, where λ is given by $(d_p/2)\sqrt{3\pi/2f}$(17), and the variation in $(\dot{\varepsilon}/D_L)(d/b)^2$ as a function of the reinforcement size, d_p, is shown in Fig. 4, where a value of $(\dot{\varepsilon}/D_L)(d/b)^2$ is determined at $(\sigma - \sigma_{th})/aG = 10^{-4}$. A value of d_p is taken to be $\sqrt{d_1 d_2}$ for the whisker-reinforced composites, where d_1 is the diameter of whiskers and d_2 is the length of whiskers. Inspection of Fig. 3 and

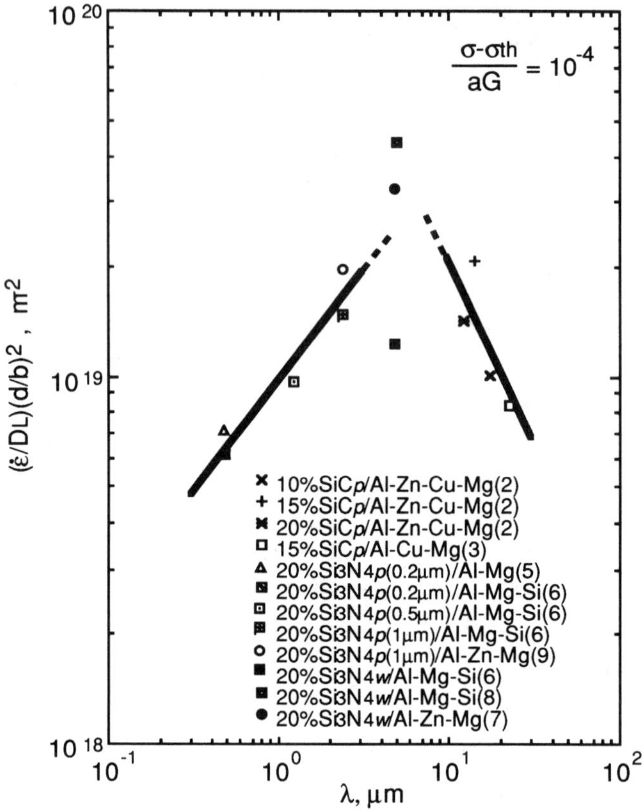

Fig. 3 The variation in $(\dot{\varepsilon}/D_L)(d/b)^2$ as a function of the reinforcement spacing λ for superplastic aluminum matrix composites.

Fig. 4 reveals that a value of $(\dot{\varepsilon}/D_L)(d/b)^2$ increases with the reinforcement spacing in a range $\lambda \leq 2$ μm and it decreases with the reinforcement size in a range $d_p \geq 5$ μm. This indicates that the dislocation-pile up mechanism affects the rate-controlling process for a small reinforcement spacing, on the other hand, the diffusional relaxation-limitation or dislocation by-pass mechanism affects the rate-controlling process for a large reinforcement size. It is suggested that the strongest strengthening effect among these processes affects the rate-controlling mechanism of superplastic flow of the composites.

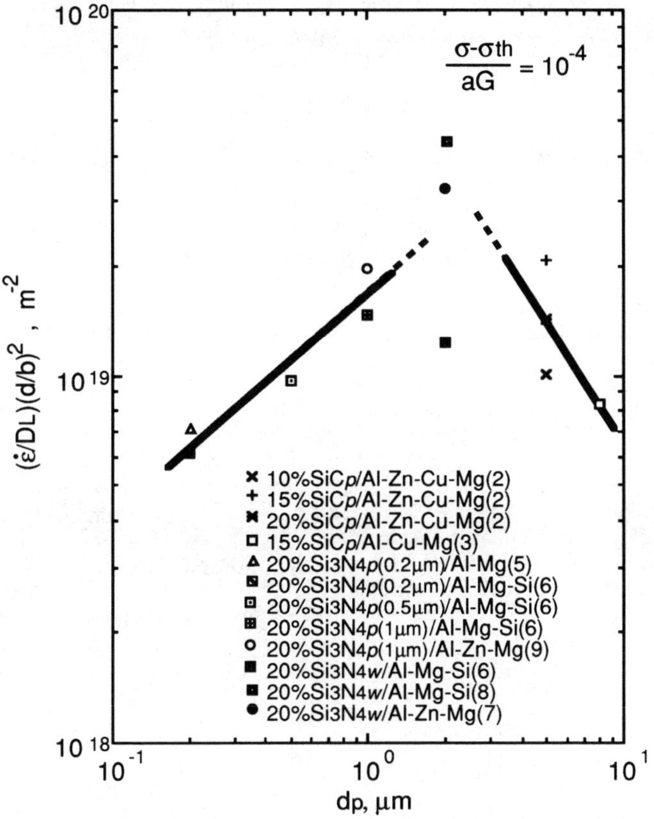

Fig. 4 The variation in $(\dot{\varepsilon}/D_L)(d/b)^2$ as a function of the reinforcement size d_p for superplastic aluminum matrix composites.

3. SUMMARY

A constitutive equation for superplasticity in a temperature range below the melting temperature was analyzed for aluminum matrix composites with discontinuous reinforcements based on the results of the tensile tests. The strengthening due to the presence of reinforcements was retained during superplastic flow and the mechanical properties of superplastic flow were affected by the reinforcement characteristics. It was suggested that there are strengthening processes of the dislocation-pileup mechanism and the diffusional relaxation-limitation or dislocation-bypass mechanism, so that the strongest strengthening effect among these processes affects the rate-controlling mechanism of superplastic flow for the composites.

REFERENCES

1. T.G.Nieh, C.A.Henshall and J.Wadsworth, Scripta Metall., 18 (1984), 1405-1408.
2. M.W.Mahoney and A.K.Ghosh, Metall. Trans. A, 18A (1987), 653-661.
3. J.Pilling, Scripta Metall., 23 (1989), 1375-1380.
4. M.W.Mahoney, A.K.Ghosh and C.C.Bampton, Proc. 6th Int. Conf. on Composite Materials, edited by F.L.Matthew, N.C.R.Buskell, J.M.Hodgkinson and J.Monton (Elesier Applied Science, London, 1986), vol. 2, 372-381.
5. M.Mabuchi and K.Higashi, Mater. Sci. Eng., A179/A180 (1994), 625-627.
6. M.Mabuchi and K.Higashi, Mater. Trans. JIM, 35 (1994), 399-405.
7. M.Mabuchi and K.Higashi, Mater. Trans. JIM, 36 (1995), 420-425.
8. M.Mabuchi and K.Higashi, to be published.
9. M.Mabuchi and K.Higashi, unpublished research.
10. M.Mabuchi and K.Higashi, Phil. Mag. Lett., 70 (1994), 1-6.
11. M.Mabuchi and K.Higashi, Scripta Mater., 34 (1996), 1893-1897.
12. O.D.Sherby and J.Wadsworth, Prog. Mater. Sci., 33 (1989), 169-221.
13. G.Bao, J.W.Hutchinson and R.M.McMeeking, Acta Metall. Mater., 39 (1991), 1871-1882.
14. J.Rosler, G.Bao and A.G.Evans, Acta Metall. Mater., 39 (1991), 2733-2738.
15. A.B.Pandey, R.S.Mishra and Y.R.Mahajan, Acta Metall. Mater., 40 (1992), 2045-2052.
16. E.Arzt, M.F.Ashby and R.A.Verrall, Acta Metall., 31 (1983), 1977-1989.
17. G.Le Roy, J.D.Embury, G.Edward and M.F.Ashby, Acta Metall., 29 (1981), 1509-1522.

Predicting High-Strain-Rate Superplasticity in UHCS-10Al

Eric M. Taleff [1], Woo-Jin Kim [2], and Oleg D. Sherby [3]

1. Assistant Professor, The University of Texas at Austin, Department of Aerospace Engineering and Engineering Mechanics, Austin, TX 78712.
2. Professor, Hong-Ik University, Department of Metallurgy and Materials Science, 72-1 Sangsu dong, Mapo ku, Seoul, Korea.
3. Professor Emeritus, Stanford University, Department of Materials Science and Engineering, Stanford, CA 94305.

Abstract

Recent studies by different investigators have produced high-strain-rate superplasticity in numerous materials. These materials have in common ultra-fine grain sizes (1 μm and less) and maximum ductilities at high strain rates (0.1 s^{-1} and greater) when deformed at elevated temperatures. One model proposed for high-strain-rate superplasticity, when incipient melting is not present, utilizes grain boundary sliding accommodated by dislocation glide across subgrains with a rate-controlling contribution from dislocation-pipe diffusion to explain observed behavior. Predictive equations for this and other mechanisms are used to construct deformation mechanism maps. Data from an ultrahigh-carbon steel are compared to these predictions. A numerical model is used to predict tensile ductility as a function of the stress exponent, n, which is controlled by the dominant deformation mechanism. At temperatures where incipient melting is not expected, these predictive relations explain the observed high-strain-rate superplastic behavior.

Modeling the Mechanical Response of
Structural Materials
Edited by E.M. Taleff and R.K. Mahidhara
The Minerals, Metals & Materials Society, 1997

Introduction

Superplasticity has been typically observed at strain rates on the order of 10^{-3} s^{-1} in standard superplastic materials [1]. Recent developments, however, have lead to superplastic behavior at strain rates of 0.1 s^{-1} and faster [2]. These observations of high-strain-rate superplasticity (HSRS) have been correlated with two characteristics, very fine grain sizes and high deformation temperatures [2]. Very fine grain sizes, on the order of 1 μm and less, enhance grain boundary sliding, which increases the strain rate at which superplasticity can be observed. For a given stress a decrease in grain size by a factor of two will increase strain rate by a factor of four to eight, depending on the diffusion mechanism accommodating grain-boundary sliding. High homologous temperatures have been associated with incipient melting, which could enhance grain-boundary sliding [3, 2]. A rheological model by Nieh *et al.* explains how incipient melting at grain boundaries could lead to HSRS, and seems applicable to many of the Al-based materials which exhibit HSRS [3, 2]. Some materials, however, exhibit HSRS at temperatures well below those at which incipient melting might occur [4, 5]. These materials rely on a fine grain size alone to produce grain-boundary sliding at high strain rates. Although such materials are a minority of those which exhibit HSRS, it is of great interest to explain their behavior.

In the present study an ultrahigh-carbon steel containing 10 wt pct Al (UHCS-10Al) and produced by powder metallurgy is investigated [4-6]. This material, processed by the mechanical attrition of powders, evidences HSRS at a temperature for which incipient melting is not expected. A deformation mechanism previously proposed for this material assumes grain-boundary sliding accommodated by both lattice and dislocation-pipe diffusion. This UHCS-10Al material exhibits a stress exponent of approximately $n = 3$ in the deformation region where HSRS is observed. Although this value is greater than the value of $n = 2$ typically observed for superplastic materials, it is sufficient to produce high ductilities. As shown in Fig. 1, high ductility can be associated with a low stress exponent. A large amount of scatter is evident in the data of Fig. 1, which can be attributed to the effects of cavitation. Cavitation generally depends on many other factors in addition to the stress exponent. By neglecting the effects of cavitation, numerical predictions for ductility of the UHCS-10Al material are made which correlate well with experimental results.

Experimental Results

Material Preparation

An ultrahigh-carbon steel alloy (UHCS-10Al) containing 1.25 wt pct C and 10 wt pct Al was prepared by the gas atomization of alloyed powders. Groups of powders were then subjected to ball-mill attrition for 0 (no milling), 40, and 80 hrs. These three groups of powders were subsequently consolidated by hot-isopressure extrusion (HIE). The chemical compositions measured for each of these three materials are presented in Table I [4]. The high carbon content of this alloy system creates a hypereutectoid composition with carbide particles present above the A_1 temperature, enhancing the formation and retention of fine grain sizes during processing. Because of the Al content, the carbides are present primarily in the form of κ-carbides (Fe_3AlC_x where $x = 0.5$ to 1). The high aluminum composition has three primary effects. First, it raises the A_1 transformation temperature to approximately 924°C. Second, it increases the oxidation resistance in air, allowing testing at high temperatures without the use of a controlled atmosphere. Third, the Al additions create a Class I solid solution with iron, affecting deformation behavior under certain conditions.

Table I: Chemistry in weight percent measured for each UHCS-10Al material.

Material	C	Si	Mn	P	S	Ni	Cr	Al	O	N
0 hr	1.26	0.041	0.58	< 0.005	< 0.001	0.007	1.61	10.16	0.0283	0.0081
40 hr	1.23	0.046	0.56	< 0.005	< 0.001	0.25	2.36	9.60	0.20	0.0193
80 hr	1.22	0.052	0.55	< 0.005	< 0.001	0.27	2.46	9.60	0.22	0.0233

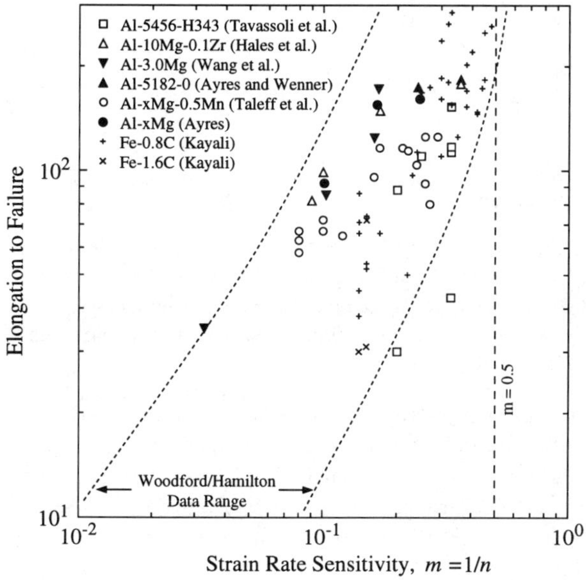

Figure 1: Elongation to failure versus strain-rate sensitivity, m, data from several investigations are shown [7-13]. Strain-rate sensitivity is the inverse of the stress exponent, $m = 1/n$. The range of data originally plotted by Woodford and Hamilton is outlined [14, 15].

Microstructure

Microstructural observations of the processed materials have been previously reported [4]. The 0 hr material exhibits a spheroidized microstructure with a mean linear intercept grain size of $L = 1.2$ μm. The 40 hr and 80 hr materials exhibit almost identical microstructures with the majority of material having a mean linear intercept grain size of $L = 0.3$ μm. These materials evidence a few pockets, less than 10% of the material, with a coarser grain size similar to that of the 0 hr material. Significant grain growth was observed in the 0 hr material during testing at elevated temperatures. The grain size of the 0 hr material was observed to be $L = 7.9$ μm after testing. The 40 hr and 80 hr materials also exhibited dynamic grain growth, with a grain size of $L = 1.1$ μm after testing.

Mechanical Testing

Each of the three materials given in Table I were subjected to tensile strain-rate change (SRC) tests. Tests were conducted at temperatures from 750 to 960°C on a screw-driven testing machine. Thermocouples attached to the testing fixtures and in contact with each sample measured the thermal gradient across each sample to be within ±4°C. Samples were pulled to an initial prestrain of approximately 10% at a moderate speed. The slowest machine speed was then selected, and subsequently faster rates then imposed on the same

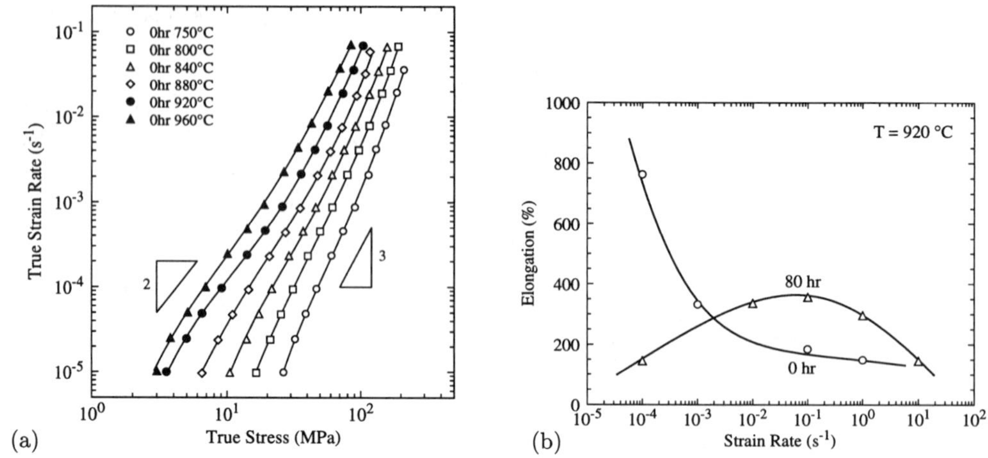

Figure 2: Data are shown in (a) for UHCS-10Al material consolidated directly from atomized powder. The slopes of two and three are representative of grain-boundary sliding deformation and solute-drag (Class I) creep, respectively. Elongation versus strain rate data are shown in (b) for the 0 hr and the 80 hr materials. The peaks in ductility are characteristic of standard superplasticity (0 hr) and HSRS (80 hr).

sample, with a minimum of 2% strain at each rate, to complete the SRC tests. In addition to SRC tests, elongation to failure tests were conducted at strain rates up to 10 s^{-1}. SRC test data for the 0 hr material are shown in Fig. 2(a), and elongation to failure data for the 0 hr and 80 hr materials are shown in Fig. 2(b). SRC test data for the 40 hr and the 80 hr materials are shown in Figs. 3(a) and 3(b), respectively.

Deformation Mechanisms

Explanation of the Data

Predictive equations for various creep mechanisms of relevance to this study are presented in Table II. The preconstants for several of these equations have been found specifically for the UHCS-10Al alloy system. Data from the 0 hr material, Fig. 2(a), can be explained by two separate mechanisms. At low strain rates and high temperatures a stress exponent of $n = 2$ is observed. At low strain rates and high temperatures, where $n = 2$, the activation energy for creep is measured to be from 224 to 287 kJ/mole, which is very close to that of iron self diffusion, $Q_{\mathrm{L}} = 270$ kJ/mole. These values indicate a grain-boundary sliding (GBS) deformation process which is rate-controlled by lattice self diffusion. The predictive equation for this mechanism is given in Table II. This mechanism is often associated with superplastic deformation, and this association is confirmed by the classical superplastic behavior evidenced in Fig. 2(b). The 0 hr material at low temperatures and high strain rates demonstrates a stress exponent near $n = 3$ and a measured activation energy for creep ranging from 140 to 185 kJ/mole. This observed range of activation energies is similar to the activation energy for Al diffusion in Fe, $Q_{\mathrm{chem}} = 155$ kJ/mole. These observations indicate a solute-drag controlled slip creep process (Class I) at low temperatures and high strain rates.

The 40 hr and 80 hr materials exhibit very similar behavior. A possible explanation for behavior of the 80 hr material has been presented previously [6]. This explanation relies on a combination of grain-boundary sliding controlled by grain-boundary diffusion and

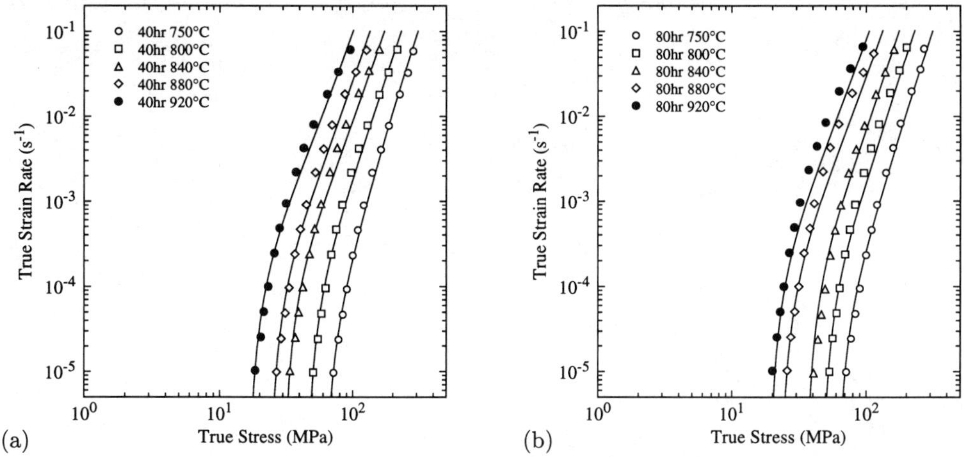

Figure 3: Data from UHCS-10Al materials attrited (a) for 40 hours and (b) for 80 hours are shown. The curves through the data are fits to a model including both lattice and dislocation-pipe diffusion as rate controlling during grain-boundary sliding.

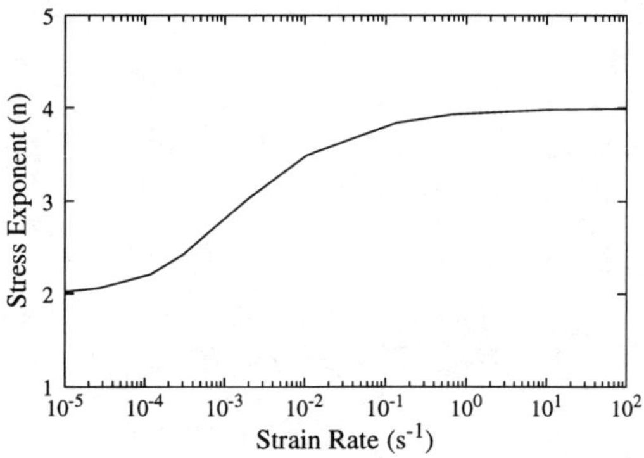

Figure 4: Stress exponent predictions for attrited UHCS-10Al at 750°C are shown.

grain-boundary sliding controlled by dislocation-pipe diffusion. The effect of pipe diffusion is highly stress dependent, and becomes an important contribution at high stresses. Pipe diffusion can raise the stress exponent observed for grain-boundary sliding from two to four. It is most likely, however, that this mechanism is additive to GBS controlled by lattice self-diffusion, resulting in a stress exponent ranging from two at low stress, and low strain rates, to three at moderate stress, and to four at high stress. This corresponds to the behavior of data presented in Fig. 3, where a value near $n = 3$ is observed. The predictive equations for these combined mechanisms are fit to the data of Fig. 3 and shown as curves. The preconstants found for these mechanisms are shown in Table II. Using these values the stress exponent of the 80 hr material is predicted in Fig. 4 at 750°C over a range of strain rates, after compensation by threshold stress.

A major complication in analyzing the data in Fig. 3, and many of the data for HSRS materials, is the presence of a strong threshold stress. The physical basis of this behavior is not well understood. In the current investigation this behavior is accounted for by simply using a threshold stress, σ_0, to compensate the flow stress data. The threshold stress is calculated in a manner described elsewhere [6]. In order to calculate activation energies

Table II: Constitutive equations used in developing deformation mechanism maps for UHCS-10Al.

Creep process	**Equation**
Diffusional flow	
Nabarro–Herring	$\dot{\epsilon} = 14\,(D_L/d^2)(E\mathbf{b}^3/kT)(\sigma/E)$
Coble	$\dot{\epsilon} = 50\,(D_{gb}\mathbf{b}/d^3)(E\mathbf{b}^3/kT)(\sigma/E)$
Slip	
Climb (pipe diffusion)	$\dot{\epsilon} = 5 \times 10^{12}\,(D_p/\mathbf{b}^2)(\sigma/E)^7$
Climb (lattice diffusion)	$\dot{\epsilon} = 10^{11}\,(D_L/\mathbf{b}^2)(\sigma/E)^5$
Glide (chemical diffusion) *	$\dot{\epsilon} = 12.15\,(D_{chem}/\mathbf{b}^2)(\sigma/E)^3$
Grain boundary sliding	
Climb (lattice diffusion) *	$\dot{\epsilon} = 3.3 \times 10^8\,(D_L/d^2)(\sigma/E)^2$
Glide (chemical diffusion) *	$\dot{\epsilon} = 5.9 \times 10^3\,(D_{chem}/d^2)(\sigma/E)$
Grain-boundary diffusion controlled *	$\dot{\epsilon} = 5.6 \times 10^8\,(D_{gb}\mathbf{b}/d^3)(\sigma/E)^2$
Pipe-diffusion controlled *	$\dot{\epsilon} = 3.2 \times 10^{11}\,(D_p/d^2)(\sigma/E)^4$

* Preconstant determined from data for UHCS-10Al.

for creep from the data of Fig. 3 the data must be compensated by threshold stress. The apparent activation energy, Q_{app}, measured without correction for threshold stress can be significantly higher than the actual activation energy for creep, Q_c. The correction provided by accounting for threshold stress is given as,

$$Q_{app} = Q_c + \underbrace{\frac{nRT^2}{\sigma - \sigma_0}\left(\frac{\sigma_0}{E}\frac{dE}{dT} - \frac{d\sigma_0}{dT}\right)}_{\text{correction}} \quad , \tag{1}$$

where R is the gas constant, E is the dynamic modulus, and T is temperature for data taken at a constant value of σ/E. It is this strong effect of threshold stress which can produce very high values of apparent activation energy for HSRS materials. After compensation for threshold stress the data of Fig. 3 provide activation energies for creep at low and high stresses of 152 and 185 kJ/mole, respectively. These values are similar to the activation energy for grain boundary diffusion, $Q_{gb} \approx 0.5\,Q_L$, and for dislocation-pipe diffusion, $Q_p \approx 0.7\,Q_L$, where $Q_L = 270$ kJ/mole. These values add further support to the proposed deformation mechanisms in the 40 hr and 80 hr materials.

Mechanism Maps

Using data from the three UHCS-10Al materials and the predictive equations of Table II enables the construction of deformation mechanism maps for UHCS-10Al. Such maps are give in Fig. 5 for temperatures of (a) 750°C and of (b) 920°C. The maps in Fig. 5 do not take into account power-law breakdown at high strain rates. These maps correspond reasonably well to the mechanisms observed in the UHCS-10Al materials. The maps of Fig. 5 point out the importance of refining grain size in order to increase the range of grain boundary sliding and superplasticity. Another important conclusion is that a rise in temperature increases the regions of grain-boundary sliding which exhibit $n = 2$, see

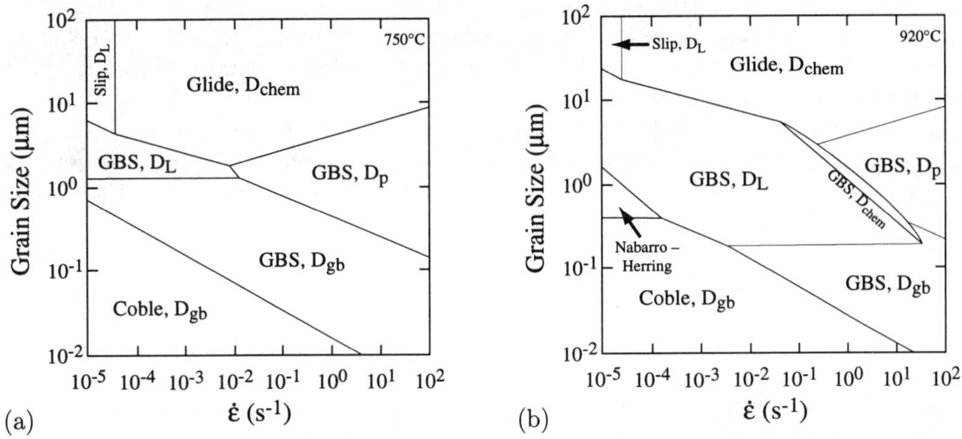

Figure 5: Deformation mechanism maps for UHCS-10Al at (a) 750°C and at (b) 920°C.

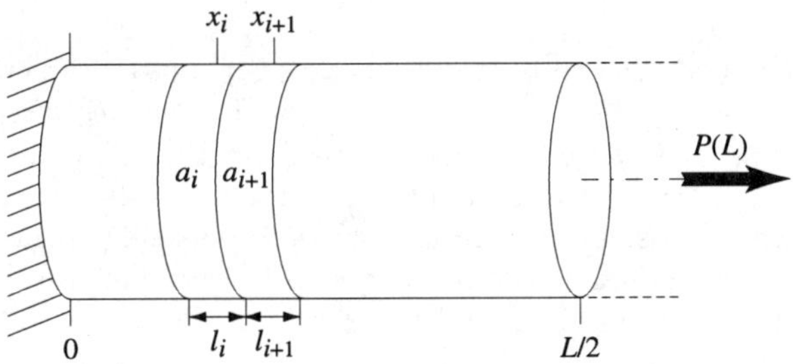

Figure 6: Schematic of the numerical model used to compute neck shapes.

Table II, and reduces the region where pipe diffusion gives $n = 4$ during GBS. Because ductility is observed to increase with decreasing n, it is of use to decrease the influence of pipe diffusion.

Predictions for Ductility

There remains one more important question to be addressed concerning the high ductilities observed for HSRS in UHCS-10Al. How are such high ductilities obtained for HSRS when the stress exponent is significantly greater than $n = 2$? To answer this question a numerical model of neck formation is used to predict tensile ductility. In this model a cylindrical specimen is considered with an initial sinusoidal perturbation in it cross-sectional area. An initial perturbation in area is necessary to initiate neck development in the numerical model. A variation in cross-sectional area which is symmetric about the middle of the sample length is considered. This makes it possible to model the sample using only half of its length. The sample half-length is divided into N sections, as shown in Fig. 6, and the length, l_i, and area, a_i, of each section is monitored during deformation.

The initial cross-sectional area is chosen to vary along the length of the sample in a sinusoidal fashion. The maximum initial perturbation is given as δ, a fraction of the initial unperturbed cross-sectional area, A_0. The initial area at any point x along the length is then given by,

$$A(x) = A_0 \left(1 - \delta \sin\left(\frac{\pi x}{\lambda}\right)\right) , \qquad (2)$$

where λ is the wavelength of the perturbation. The longest wavelength which is symmetric about the middle of the sample is $\lambda = L$, and this is the value used in the model. This so-called long wavelength assumption has two advantages. First, it provides for a slightly varying cross-sectional area. Under these conditions non-axial stresses are small, and only the axial normal stresses need be accounted for in the model for reasonable accuracy. Second, Hutchison and Obrecht have shown the long wavelength condition to provide the most rapid neck growth, thus giving a conservative estimate of elongation [16]. The loading condition used in the model is that of a constant stress creep test, where load, P, is varied with specimen elongation in the manner,

$$P(L) = \sigma_0 \, A_0 \, \frac{L_0}{L} \ , \tag{3}$$

where σ_0 is a "constant" stress of 10 MPa, L_0 is the initial sample length, and L is the sample length at any time t.

The constitutive relation used in the numerical model for material behavior is that of a simple power-law creeping material,

$$\dot{\epsilon} = \alpha \sigma^n \ , \tag{4}$$

where the preconstant is chosen to be a value typical for solute-drag creep of aluminum-magnesium materials, $\alpha = 4.89 \times 10^{-8}$ $\text{MPa}^{-n} \cdot \text{s}^{-1}$. The shapes of sample necks after deformation were computed using this model with $N = 100$. Time steps were taken at $\Delta t = 0.01\, t_c$, where t_c is the critical time for failure found by Hutchison and Obrecht for a constant sample load and sinusoidal initial variation in sample radius. Failure is taken as the point at which the minimum sample area goes to zero numerically. This numerical model is very similar to that used by Burke and Nix to predict the effect of necking during creep tests [17].

The results of the numerical model are given in Figs. 7 and 8 as sample radius plotted against length. Fig. 7 shows the original shape of the sample half-length along with the shapes at failure for stress exponents of 5, 3, and 2. The original shape of the full-length sample is shown in true perspective in the inset to the figure. The non-dimensional length and radius of the undeformed sample are given in the figure. An initial perturbation in area of $\delta = 0.01$ is used in all calculations. The top horizontal scale indicates the elongation achieved for each stress exponent. Of note in Fig. 7 is that an elongation of 500% is achieved for a stress exponent of three, while an elongation of over 2000% is achieved for a stress exponent of two. Fig. 8 shows results of the numerical model for stress exponents of 3.0, 2.75, 2.5, 2.25, and 2.0. These results indicate that even a stress exponent of $n = 3$ can produce significant ductilities, such as those found in the UHCS-10Al during HSRS.

Conclusions

1. High-strain-rate superplasticity is observed in UHCS-10Al material processed to achieve a grain size on the order of 1 μm.

2. Predictive equations for creep can be used to provide a reasonable description of high temperature deformation in the UHCS-10Al materials.

3. These same equations produce deformation mechanism maps which are consistent with the observed deformation behavior.

4. Dislocation-pipe diffusion plays an important role in HSRS in the UHCS-10Al materials and increases the stress exponent from 2 to a value near 3.

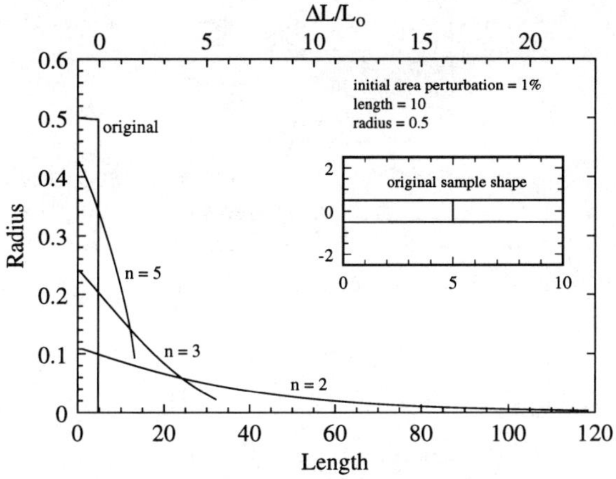

Figure 7: Predicted neck shapes for various values of the stress exponent.

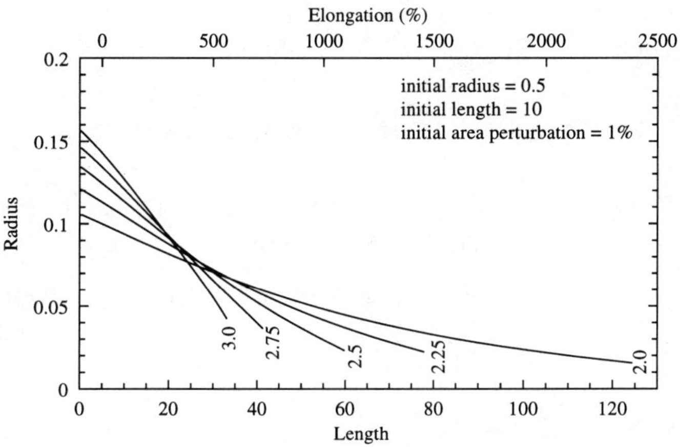

Figure 8: Predicted neck shapes for small values of stress exponent.

5. A numerical model predicts high ductilities even when the stress exponent is higher than that of typical superplastic materials.

Acknowledgments

The authors would like to extend their gratitude to Prof. Kenji Higashi of Osaka Prefecture University for his important contributions to this work. The authors thank Dr. George Yoder of the U.S. Office of Naval Research, under whom a portion of this work was supported through contract N-00014-91-J-1197.

References

[1] O. D. Sherby and J. Wadsworth. *Prog. Mater. Sci.*, **33** (1989) pp. 169–221.

[2] T. G. Nieh, J. Wadsworth, and O. D. Sherby. *Superplasticity in Metals and Ceramics*, pp. 154–185. Cambridge University Press, New York, NY (1997).

[3] T. G. Nieh, J. Wadsworth, and T. Imai. *Scripta Metall. Mater.*, **26** (1992) pp. 703–708.

[4] E. M. Taleff, M. Nagao, K. Higashi, and O. D. Sherby. *Scripta Metall. Mater.*, **34** (1996) pp. 1919–1923.

[5] E. M. Taleff and W.-J. Kim. In J. C. Earthman, ed., *Seventh International Conference on Creep and Fracture of Engineering Materials and Structures*, pp. 215–226. TMS, Warrendale, PA (1997).

[6] E. M. Taleff, M. Nagao, Y. Ashida, and O. D. Sherby. *J. Mater. Res.*, **11** (1996) pp. 2725–2730.

[7] A. A. Tavassoli, S. E. Razavi, and N. M. Fallah. *Metall. Trans.*, **6A** (1975) pp. 591–594.

[8] S. J. Hales, T. R. McNelley, and H. J. McQueen. *Metall. Trans.*, **22A** (1991) pp. 1037–1047.

[9] J. Wang, Z. Horita, M. Furukawa, M. Nemoto, N. K. Tsenev, R. Z. Valiev, Y. Ma, and T. G. Langdon. *J. Mater. Res.*, **8** (1993) pp. 2810–2818.

[10] R. A. Ayres and M. L. Wenner. *Metall. Trans.*, **10A** (1979) pp. 41–46.

[11] E. M. Taleff, D. R. Lesuer, and J. Wadsworth. *Metall. Mater. Trans.*, **27A** (1996) pp. 343–352.

[12] R. A. Ayres. *Metall. Trans.*, **10A** (1979) pp. 849–854.

[13] E. S. Kayali, H. Sunada, T. Oyama, J. Wadsworth, and O. D. Sherby. *J. Mater. Sci.*, **14** (1979) pp. 2688–2692.

[14] D. A. Woodford. *Trans. Am. Soc. Metals*, **62** (1969) pp. 291–293.

[15] C. H. Hamilton. In *Superplasticity.* Nation Technical Information Services, Springfield, Va, (1989).

[16] J. W. Hutchinson and H. Obrecht. *Fracture*, **1** (1977) pp. 101–116.

[17] M. A. Burke and W. D. Nix. *Acta Metall.*, **23** (1975) pp. 793–798.

Fracture Processes

INVESTIGATION OF TEXTURE AND FRACTURE TOUGHNESS IN AA 7050 PLATE

Armand J. Beaudoin
University of Illinois at Urbana-Champaign
354 Mechanical Engineering Building, MC-244
1206 West Green Street
Urbana, Il 61801

William A. Cassada
Reynolds Metals Company
13203 N. Enon Church Rd.
Chester, VA 23836

Abstract

Aluminum alloy 7050 was rolled using an experimental regime designed to vary texture development at the T/4 (quarter-) plane. Samples of 5" gauge plate were finished-processed to a T7451 temper. Samples were prepared for evaluation of plane strain fracture toughness (K_{Ic}) using compact specimens. Tests were carried out with normal to the crack in the transverse direction (T-L) as well as in the longitudinal direction (L-T). Experimental pole figures were measured and quantitative texture analysis was performed to develop the orientation distribution function. Yield surfaces were derived for each sample using polycrystal plasticity theory. Metrics developed through the quantitative texture analysis demonstrated a correlation with the difference in fracture toughness measured in the T-L and L-T samples. Specifically, the difference in the L-T and T-L fracture toughness was related to the ratio of the 111 and 100 intensities, read from the center of the corresponding pole figures. As such, the texture is providing a marker for the process history and resulting directionality in toughness response.

Modeling the Mechanical Response of
Structural Materials
Edited by E.M. Taleff and R.K. Mahidhara
The Minerals, Metals & Materials Society, 1997

Introduction

Thick heat-treatable plate, produced through the rolling process, offers a cost-effective option for the manufacture of structural components for use in airframe applications. Manufacture through rolling offers both high productivity and product consistency. A natural consequence of rolling is the variation of the thermomechanical processing conditions for different streamlines – as viewed through the thickness of the plate. As properties follow from the thermomechanical history, these processing gradients in turn lead to through-thickness variations in the plate mechanical properties. In applications in which milling operations remove a considerable amount of material, the result is a component that has different material properties at different points in the part. Specifically, if a property such as fracture toughness varies with spatial orientation in the plate blank, these spatial variations will carry over into the milled part. An understanding of the underlying metallurgical features that contribute to material properties will be helpful in assessing the effect of gradients in material properties.

Crystallographic texture plays a role in the performance of metals in many applications. With regard to sheet materials, anisotropy following from crystallographic texture can alter the strain path in a target forming operation. Examples are the development of ears in cup drawing and thinning in sheet forming. Research on the effects of texture in the formability of sheet materials has been a subject of intense study. Relatively speaking, less information is available on the analysis of texture in thick plate materials. In part, this is because the processing imparts gradients in the crystallographic texture through the plate thickness. Characterization of the performance must consider the spatial location within the plate; the analysis is more complex.

For the rolling of plate, in contrast to the finish passes for sheet materials, roll gap geometry is such that the resulting deformation field is quite non-uniform. The arc of contact is less than, or on the order of, the plate thickness. Frictional effects - induced by the rolls - further contribute to the deformation inhomogeneity. As a consequence, shear deformations are developed at the quarter-plane of the plate. Simple shear leads to the formation of orientations of the form {111}<*uvw*>, where the first index refers to the plane of shear and the second refers to the shear direction [1]. Studies of rolling with high reduction ratio also indicate the 45° ND-rotated cube as a dominant component of the texture [2].

In this work, an analysis of texture for 7050-T7451 plate is conducted. The texture analysis is then related to fracture toughness data. Yield surfaces are developed for selected samples to highlight the magnitude of the effect of crystallographic texture on mechanical properties.

Materials and Method

AA7050 ingots were experimentally hot rolled on a reversing mill to a slab gauge of 5" thickness. Rolling temperature, reduction schedule and the degree of cross-rolling were varied so as to produce lot-to-lot differences in the fracture toughness response. The 5" gauge plates were subsequently processed to develop T7451 temper. Samples for mechanical testing and texture measurement were taken from the quarter thickness locations of each plate. L-T and T-L plane strain fracture toughness tests (K_{Ic}) were performed using conventional sample coordinates to identify the test specimen directions (per ASTM E399). Sample directions were identified as L, T, and N being the rolling direction, transverse to the rolling direction, and normal to the plate surface, repectively. In addition to fracture toughness testing, routine tensile tests were performed for the samples. The yield strength was found to be independent of sample orientation in the L-T plane at the quarter thickness.

Pole figures were measured with the normal to the sample in the rolling direction. Because of the aspect ratio of the microstructure, this provided for a maximum number of distinct grains to be included in the measurement. 111, 100, 110, and 311 pole figures were collected. Texture analysis was performed using the popLA (preferred orientation package – Los Alamos) software package. The following procedure was employed:

1. Pole figures were normalized;
2. Orientation Distribution Function (ODF) analysis was performed without assuming any sample symmetry and pole figures were recomputed;
3. The recomputed pole figures were tilted about the transverse direction such that the normal to the plate was in the pole figure center;
4. Minor rotations were performed to optimize symmetry;
5. A second ODF analysis was performed, assuming orthotropic sample symmetry.

Upon completion of the quantitative texture analysis, yield surface sections were developed using the LApp (Los Alamos code for polycrystal plasticity) program. A discrete collection of 576 weighted grains was generated using the ODF produced through step 5, above, as input. The orientations of the weighted grains provide a complete coverage of the orientation space. The weighted grains were then used as input to the LApp program to develop the yield surface section. Use of the LApp program provided a means of condensing the details of the ODF into a single measure - here taken as a position on the yield surface section.

Results

Listed in the Table I are the 111 and 100 intensities taken from pole figures recalculated from the first ODF analysis (assuming no sample symmetry). The intensities denoted I_{111} and I_{100} were read from the center of the 111 and 100 pole figures, respectively. These two intensities provide a measure of number of {111} or {100} planes having the normal of the crystallographic plane aligned with normal to the plate surface. Pole figures for two of the specimens, one with relatively high 111 intensity (sample 106) and one with low 111 intensity (sample 418), are shown in Figure 1. A well-defined peak is notable at the center of the111 pole figure of Figure 1A.

As a point of reference, pole figures measured from the T/2 (center-) plane are given in Figure 2. These pole figures show a well-developed rolling texture, with some evidence of a cube texture component. There is no indication of {111}<*uvw*> fiber development in the pole figure; the density in the center of the 111 pole figure is considerably less than random.

Fracture toughness measurements for the samples are also listed in Table 1. There is a difference in fracture toughness between specimens taken in the L-T and T-L direction, with the L-T measurement being of greater magnitude. ΔK_{Ic} is defined as the difference in fracture toughness between the L-T and T-L specimens. There exists a relationship between the 111 intensity – read from the center of the 111 pole figures and the difference measure ΔK_{Ic}. Specifically, there exists a relation between the texture intensity ratio I_{111}/I_{100} and ΔK_{Ic}; this relationship is shown in Figure 3. A similar relationship between K_{Ic} - taken in either the L-T or T-L orientation and pole figure intensity was not found in the current study.

Table I

Sample ID	111 Intensity	100 Intensity	I_{111}/I_{100}	K_{Ic} (L-T) $[ksi\sqrt{in}]$	K_{Ic} (T-L) $[ksi\sqrt{in}]$	ΔK_{Ic}
105	1.84	3.41	0.54	29.4	25.9	3.5
106	2.02	3.08	0.66	29.8	26.8	3.0
348	1.01	2.11	0.48	30.1	25.8	4.3
350	0.87	3.12	0.28	32.0	26.1	5.9
352	1.32	3.10	0.43	28.9	23.7	5.2
354	1.52	2.11	0.72	21.0	20.9	0.1
416	1.02	2.50	0.41	32.7	26.1	6.6
418	1.00	2.75	0.36	32.0	25.3	6.7
420	0.87	2.08	0.42	31.7	25.5	6.2

Greater detail of crystallographic orientation can be developed through examination of the ODF. Sections of the ODF that contain the {111}<*uvw*> fiber are plotted in Figure 4. The angle convention selected, Kocks' angles, uses the angles (ψ,θ,ϕ) [3]. The {111}<*uvw*> fiber is located at an angle θ=54° in the section with ϕ=45°. For the textures examined in this study, the fiber was more developed in the ϕ=50° section; as such, this is the section plotted. The approximate location of θ=54° is indicated by the dashed white line. Two sections are shown which correspond to the pole figures of Figure 1. The sample with higher 111 intensity shows a well-developed fiber with peaks near the {111}<011> and {111} <211> orientations.

To detail orientations of the cube variety, it is necessary to modify the angle description. Cube orientations represent a point of degeneracy in the ODF; as $\theta = 0°$, only one angle is needed to characterize rotations about the <100> axis. We follow the Kocks convention and use the definition [3]

$$\nu = \psi - \phi$$

such that ν represents a rotation about the <100> axis. The {100}<001> (cube) orientation is then identified by $\nu = 0°$ and the {100}<011> (45° ND-rotated cube) is denoted by $\nu = 45°$. Given in Figure 5 are the intensities of ODF orientations with $\theta = 0°$. Samples taken from the T/4 plane show an intensity peak at the 45° ND-rotated cube position – as would be expected based on the shear deformation history for this location of the plate. In contrast, a sample from the (T/2) plane exhibits a retained texture component of the cube variety.

Yield surface sections developed for the two samples shown in Figures 1 and 4 are plotted in Figure 6. The section is plotted using the Taylor factor, because it provides a comparison between the anisotropy from the two textures independent of the hardening behavior. In this particular section, the out-of-plane stress component is eliminated through iteration in the LApp program. This particular section was selected as it highlights anisotropy in the L-T plane (for that reason, it is commonly used to characterize sheet materials). However, the stress states presented in this section - having the out-of-plane Cauchy stress equal to zero - are not truly representative of those developed in the plane strain fracture toughness test. Two normal stresses are represented in the section. The horizontal axis represents the normal stress in the L direction. The vertical axis represents the normal stress in the T direction.

The largest difference between stress prediction for the samples presented in Figure 6 is along

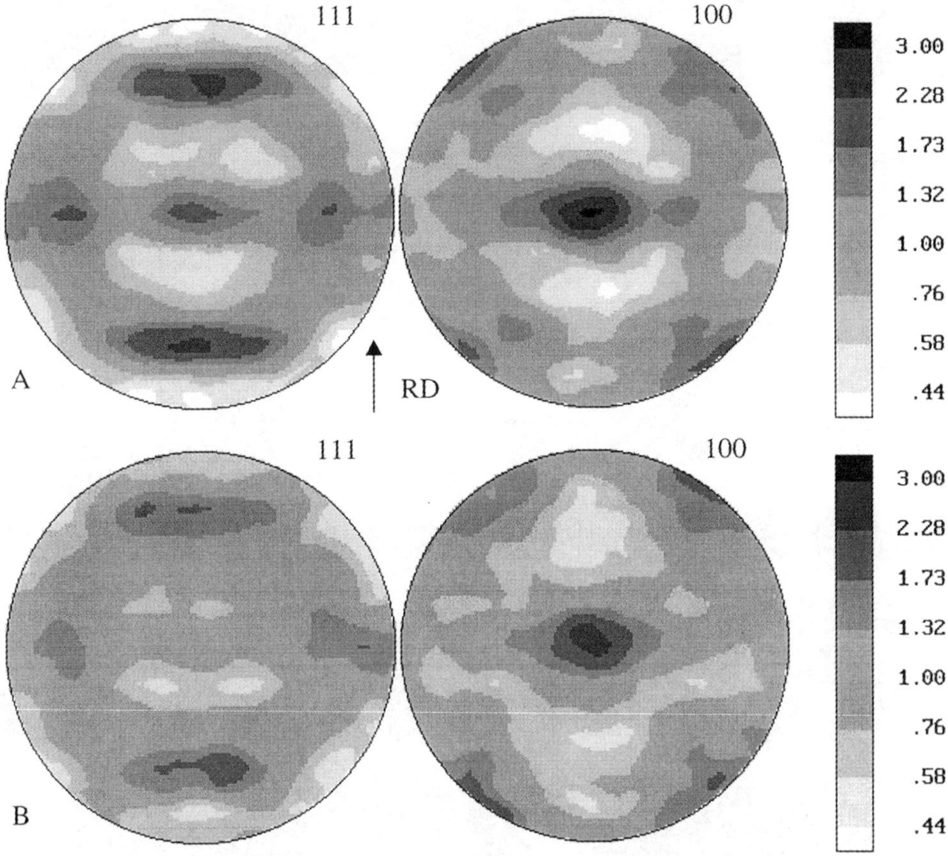

Figure 1 – Pole figures of samples taken from the T/4 plane with A) 111 intensity of 2.02 and B) 111 intensity of 1.0.

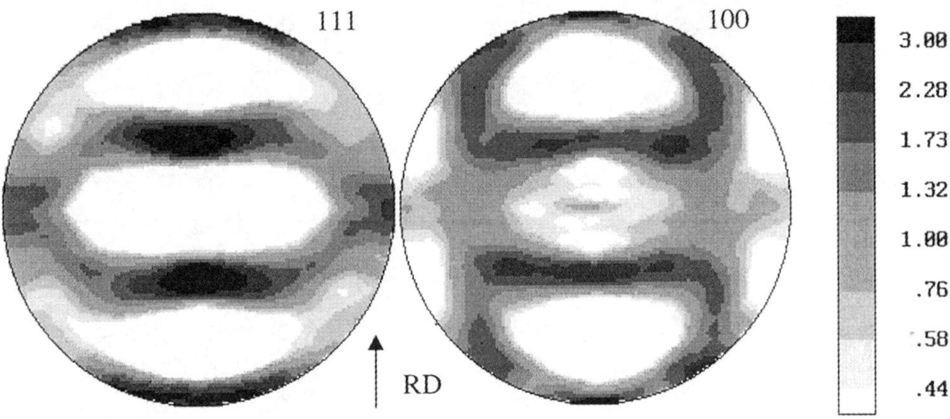

Figure 2 – Pole figure of sample taken from the T/2 plane.

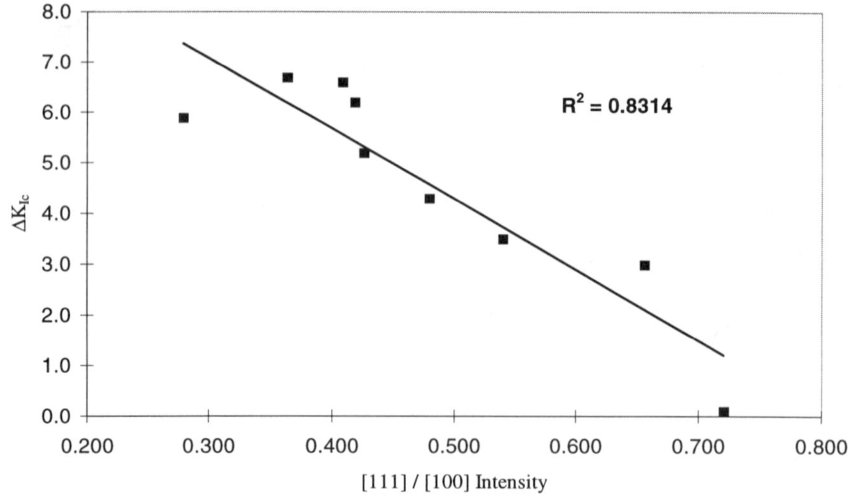

Figure 3 – Relationship between the toughness difference, ΔK_{Ic}, and (111)/(100) ratio.

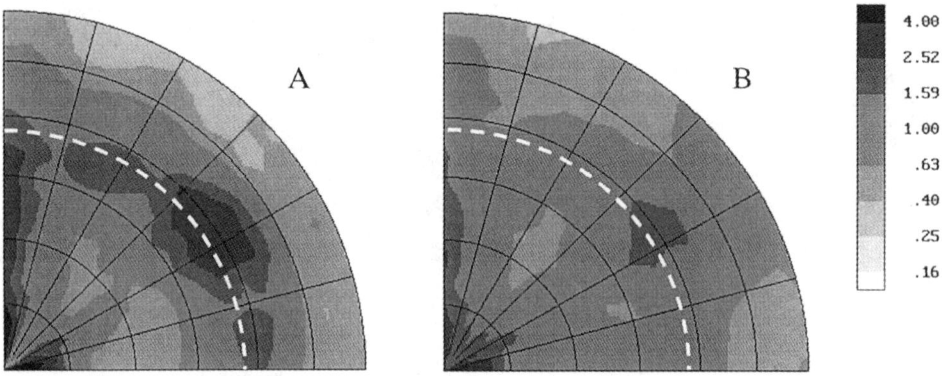

Figure 4 – ODF section containing the {111}<*uvw*> fiber for samples A and B of Figure 1.

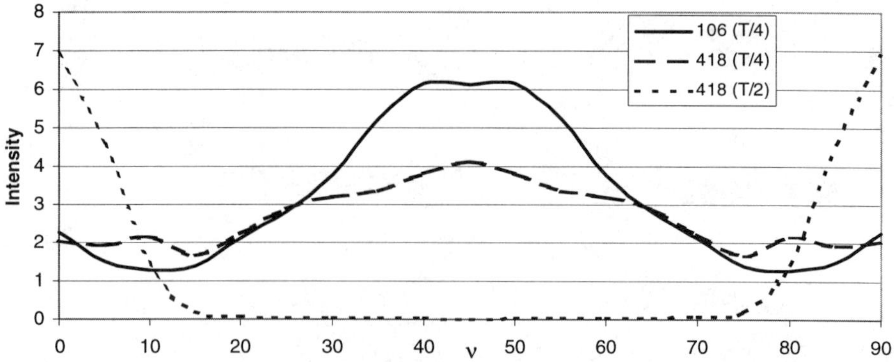

Figure 5 – Intensity as a function of the orientation angle ν.

the T (vertical) axis. Yield surface sections, and therefore the normal stress in the T direction, were developed for all samples considered in the study. The Taylor factor in the T direction is plotted along with the ΔK_{Ic} measurement and shown in Figure 7. Numerous yield surface sections and projections (including the π-plane) were examined in the study. Attempts were also made to correlate differences in yield stress prediction with $\Delta K_{Ic.}$ While some of these attempts met with success, the correlation shown in Figure 7 is the strongest relation between a yield stress and the difference between L-T and T-L fracture toughness.

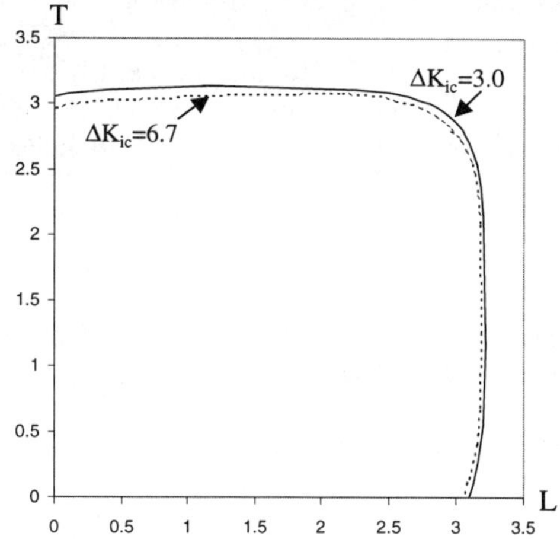

Figure 6 - Predictions of Yield Surface Sections

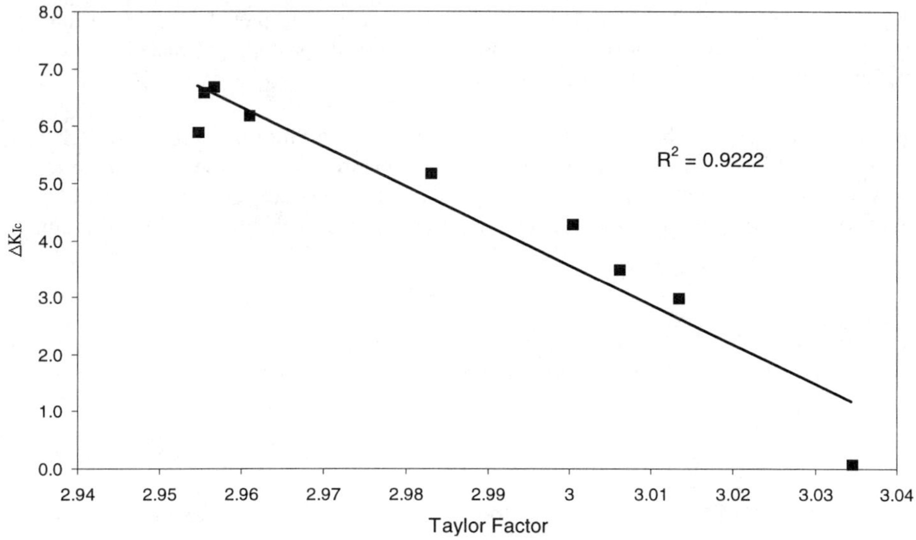

Figure 7 - Relationship between yield surface section Taylor factor in T direction and ΔK_{Ic}

Discussion

There are several metallurgical features that affect the fracture toughness of 7XXX alloys. Examples are chemical composition, distribution of the second phase, and the degree of recrystallization. A relationship between the magnitude of fracture toughness and crystallographic texture was not found in this work. However, for the samples considered in this study, the directionality of the fracture toughness does correlate with experimental texture measurement at the (T/4) plane of the plate. Reduction of the difference of K_{Ic} as measured in the L-T and T-L directions is associated with a greater density of {111} crystallographic planes sharing the normal direction to the plate.

The texture measure adopted in this work, I_{111}/I_{100}, is a heuristic choice. Mechanical properties have been related to measures of texture similar to the one employed herein. O'Brien et al. [4] related the R-value of fiber-textured tantalum to the parameter

$$P_{111} = \frac{I_{111}}{I_{111} + I_{100}}$$

Applying this ratio for the data of Table 1 leads to a similar correlation between texture and toughness difference, with R^2=0.78.

The correlation between ΔK_{Ic} and the intensity ratio I_{111}/I_{100} indicates a possible relationship between material anisotropy and plane strain fracture toughness. The further correlation of ΔK_{Ic} with a normal stress taken in the T direction (from the L-T yield surface section) suggests that anisotopy is modifying the plastic zone at the crack tip. However, there is relatively little difference in the yield surface sections shown in Figure 6. Correspondingly, the range in the Taylor factor shown in Figure 7 is quite small - on the order of one percent. Hence, it is not possible to definitively state that material anisotropy has a significant effect on the orientation dependence of fracture toughness. There is the possibility that the texture is simply reflecting a microstructural feature that follows from processing - for example, the distribution of the second phase or grain aspect ratio. In any case, the texture is providing a marker for a process history that led to a particular microstructure.

Future research efforts will be directed at characterization of texture development at the quarter plane. Current finite element implementations with constitutive response derived from crystal plasticity are capable of addressing this problem. Results from such simulations would provide insight as to the processing path indicated by the experimental textures considered herein.

Acknowledgements

Conversations with Dr. U. F. Kocks on the application of the popLA and LApp codes were most helpful in carrying out the analysis. Comments on the manuscript by J. A. Dantzig and U. F. Kocks are sincerely appreciated.

References

1. G. R. Canova, U. F. Kocks, and J. J. Jonas, "Theory of Torsion Texture Development," Acta Metallurgica, 36 (1984), 211-226.
2. A. I. Fedoseev and G. Gottstein, "Inhomogeneous Deformation and Texture during Rolling," in Proceedings of the 1st International Conference on Modelling of Metal Rolling Processes (London, The Institute of Metals, 1993), 296-308.
3. U. F. Kocks, "A Symmetric Set of Euler Angles and Oblique Orientation Sapce Sections, in Eighth In. Conf on Textures of Materials, J. S. Kallend & G. Gottstein, eds. (Warrendale, PA, The Metallurgical Society), pp. 31-36.
4. J. O'Brien, R. Logan, and W. Hosford, "Anisotropy of Fiber-Textured Tantalum," in Physics and Mechanics of Finite Plastic and Viscoplastic Deformation, Proceedings of Plasticity '97 (Fulton, Maryland, Neat Press, 1997), pp. 73-74.

ANISOTROPY OF STRUCTURE AND FRACTURE TOUGHNESS

OF ROLLED BCC-METALS

M.Danylenko

Frantsevich Institute for Problems
of Materials Science, 3, Krzhizhanovsky St.,
Kiev 252680,
Ukraine

Abstract

Effect of dislocation structure on fracture toughness, yield stress and mechanisms of fracture of Mo-, Cr-, and Fe-based alloys have been studied. Fracture toughness was determined by bending of specimens with cracks introduced into the plane perpendicular (direction 1) and parallel (direction 2) to the plane of rolling. Tests were carried out at range of temperatures below ductile-brittle transition temperature.

Non-monotonous dependence of fracture toughness against strain connected with unstability of dislocation cell structure for low prestrained materials (for both directions). The increasing of fracture toughness in direction 1 and decreasing of fracture toughness in direction 2 for high strained materials connected with changing of mechanism of fracture.

On basis of this investigations the explanation of some phenomena which take place during tension of samples were presented.

Modeling the Mechanical Response of
Structural Materials
Edited by E.M. Taleff and R.K. Mahidhara
The Minerals, Metals & Materials Society, 1997

Introduction

A lot of publications have been devoted to investigation of regularities of structure formation under plastic deformation. In papers of V.Trefilov at al.[1] results obtained for bcc-metals have been presented in generalized schemes (fig.1) which characterized structure evolution during plastic deformation within wide temperatures and deformation degree. Analysis if such diagrams for various bcc-metals permits to conclude that there are some general mechanisms of structure transformation under deformation.

In papers of V.Moiseev at al. [2] it was shown that for certain state of structure deformation curve in terms "S-$e^{1/2}$ " (here S - true deformation stress, e- true strain) consisted from straight lines sections, whereas bends were corresponded to structure state changes.

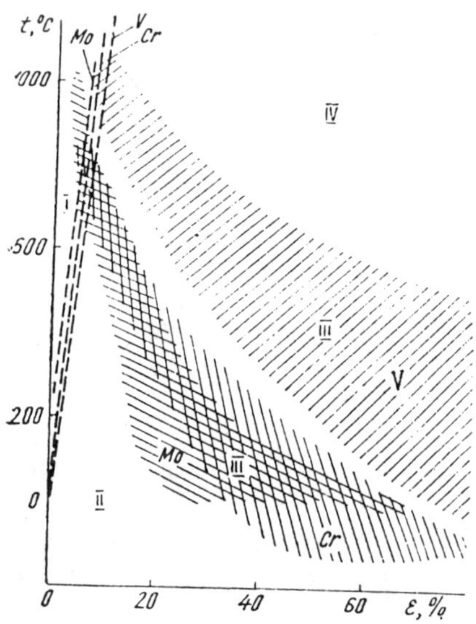

Figure 1 - Diagrams of structural states of bcc-metals, I - single dislocations and pile-up, II - chaotic distribution of dislocations III - intermediate states, IV - cell structure.

The non-monotonous dependence of ductile-brittle transition temperature on previous strain was observed in papers of V.Trefilov at al. [1] It was discovered that peculiarities of this dependence were caused by structure state changes present work has been devoted to investigation of structure influence on the fracture toughness of strained bcc-metals.

Materials and methods

The investigations have been carried out with the specimens of strained alloys of molybdenum, chromium and iron. Mechanical properties of the alloys, their yield stress, fracture toughness, were tested under temperatures 77-373K. For studying the anisotropy of the fracture toughness cracks were introduced into planes perpendicular (direction 1) and parallel (direction 2) to the plane of rolling. Special grippes for testing microspecimens with cracks within the plane of rolling were made. Fine structure of deformed materials was studied by transmission electron microscopy. Fractographic studies were carried out by scanning electron microscopy.

Results and discussion

The dependence of the fracture toughness of bcc-metals on previous strain within the range of D-B-T temperatures is shown in Fig.2 [3, 4]. It is evident that availability of single dislocations and pile-ups of dislocations within the range of small strains causes the decrease of fracture toughness.

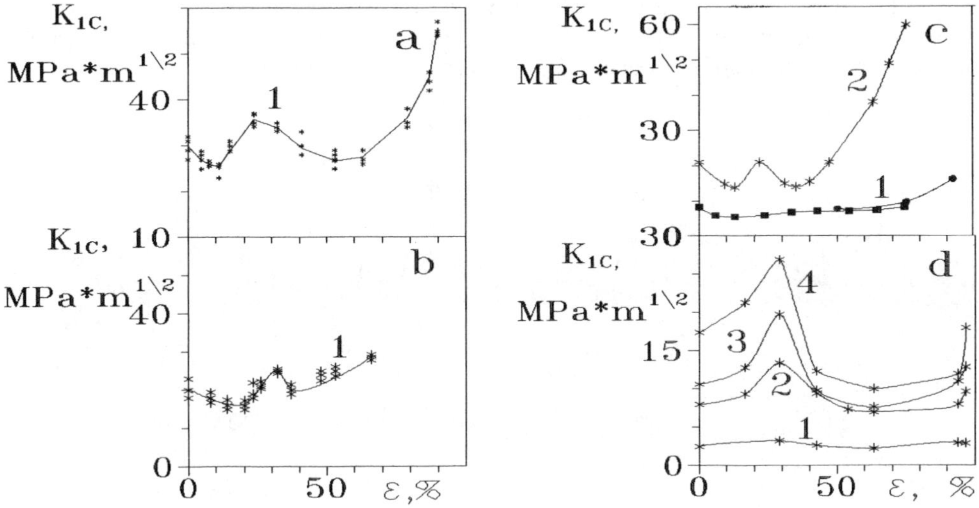

Figure 2 - Fracture toughness of strained bcc-metals for brittle failure (a,b -iron, c - molybdenum, d - chromium, (direction 1), (test temperature 1 -77K, 2 - 293K, 3 - 393K).

The decrease of the fracture energy at the early stages of deformation was theoretically predicted in a number of works [5, 6]. Gilman [5] analyzed the effect of stress fields originating near single dislocations on the effective specific energy of fracture by microcleavage mechanism under conditions when the crack absorbed elastic energy of dislocations arranged in the cleavage plane. Estimations have shown that when the dislocation density was adopted as 10^{12} cm^{-3} the decrease of the effective specific surface energy amounted to 5-10% of the true specific surface energy.

S.A.Firstov et al [6] proposed an expression for the diminution of the specific surface energy $\Delta\gamma$ at the moment of fracture in the presence of dislocation substructure

$$\Delta\gamma = Gb\rho^{1/2} / 4\pi K_s \qquad (1)$$

here G - the shear modules, b - the Burgers vector, ρ - the dislocation density, K_s =0,8 - a parameter of the model. The authors have estimated surface energy diminution for Armco-iron when r was adopted 10^{12} cm^{-3} The $\Delta\gamma$ value obtained was amounted to 10% of the true specific surface energy.

The emergence of slightly misoriented cell structures is accompanied by a sharp increase of K_{1C} with elevation of test temperature. Maxima on the curves correspond to the deformation for which formation of slightly misoriented cells structure was observed. It was found that maximum corresponding with deformation 20-30% has grown due to test temperature increase. Peak fracture toughens is not observed at low test temperatures Mo- and Cr-alloys. Maximum of fracture toughness corresponding with deformation 20-30% is stipulated by interaction of dislocations being emitted from the crack tip with boundaries of slightly misoriented cells. The dissolution of cellular structure under deformation have been observed by Ohr [7], Bassim [8] in situ HVEM. The mobile dislocations, which have arisen due to dissolution of cell structure, cause the blunting of crack and increasing of fracture toughness. The non-monotonous dependences of yield stress against strain was observed too (fig.3)

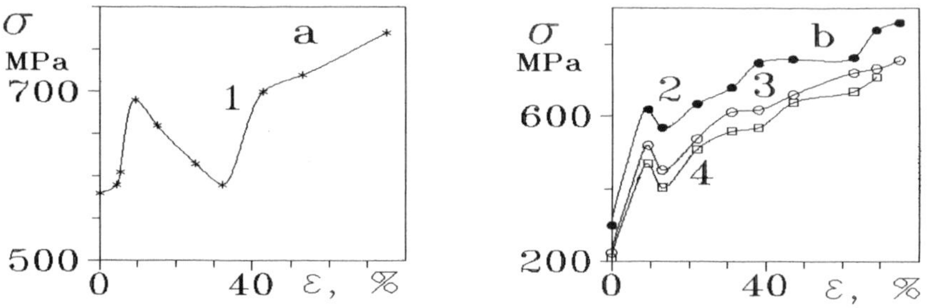

Figure 3 - Yield stress of strained iron (a) and molybdenum (b), (test temperature 1 - 98K, 2 - 293K, 3 - 373K, 4- 473K)

Fracture toughness growth being observed for high deformed materials is only stipulated by increasing of number of high angle boundaries. This boundaries being effective barriers for cleavage crack, stimulate the fracture toughness enhancement in this plane.

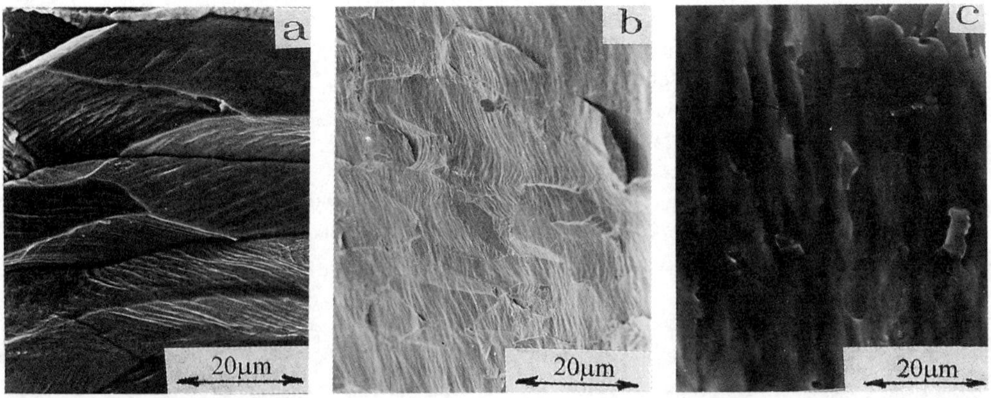

Figure 4 - Failure of the strained Mo (a) - $\varepsilon = 50\%$, (b) - $\varepsilon = 74\%$, (c) - 92%, direction 2.

At large deformation (over 75%), the presence of highly misoriented cells favors intercellular fracture that leads to decrease of fracture toughness in the plane of rolling (direction 2).
The effect of plastic deformation on the fracture toughness of Mo, Cr and Fe alloys in the plane of rolling is shown in Fig. 5 [9]. The alloys of Cr and Fe in their starting state show transcrystalline failure, and delamination has been observed at cell boundaries under high strains (ε>75% for Fe and ε>85% for Cr). Molybdenum alloys are usually susceptible of intergranual failure, and the dislocation structure of grains does not influence the structural preparation of fracture for cracks in direction 2 (Fig. 4). A monotonous decrease of fracture toughness at ε>20% has been observed.
This phenomenon can be explained by the peculiarities of the crack propagation in a deformed material. For brittle intergranular failure fracture toughness (K_{1C}) depends on angle of rotation of crack φ

$$K_{1C}= K^{b}_{1C} *f(\varphi) \qquad (2)$$

where - K^{b}_{1C} fracture toughness of the boundary parallel to the plane of macrocrack propagation, and $f(\varphi)$ - is angle distribution of stress at the crack tip [10].

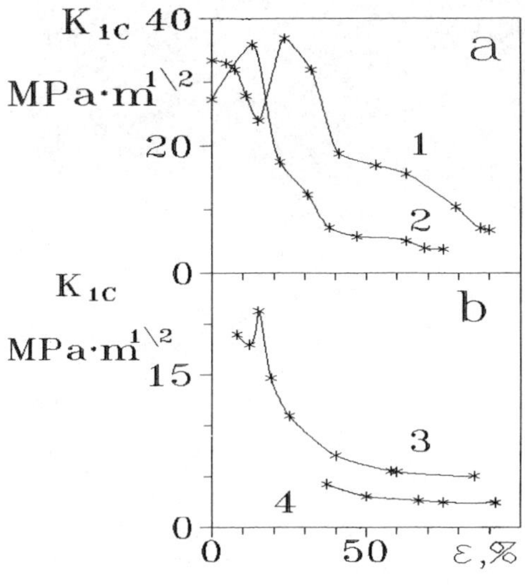

Figure 5 - Fracture toughness of strained alloys of Fe (1), Mo (2, 4) and Cr (3), (direction 2).

Energy loss onbypassing the boundary in direction 1 is very large and the straight propagation of the crack through the grain body is more advantageous from the point of view of energy saving. In direction 2 grain boundaries and plane of crack propagation almost the same, and failure takes place along the boundary (fig.4). Effect of direct microalloying of W-based alloys for improving grain body and grain (cell) boundaries was presented in paper [9].
On basis of this investigations the explanation of out-neck failure which take place during tension of samples may be given. After a neck formation during tension the fracture takes place not on the minimal cross section (fig. 6) According to dependency $K_{1C}=f(\varepsilon)$ (fig.2)

fracture toughness is increasing with growing of deformation (for ε>50%). This increasing is not be compensated by decreasing of cross-section. According to criteria of quasi-cleavage:

$$\sigma_{cr} = K_{1C}*(\pi*c_{cr})^{1/2} \quad (3)$$

the section where crack will propagate may be determined. In the case of neck formation for each section $\sigma_k = P/S_k$ and

$$P_{cr} = S_{cr}* K_{1C}*(\pi*c_{cr})^{1/2} \quad (4)$$

The critical size of defect c_{cr} is a weak function of deformation degree. The size cross-section depends on deformation as:

$$S_k = S_o/\exp \varepsilon , \quad (5)$$

and dependence K_{1C}=f(ε) is shown at fig. 2. The section where the fracture will take place may be founded from minimum of the product of $S_k(\varepsilon)*K_{1C}f(\varepsilon)$.

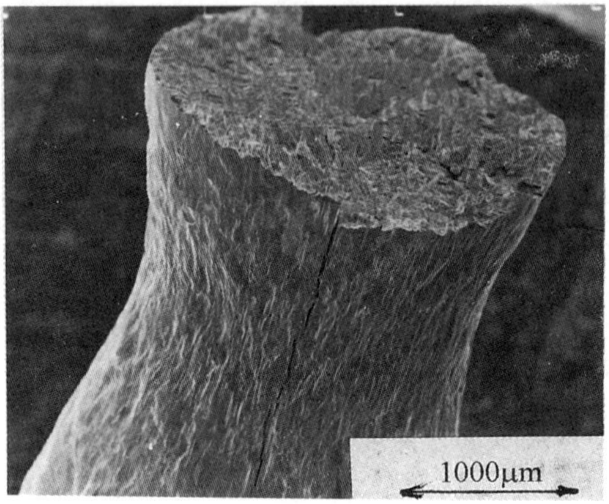

Figure 6 - Out-neck failure of specimen.

Conclusion

The non-monotonous dependence of fracture toughness of bcc-metals against strain caused by changing of structural stages of materials. For low test temperatures fracture toughness is weak function (fig. 2, curves 1 for Mo- and Cr-based alloys). Peak of fracture toughness (ε=20-30%) is growing with test temperature increasing. Fracture toughness growth of high strained materials connected with formation of high angle cell structure, which are effective barriers for crack propagation in the direction 1. In the plane of rolling (direction 2) inter cellular failure

References

1. V.I.Trefilov, Yu.V.Milman, and S.A.Firstov, Physical Principles of the Strength of Refractory Metals in Russian, (Naukova Dumka, Kiev, 1975), 315.

2. V.Trefilov et al., Deformation strengthening and fracture of polycrystaline metals, (Kkiev, Naukova Dumka, 1989), 256.

3. N.I.Danilenko, A.N.Demidik, Yu.N.Podrezov, and Rachek, "Effect of Plastic Deformation on the Fracture Energy of Powder and Cast Iron," Soviet Powder Metallurgy and Metal Ceramics, 30 (1991), 780-784.

4. N.I.Danilenko, Yu.N.Podrezov, The Influence of Plastic Deformation on Fracture toughness of Molybdenum, Chromium and Iron", Fracture Mechanicics: Successes and Problems, vol.2 (Lviv, Collection of Abstracts, 1993), 423.

5. D.C.Drucker, and J.J.Gilman, ed., Fracture of Solids (Maple, Valley, Washington, 1962), 492.

6.Yu.Cornjushin, V.Trfilov, and S.Firstov, "Effect of dislocation Structure on the Crack's Propagation", Problems of Strength, 9 (1976), 94-98

7. S.M.Ohr, "An Electron Microscope Study of Crack Tip Deformation and Impact on the Dislocation Theory of Fracture", Mater. Sci. Eng. 72, (1985), 1-35.

8. M.Nabil, M.N.Bassim, "The Role of Dislocations in the Fracture of Prestrained AISI Steel", J.Mat.Sci. and Eng.,95, (1987), 199-207.

9.N.Danilenko et al., "Mechanical Properties of W-Re Made by the Powder Metalluirgy Method", Rhenium and Rhenium Alloys, ed. B.D.Bryskin (Rhenium Alloys, Inc. Erilia, Ohio, 1997), 391-395.

10. Yu.N.Podrezov, A.S.Drachinsky, and V.I.Trefilov, "Effect of Structural Elements on the Energy of Itergranular Failure", Phys. Met. Metalloved., 55, 1, (1983), 157-164, (in Russian).

MODELING OF CRACKING AND CAVITATION DURING HOT WORKING OF GAMMA TITANIUM ALUMINIDES

V. Seetharaman* and S.L. Semiatin

Materials and Manufacturing Directorate, Air Force Research Laboratory,
Wright-Patterson AFB, OH 45433-7817.

*Materials and Processes Division, UES, Inc., Dayton, OH 45432-1894.

Abstract

Hot workability of near gamma titanium aluminides was assessed using hot compression and hot tension tests. A simple model for the brittle intergranular failure of these alloys during hot working was developed to seek correlations between flow properties, microstructural characteristics, and the propensity for cracking. The onset of wedge cracking was predicted using a criterion based on a critical value of the product of the peak flow stress and the square root of the grain size. In contrast, the occurrence of ductile fracture of the workpiece in regions subject to secondary tensile stresses is controlled by the kinetics of nucleation, growth, and coalescence of cavities. This type of cavitation damage could be predicted using the maximum tensile work criterion proposed originally by Cockcroft and Latham. Failures resulting from severe flow localization were analyzed in terms of the relative magnitudes of the normalized flow softening rate and the strain rate sensitivity. The applicability of the preceding criteria for the onset of brittle and ductile fracture during deformation processing of near gamma titanium aluminide alloys is discussed.

Modeling the Mechanical Response of
Structural Materials
Edited by E.M. Taleff and R.K. Mahidhara
The Minerals, Metals & Materials Society, 1997

Introduction

Two-phase gamma titanium aluminide alloys containing gamma (TiAl) and alpha-two (Ti_3Al) phases have emerged as potential engineering materials for high temperature applications in gas turbine engines and in the automobile industry. These alloys possess attractive properties such as low density, resistance to oxidation and burning, and high strength at elevated temperatures; however, the tensile ductility and fracture toughness of these alloys at ambient temperatures are rather low [1]. Improvements and optimization of mechanical properties are generally sought through control of microstructural characteristics, viz., alpha grain size, interlamellar spacing, and the fraction of lamellar grains. Thermomechanical processing sequences employing different hot working and heat treatment operations are used to control the microstructure development in these alloys [2]. Considerable success has been achieved in processing of the near-gamma titanium aluminide alloys via techniques such as isothermal forging, conventional (canned) hot forging, conventional (canned) hot extrusion, and pack rolling [3-5]. Nevertheless, processing parameters to avoid fracture during ingot breakdown or secondary processing are still selected to some degree by trial-and-error techniques or by recourse to an expanding experience base for a variety of binary, ternary, and higher-order gamma plus alpha-two alloys of various compositions and cast grain sizes.

Deformation processing of near gamma titanium aluminides constitutes a crucial step in the successful development and application of these materials. High strength and resistance to plastic deformation exhibited by gamma titanium aluminides at high homologous temperatures make them very attractive for high temperature service, but drastically reduce their hot workability. In general, the hot working regime in which the material can be processed with adequate control over both microstructure evolution and nucleation and propagation of cracks is quite narrow. This is especially true for cast materials with relatively coarse grain sizes, which are prone to cracking and fracture under the influence of tensile stresses. Even though primary hot working processes such as forging or extrusion used for the breakdown of the ingot structure involve nominally compressive states of loading, secondary tensile stresses can be generated due to geometrical, frictional, or thermal effects. Examples of failures induced by tensile stresses include free surface bulging and fracture in open die forging, edge cracking during rolling, and nose fracture during canned extrusion [2,6,7]. While ductile failures are caused by the nucleation, growth, and coalescence of cavities at the grain boundaries, brittle failures are invariably the result of catastrophic propagation of wedge cracks. Thus, the development of models for the initiation and propagation of defects during hot working can be very helpful in the selection of processing conditions in order to avoid extensive and expensive laboratory experimentation.

The objective of the present work was to begin the development of models for the hot workability of near-gamma titanium aluminide alloys under high-rate deformation conditions that characterize conventional metalworking processes. For this purpose, isothermal, hot compression and hot tension test results for a few near-gamma titanium aluminides were analyzed to formulate suitable criteria for the onset of both brittle and ductile failures.

Brittle Fracture During Ingot Breakdown

Hot workability is generally evaluated through the use of compression, tension, or torsion tests. Of these, the hot compression test is useful for determining the resistance of a given material to plastic deformation at different temperatures and strain rates without the complication of concurrent fracture related events. By contrast, torsion and tension tests provide valuable information on the hot ductility of the alloy. Hot ductility is a reliable and accurate measure of the intrinsic hot workability and is affected by dynamic structural changes and by the occurrence of cavitation and wedge cracking phenomena.

Seetharaman, et al. [6] investigated the hot tension behavior of a model alloy, Ti-49.5Al-2.5Nb-1.1Mn, in both the cast plus HIP'ed condition and the wrought condition. Both material conditions exhibited a complex relationship between reduction in area, test temperature, and strain rate (Figure 1). However, in broad terms, the data revealed transitions from brittle behavior (in which wedge cracks grew and led to catastrophic failure at very low strains) to ductile behavior (in which microvoid initiation and growth was gradual) over a rather narrow temperature. For each material condition, the brittle-to-ductile transition temperature increased with increasing strain rate (Figure 2). Furthermore, the transition temperatures for a given strain rate were higher for the coarse-grained, cast material than for the finer-grained, wrought material. An Arrhenius-type analysis of the transition-temperature data yielded values of activation energy comparable to those that describe the dynamic recrystallization of gamma titanium aluminide alloys during hot compression testing [8]. Furthermore, detailed metallographic studies [6,9] conducted on the longitudinal sections of the test specimens confirmed the crucial role of dynamic recrystallization in suppressing the brittle fracture. The

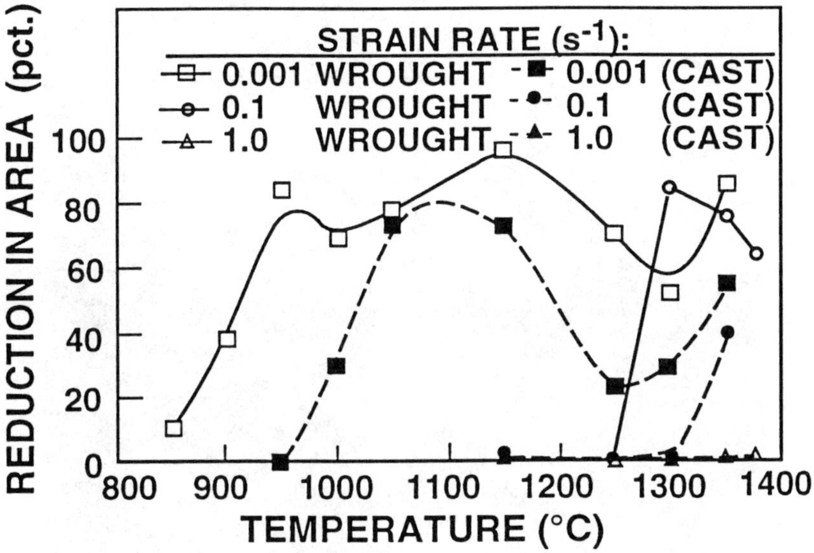

Figure 1 Dependence of reduction in area on test temperature and strain rate for hot tension tests on cast or wrought (extruded + recrystallization heat treated) samples of Ti-49.5Al-2.5Nb-1.1Mn.

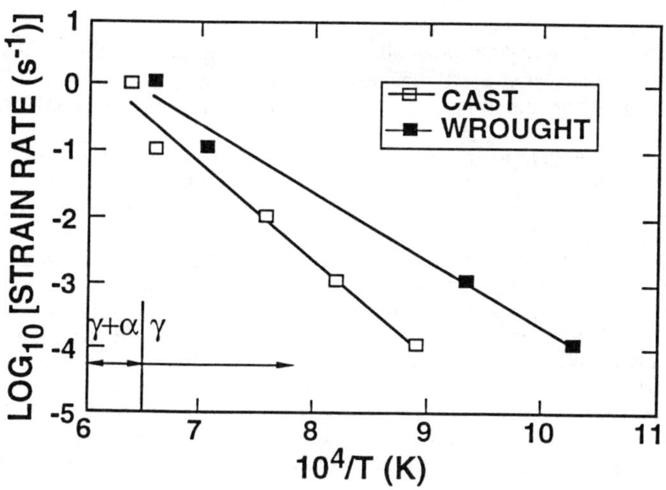

Figure 2 Relationship between strain rate and inverse of the ductile-to-brittle transition temperature for the hot tension behavior of cast and wrought (extruded + recrystallization heat treated) Ti-49.5Al-2.5Nb-1.1Mn.

interaction of recrystallization and cavitation is illustrated in Figure 3 for a cast + HIP'ed tension specimen pulled to failure at 1050°C and a nominal strain rate of 0.001 s^{-1}; this sample had a reduction in area of approximately 70 percent. The micrographs reveal a fairly high density of cracks/cavities on grain boundaries normal to the stress axis and wedge cracks which are invariably associated with boundaries between lamellar grains and gamma grains. In addition, dynamically recrystallized, fine grains, approximately 10-15 μm in diameter and forming a necklace-type structure, are seen along the original gamma grain boundaries.

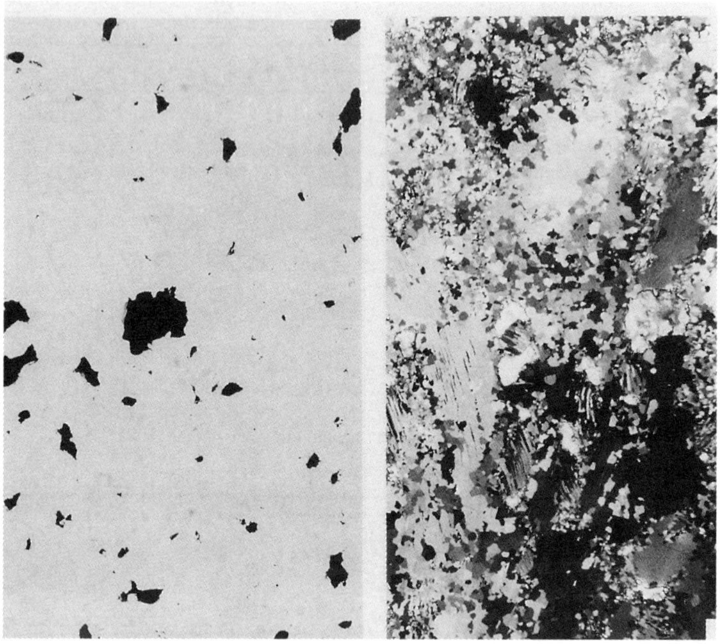

(a) (b)

Figure 3 Optical microstructures observed in a cast plus HIP'ed Ti-49.5Al-2.5Nb-1.1Mn specimen tested at 1050°C and a nominal strain rate of 0.001 s^{-1} : (a) bright field micrograph and (b) polarized-light micrograph of the same field shown in (a).

An attempt to develop a criterion for the initiation of wedge cracking and brittle failure during hot working of gamma titanium aluminide alloys was made recently by Semiatin and Seetharaman [10]. In this work, the occurrence of wedge cracks during isothermal hot compression testing of the *wrought* alloy Ti-46.6Al-2.7Nb-0.3Ta was investigated over a wide range of strain rates and temperatures. Test temperatures were in both the two phase (alpha + gamma) field as well as the single-phase alpha field. Wedge cracks were found both at the center of the test samples, at which the stress state was essentially compressive, as well as at and near the bulged free surfaces, at which a component of secondary tensile stress had developed. Two possible fracture criteria that may be hypothesized are those based on (i) a maximum stress or (ii) a critical combination of stress and some microstructural feature (e.g., grain size because the failure is intergranular). The incidence of microcracking increases with stress. However, the stresses at which wedge cracks were observed are quite a bit higher for subtransus temperatures (σ_p > 250 MPa) than for supertransus temperatures (σ_p > 60 MPa). Thus, the second possibility, a criterion incorporating stress and some feature of the microstructural scale, was examined to see if all of the observations could be explained in a unified manner.

The form of a simple microstructure-based wedge cracking criterion may be inferred from previous models such as the classical relations for fracture of brittle metallic materials [12] or cavity nucleation at grain boundaries due to dislocation pileups [13]. The former relation, derived by Griffith and expanded by Orowan, is

$$\sigma_f = \sqrt{2E\gamma_p/\pi c}, \tag{1}$$

in which σ_f is the (far-field) axial stress, E is Young's modulus, γ_p is the plastic work expended per unit of crack extension, and 2c is the size of the assumed elliptical, pre-existing flaw. Because the opening of wedge cracks is known to occur as a result of grain boundary sliding which leads to stress concentrations which cannot be relieved quickly enough by diffusional processes or matrix slip, a criterion analogous to the above relations may be proposed for wedge cracking. Xinggang, et al. [13] derived a similar relation for the stress σ_{cn} for cavity nucleation at grain boundaries due to dislocation pileups:

$$\sigma_{cn} = \sqrt{3E\gamma_s/d}, \tag{2}$$

in which γ_s is the surface energy associated with intergranular cavities, and d denotes the grain size.

The preceding equations suggest that a criterion based on a threshold value of the product of the peak flow stress and the square root of the gamma (or alpha) grain size d should predict the occurrence of wedge cracks. Examination of the experimental results in light of this criterion (Figure 4) did indeed suggest that such an approach may be promising. For the wrought Ti-46.6Al-2.7Nb-0.3Ta alloy, the critical value of $\sigma_p \sqrt{d}$ was found to be approximately 1.2 MPa $\sqrt{m}$.

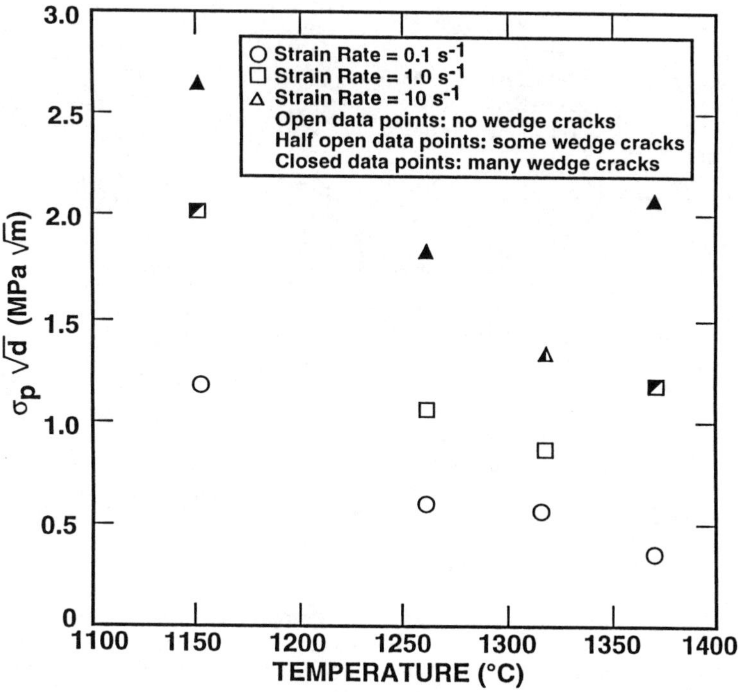

Figure 4 Plot of ($\sigma_p \sqrt{d}$) versus temperature indicating the occurrence of wedge cracking during hot compression testing of Ti-48Al-2.5Nb-0.3Ta.

In a parallel effort, Seetharaman and Semiatin [11] probed the efficacy of the $\sigma_p \sqrt{d}$ criterion to explain the strain rate-temperature conditions under which brittle fracture occurred during tension testing of the Ti-49.5Al-2.5Nb-1.1Mn in the cast + HIP'ed condition and the wrought (extruded + recrystallization heat treated) condition. The wrought material results suggested a critical value of $\sigma_p \sqrt{d}$ for brittle fracture, i.e., 1.0 MPa $\sqrt{m}$, which was comparable to that found for wrought Ti-46.6Al-2.7Nb-0.3Ta. By contrast, the cast Ti-49.5Al-2.5Nb-1.1Mn material exhibited critical values of $\sigma_p \sqrt{d}$ that varied from approximately 1.5 to 3.1 MPa $\sqrt{m}$ over the temperature range investigated (Figure 5).

It is of interest to compare the brittle fracture results for cast Ti-49.5Al-2.5Nb-1.1Mn to those for the *cast* alloys studied by Nobuki and Tsuijimoto[14] and Shih and Scarr [15] in the context of the $\sigma_p \sqrt{d}$ criterion. This comparison provides only a very qualitative insight because the failure observations reported by Nobuki, *et al.* and by Shih and Scarr were based essentially on the occurrence of free surface cracking at rather large strains rather than the initiation of wedge cracks per se. Keeping this limitation in mind, values of $\sigma_p \sqrt{d}$ were estimated from the reported critical peak stresses and grain/colony sizes. These calculations (Table I) give critical values of $\sigma_p \sqrt{d}$ between 3.1 and 4.9, or levels comparable to the higher numbers reported by Seetharaman and Semiatin [11]. These findings certainly warrant further experimental and theoretical work. Regardless the magnitude of $\sigma_p \sqrt{d}$, the fact that this parameter remained constant for a given microstructure is encouraging. This approach may also be used to select processing conditions for materials with different grain sizes.

Table I. Fracture Data for Cast Gamma Titanium Aluminide Alloys

Investigator	Alloy	Grain Size, d (μm)	Critical Stress, σ_p (MPa)	$\sigma_p \sqrt{d}$ (MPa $\sqrt{m}$)
Nobuki, *et al.*	Ti-43.8Al	560	150	3.55
Nobuki, *et al.*	Ti-44.9Al	530	150	3.45
Nobuki, *et al.*	Ti-48.2Al	130	275	3.14
Nobuki, *et al.*	Ti-49.5Al	300	275	4.76
Nobuki, *et al.*	Ti-50.2Al	320	275	4.92
Shih and Scarr	Ti-48Al-2Cr-2Nb	~100	475	4.75

Ductile Fracture During Isothermal Forging

Ductile behavior during hot working of gamma titanium aluminide alloys is associated with strain rate-temperature conditions for which void initiation occurs at large strains and/or void growth is slow and somewhat stable. In these cases, fracture is a damage-propagation, rather than initiation, controlled process in which secondary tensile stresses play an important role in the kinetics of failure. An example of such a failure mode was described in the work of Seetharaman, *et al.* [6] who investigated free surface cracking of Ti-49.5Al-2.5Nb-1.1Mn during pancake forging of cylindrical mults. Metallographic sections of forged mults revealed that cavity formation, growth, and coalescence, i.e., ductile fracture, controlled the formation of the surface cracks. During pancake forging, the degree of bulge and thus the level of the secondary tensile stresses is a function of the instantaneous aspect ratio of the workpiece and the die-workpiece interface friction conditions.

For the specific geometry and interface friction involved in the experiments of Seetharaman, *et al.*, it was found that the free-surface fracture could be predicted using the maximum tensile work criterion first proposed by Cockcroft and Latham [16] (for ductile fracture under cold-working conditions). To this end, the "critical" tensile work to fracture, C*,

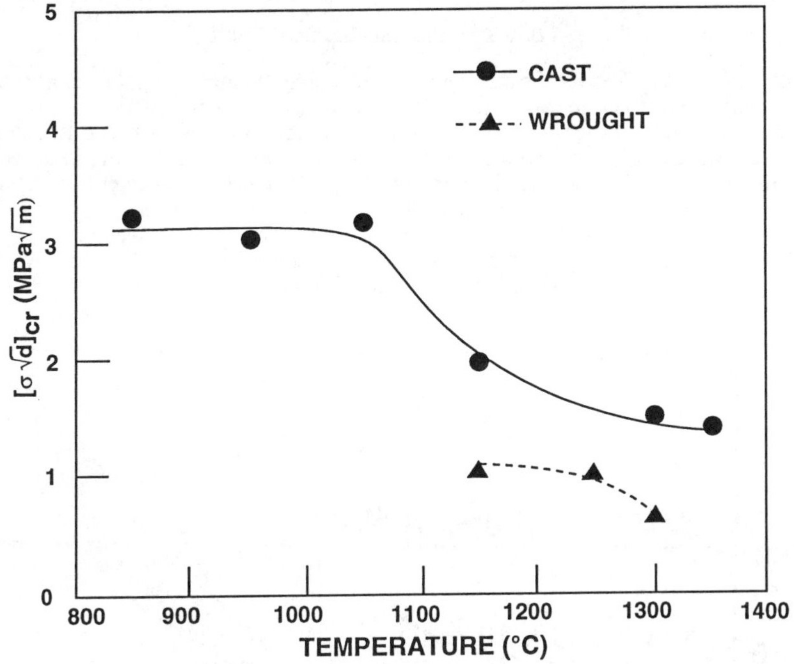

Figure 5 Variation in the parameter $(\sigma_p \sqrt{d})_{cr}$ with temperature for cast and wrought materials.

was measured in uniaxial tension tests and compared to the work C done through the maximum tensile stress at the equator of the bulge developed during isothermal pancake forging, as estimated using a finite element modeling (FEM) technique. This comparison is illustrated in Figure 6. The broken line corresponds to the critical value C* for strain rates $\dot{\varepsilon}$ between 10^{-3} and 10^{-2} s^{-1} at a temperature of 1150°C. The shaded bands are the FEM - predicted values for C as a function of height strain e for friction factors between 0.2 and 0.4, or the typical range of measured friction factors. For $\dot{\varepsilon} = 10^{-3}$ s^{-1}, the values of C ($\bar{\varepsilon}$) are well below the critical damage factor C*; thus free surface fracture would not be expected. Experimentally, it was found that for reductions as high as 6:1 at this strain rate and a temperature of 1150°C, no free surface fractures were observed, thus validating the predictions. On the other hand, for $\dot{\varepsilon} = 3 \times 10^{-2}$ s^{-1}, C is below C* for a 2:1 reduction, but becomes comparable to C* at a 4:1 reduction. Thus, a tensile type of fracture is predicted at deformations near the latter reduction, as indeed was observed.

Limits on the hot workability of gamma titanium aluminide alloys due to the generation of secondary tensile stresses and thus cavitation can be extended by superimposing high levels of hydrostatic pressure. Watkins, et al. [17] studied the isothermal, upset forging of cast plus HIP'ed specimens at 1050°C both with and without a superimposed pressure of 260 MPa. Pressure was found to have a significant effect on the free surface fracture behavior of these specimens. For instance, under ambient pressure conditions, specimens deformed at 1050°C and a strain rate of 1 s^{-1} up to average height strains as low as 0.6 exhibited noticeable cracks on the bulged free surface as well as subsurface voids. On the other hand, specimens deformed at the same temperature and strain rate but with a superimposed pressure of 260 MPa showed neither cracks at the free surface nor subsurface voids, even at average height strains equal to 1.6. Such improvements in ductility are analogous to the well-documented effects of pressure on the ductility of a number of conventional metallic alloys.

<u>Failure During Shear Deformation</u>

Developed in the former Soviet Union, the equal channel angular extrusion (ECAE) process offers an attractive alternative for the primary breakdown of conventional and advanced alloys. In the process, an ingot (or prior worked billet) is extruded through a channel consisting of two continuous sections situated at an angle to each other; the imposed strain is a function of the channel angle and the number of passes that the workpiece is put through the tooling. The ability to impart large deformations without a change in the cross-sectional area of the workpiece is the principal advantage of the ECAE process. However, the application of this method to titanium alloys and titanium aluminides appears to be limited by the frequent occurrence of severe shear localization and shear cracking.

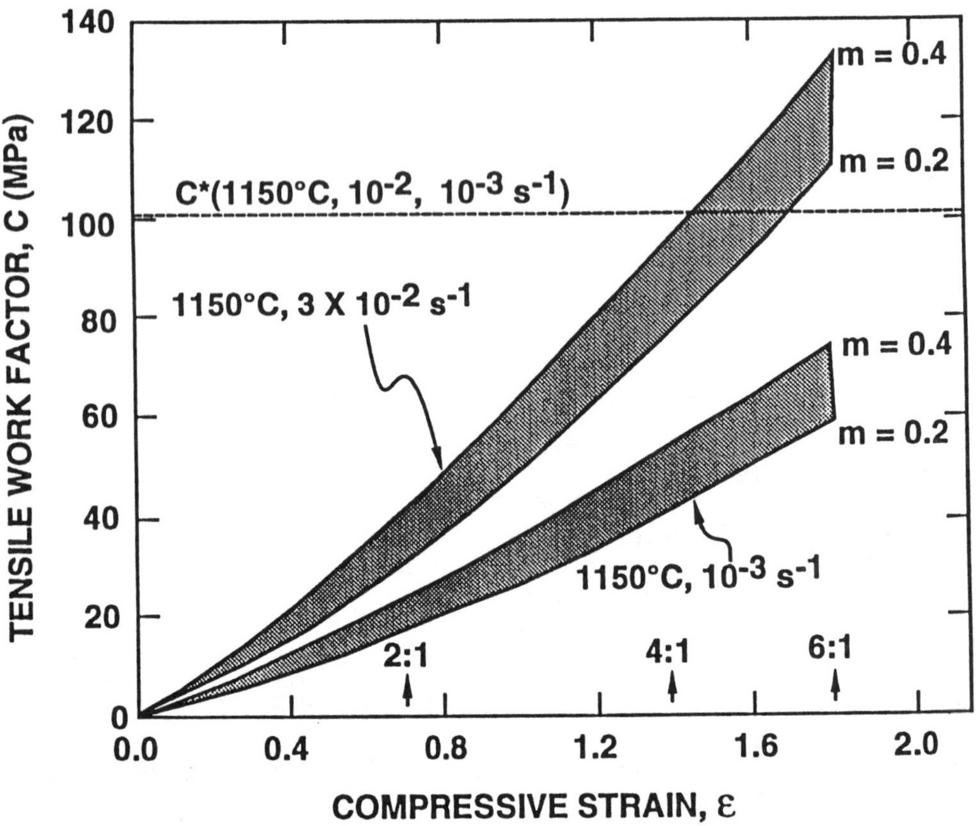

Figure 6 Comparison of the tensile work factor C during isothermal forging of cast + HIP'ed Ti-49.5Al-2.5Nb-1.1Mn pancakes and the critical tensile work for fracture, C*.

Semiatin, *et al.* [19] investigated the workability of the near-gamma alloy Ti-45.5Al-2Cr-2Nb during ECAE. They showed that the tendency for nonuniform deformation increased rapidly as the preheat temperature decreased. For example, a relatively sound product was produced by extrusion at 1250°C of cast plus HIP'ed Ti-45.5Al-2Cr-2Nb canned in type 304 stainless steel, but shear bands and shear cracks were developed when extrusion was attempted at 1150°C (Figure 7). The propensity for nonuniform flow during ECAE of the near gamma titanium aluminides can be interpreted using the concept of the flow localization parameter, α [19]. It provides a first order gage of the competition between the destabilizing influence of material flow softening (due to deformation heating, microstructure changes, etc.) and the stabilizing effect of a moderate-to-high strain rate sensitivity of the flow stress on the development of

3. S.L. Semiatin, V. Seetharaman, and V.K. Jain, "Microstructure Development During Conventional and Isothermal Hot Forging of a Near-Gamma Titanium Aluminide", *Metall. and Mater. Trans. A*, 25A (1994), 2753-2768.
4. H. Clemens, P. Schretter, K. Wurzwallner, A. Bartels, and C. Koeppe, "Forging and Rolling of Ti-48Al-2Cr on Industrial Scale", in R. Darolia, *et al.*, eds., *Structural Intermetallics* (Warrendale, PA: TMS, 1993), 205-214.
5. S.L. Semiatin and V. Seetharaman, "Deformation and Microstructure Development During Hot-Pack Rolling of a Near-Gamma Titanium Aluminide Alloy", *Metall. And Mater. Trans. A*, 26A (1995), 371-381.
6. V. Seetharaman, R.L. Goetz, and S.L. Semiatin, "Tensile Fracture Behavior of a Cast Gamma-Titanium Aluminide", in L.A. Johnson, *et al.*, eds., *High-Temperature Ordered Intermetallic Alloys IV* (Pittsburgh, PA: MRS, 1991), 895-900.
7. V. Seetharaman, S.L. Semiatin, C.M. Lombard, and N.D. Frey, "Deformation and Fracture Characteristics of a Gamma Titanium Aluminide at High Temperatures", in I. Baker, *et al.*, eds., *High-Temperature Ordered Intermetallic Alloys V* (Pittsburgh, PA: MRS, 1993), 513-518.
8. V. Seetharaman and C.M. Lombard, "Plastic Flow Behavior of a Ti-Al-Nb-Mn Alloy at High Temperatures", in Y-W. Kim and R.R. Boyer, eds., *Microstructure/Property Relationships in Titanium Aluminides and Alloys* (Warrendale, PA: TMS, 1991), 237-251.
9. V. Seetharaman and S.L. Semiatin, "Plastic Flow and Microstructure and Microstructure Evolution During Hot Deformation of a Gamma Titanium Aluminide", *Metall. and Mater. Trans. A*, in press.
10. S.L. Semiatin and V. Seetharaman, "A Criterion for Intergranular Failure During Hot Working of a Near-Gamma Titanium Aluminide Alloy", *Scripta Mater.*, 36 (1997), 291-297.
11. V. Seetharaman and S.L. Semiatin, "Intergranular Fracture During Hot Working of a Near Gamma Titanium Aluminide Alloy", *Metall. and Mater. Trans. A*, in press.
12. E. Orowan, *Fatigue and Fracture of Metals*, (New York, NY: John Wiley, 1950).
13. J. Xinggang, C. Jianzhong, and M. Longxiang, " A Cavity Nucleation Model During High Temperature Creep Deformation of Metals", *Acta Metall. et Mater.*, 41, (1993) 539-542.
14. M. Nobuki and T. Tsujimoto, "Influence of Alloy Composition on Hot Deformation Properties of Ti-Al Binary Alloys", *ISIJ International*, 31 (1991) 931-937.
15. D.S. Shih and G.K. Scarr, "High Temperature Deformation Behavior of the γ Alloy Ti-48Al-2Cr-2Nb", in L.A. Johnson, D.P. Pope, and J.O. Stiegler (eds.), *High-Temperature Ordered Intermetallic Alloys IV*, (Pittsburgh, PA: Materials Research Society 1991), 727-732.
16. M.G. Cockcroft and D.J. Latham, "A Simple Criterion of Fracture for Ductile Metals", N.E.L. Report No. 240, National Engineering Laboratory, East Kilbride, Glasgow, Scotland, 1966.
17. D. Watkins, H.R. Piehler, V. Seetharaman, C.M. Lombard, and S.L. Semiatin, "Effect of Hydrostatic Pressure on the Hot-Working Behavior of a Gamma Titanium Aluminide", *Metall. Trans. A*, 23A (1992), 2669-2672.
18. S.L. Semiatin, V.M. Segal, R.L. Goetz, R.E. Goforth, and T. Hartwig, "Workability of a Gamma Titanium Aluminide Alloy During Equal Channel Angular Extrusion", *Scripta Metall. et Mater.*, 33 (1995), 535-540.
19. S.L. Semiatin and J.J. Jonas, *Formability and Workability of Metals* (Materials Park, OH: ASM International, 1984).

ON THE TEMPERATURE DEPENDENCE OF FRACTURE TOUGHNESS OF DEFORMED AND RECRYSTALLIZED LOW-ALLOYED CHROMIUM

Anatoly V. Sameljuk, Alexandr D.Vasilev and Sergey A. Firstov

Frantsevich Institute for Problem of Materials Science,
3 Krzchizchanovskogo St., 252680 Kiev , Ukraine

Abstract

The temperature dependences of fracture toughness as well as sequences of fracture mechanisms changes of deformed and recrystallized chromium alloyed with 0,5% (w/w) La_2O_3 were studied in a wide temperature range. It was found, that the fracture toughness is nonmonotonously depended on temperature. As temperature is raised cleavage fracture toughness increases up to 500° C for deformed state and up to 600° C for recrystallized one. The dynamic deformation ageing is responsible for appearance of brittle intergranular fracture and decreasing the fracture toughness of recrystallized material at temperature above 600° C. Sharp increasing of the dependence of deformed material above 500° C is due to plane-strain to plane-stress transition. The decreasing fracture toughness of deformed material at higher temperature results from changing of fracture mechanisms from cleavage to pore coalescence.

Modeling the Mechanical Response of
Structural Materials
Edited by E.M. Taleff and R.K. Mahidhara
The Minerals, Metals & Materials Society, 1997

Introduction

Most high-strength material are susceptible to brittle fracture. Chromium fails by cleavage over an unusually wide range of structural states [1]. In contrast of other BCC metals, chromium has limited number of fracture mechanism and cleavage is commonly changed by pore coalescence with temperature [2]. This behaviour of chromium may be due to a high crack velocity which result also in cleavage crack branching polycrystalls material [3] as well as monocrystal one [4]. Such high crack velocity are due to the extremely high strength and the relatively low effective surface energy for fracture [5]. Chromium is also extremely notch-sensitive [6]. Using the big samples with sharp notch allows to study cleavage fracture toughness of chromium in wide temperature range.

The purpose of this paper is to study of fracture toughness temperature dependences and fracture mechanism changes of deformed and recrystallized low alloyed chromium in a wide temperature range.

Experimental procedure

The cylindrical specimens (diameter is 16 mm) with a sharp circular notch (diameter after notching is 10 mm) was used for fracture toughness determination of deformed and recrystallized chromium alloyed with 0,5% (w/w) La_2O_3. The extrusion deformation was performed at 1100° C with 48% deformation degree. The recrystallized samples was obtained by annealing for 4 hr at 1100° C in vacuum. The material contains lanthanum oxide particles, about 2-3 μm in size in both states. The grain size was about 20 μm for recrystallized material and 2 μm for deformed one. Tensile tests were carry out on electro-mechanical machine at a constant strain rate of 1 mm/min in temperature range from 20° C to 750° C. Fractographical investigation and chemical composition of secondary phases determination were carry out with "Superprobe-733" (JEOL) X-ray microanalyzer.

Results and discussion

The fracture toughness behaviour of deformed and recrystallized chromium over a wide range of temperature is shown in Fig. 1. It is observed a monotonous increasing of fracture toughness of recrystallized material in temperature range 20-600° C and deformed material in temperature range 20-500° C. The dependences have a different behaviour at higher temperatures. The fracture toughness of recrystallized material decreases above 600° C. In contrast of this, the fracture toughness of deformed material increases sharp above 500° C and decreases at temperature above 700° C.

Fractographical investigation of the samples tested allows to reveal the fracture features of both states of the chromium alloy. The material fractures by cleavage in temperature ranges 20-600° C for recrystallized state (Fig. 2,,a) and 20-500° C for deformed state (Fig. 3,a). The fracture surfaces of this samples is located in a notch plane. As the temperature exceeds 600°C, the brittle intergranular fracture of recrystallized material (Fig. 2,b) preceded the cleavage. As was mention above this fracture mechanisms is rarely observed in chromium. It is worth to note, that the area of the intergranular fracture increases with temperature. The fracture surfaces contained the area of brittle intergranular fracture is also located in the notch plane. Fracture surfaces of deformed material contains the area failed by pore coalescence (Fig. 3,b) and deviates from the notch plain at temperature above 500° C. The area increases with temperature also.

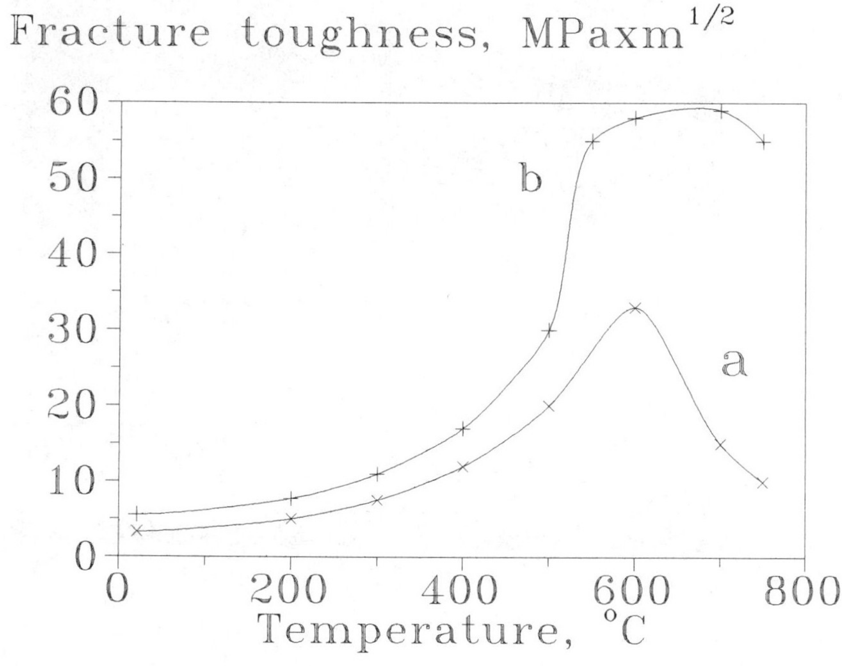

Figure 1 - Effect of temperature on fracture toughness of recrystallized (a) and deformed (b) chromium alloyed with 0,5% (w/w) La_2O_3.

The monotonous increasing of fracture toughness of recrystallized material in temperature range 20-600° C and deformed one in temperature range 20-500° C is due to the decreasing of the yield strength. The decreasing results in the crack tip blunting and plastic zone size increasing. To compensate this, it is necessary to increase the local tensile stress at the crack tip.

It is known, that chromium and its alloys are susceptible to dynamic deformation ageing in temperature range 600-800° C. The dynamic deformation ageing is responsible for fracture toughness decreasing of recrystallized material at temperature above 600° C. The effect results in strengthening of the grain body. The brittle intergranular fracture appears when the grain body strength exceeds the grain boundary strength. The final fracture by cleavage starts from this area of intergranular fracture. The increasing of the area results in decreasing the fracture toughness of recrystallized material with temperature.

The excess of the plastic zone over the specimen size lead to the sharp increasing of fracture toughness of deformed material at temperature above 500° C. The deviation of fracture surfaces from the notch plane indicates a plane-strain to plane-stress transition. The decreasing of fracture toughness of deformed material at higher temperature is due to changing fracture mechanisms from cleavage to pore coalescence.

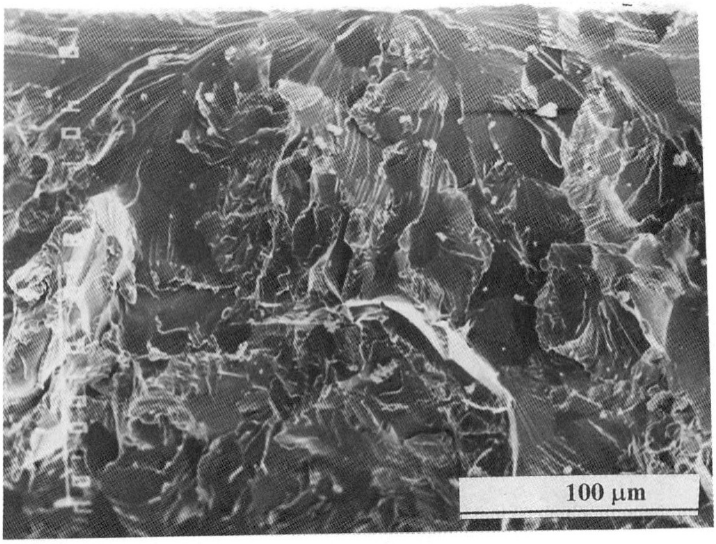

a

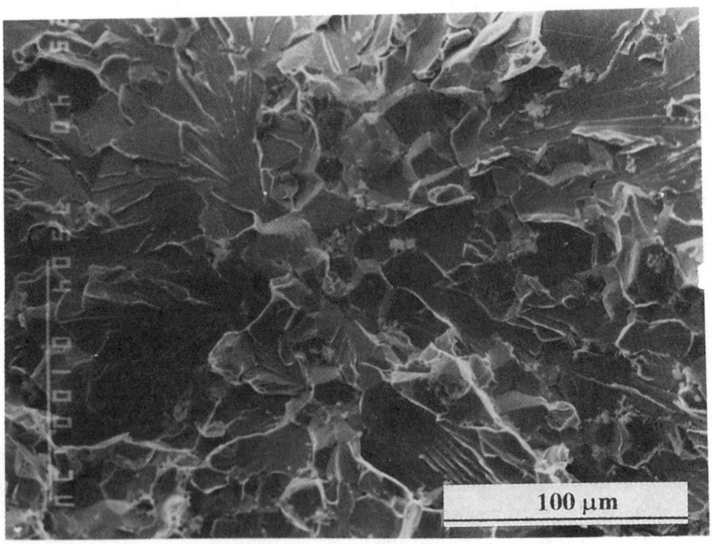

b

Figure 2 - Fracture surfaces of recrystallized chromium alloy. Cleavage fracture (a, 200° C) and the area of brittle intergranular fracture preceded the cleavage (b, 600° C).

THE ROLE OF DELAMINATION ON THE DUCTILE FRACTURE

BEHAVIOR OF LAMINATED COMPOSITES

S.B. BINER

Ames Laboratory,
Iowa State University
Ames, IA 50011, U.S.A.

ABSTRACT

In this study , the failure of the ductile layers from collinear and delaminating cracks that occur in laminated composite systems was studied using a constitutive relationship that accounts for strength degradation resulting from the nucleation and growth of voids. The results indicate that in laminated composites, void nucleation and growth ahead of the cracks occur at a much faster rate due to evolution of much higher stress values at the interface region. For delaminating crack cases, the fracture behavior is strongly influenced by the delamination length. The resistance of the ductile layers to crack extension can be significantly reduced by short delamination lengths.

Modeling the Mechanical Response of
Structural Materials
Edited by E.M. Taleff and R.K. Mahidhara
The Minerals, Metals & Materials Society, 1997

INTRODUCTION

Laminated composites containing ductile layers are currently under development not only to improve the fracture toughness of inherently brittle intermetallics and ceramics [1-9], but also to increase the relatively low fracture toughness of metal matrix composite systems[10,11]. The recent reviews treating the contribution of the ductile phase to the composite toughening have indicated that this contribution depends on the strength, volume fraction and failure characteristics of the ductile reinforcing phase. It is also apparent that the failure of the ductile layers is strongly affected by the degree of the constraint that is largely controlled by the interfacial behavior of the layers[12-17].

As schematically illustrated in Fig.1, several failure modes of the brittle layers such as only formation of collinear cracks, with multiple cracking and with delamination in various lengths are usually seen in most of the experimental studies[1-11]. In this study, the growth of such cracks into the ductile layers are studied by using a constitutive relation that accounts for strength degradation resulting from the nucleation and growth of micro-voids in the ductile layers. The results obtained for the laminated composite cases are compared with the growth behavior of the cracks in the same ductile material as a homogenous isotropic case.

NUMERICAL ANALYSIS

The basis for the constitutive model for the ductile layers is a flow potential introduced by Gurson[18,19], in which voids are represented in terms of a single internal variable, f, the void volume fraction.

$$\phi = \frac{\sigma_e^2}{\sigma_o^2} + 2f^* q_1 \cosh\left(\frac{q_2 \sigma_h}{2\sigma_o}\right) - 1 - q_1^2 f^{*2} = 0 \qquad (1)$$

where σ_o is the flow strength of the ductile layer, σ_e is the equivalent stress and σ_h is the hydrostatic stress. The parameters q_1 and q_2 were introduced in order to provide a better relationship between unit-cell analysis and eq.1[20,21]. Later, he function f^* was proposed to account for the effect of rapid void coalescence at failure[22]. Initially $f^*=f$, but at some critical void fraction, f_c, the dependence of f^* on f is changed. This function is expressed by

$$f^* = \begin{cases} f & f \le f_c \\ f_c + \dfrac{f_u^* - f_c}{f_f - f_c}(f - f_c) & f \ge f_c \end{cases} \qquad (2)$$

The constant f_u^* is the value of f^* at zero stress in eq.1 (i.e. $f_u^*=1/q_1$) and f_f is the void volume fraction at fracture. As $f \to f_f$, $f^* \to f_u^*$ and the material loses all stress carrying capacity.
The increase in void volume fraction f arises from the growth of existing voids and from the nucleation of new voids. The growth rate of existing voids is related to the macroscopic dilation rate by

$$(\dot{f})_{growth} = (1 - f)\delta^{ij} \dot{\eta}_{ij}^p \qquad (3)$$

where $\dot{\eta}_{ij}^p$ is the plastic part of the deformation.

The increase in the volume fraction due to the nucleation process is assumed to occur with strain controlled nucleation and follow a normal distribution as suggested by Chu et al.[23].

$$(\dot{f})_{nucleation} = \left[\frac{f_N}{s_N (2\pi)^{1/2}} \exp\left(-\frac{1}{2}\left(\frac{\varepsilon^p - \varepsilon_N}{s_N}\right)^2\right)\right]\dot{\varepsilon}^p \qquad (4)$$

where f_N is the volume fraction of void nucleating particles, ε_N is the mean strain for nucleation, ε^p is the current value of effective plastic strain and s_N is the corresponding standard deviation.

In the present investigation a rate-sensitive version of Gurson's model was employed. In the ductile layers, the effective plastic strain rate $\dot{\varepsilon}^p$ is represented by the power law relation.

$$\dot{\varepsilon}^p = \dot{\varepsilon}_o \left[\frac{\sigma_m}{g(\varepsilon^p)} \right]^{\frac{1}{m}} \tag{5}$$

where m is the strain rate sensitivity exponent, $\dot{\varepsilon}_0$ is a reference strain rate and ε^p is the current value of the effective plastic strain representing the actual microscopic strain state. The function $g(\varepsilon^p)$ represents the effective tensile flow stress in a tensile test carried out at strain rate that is equal to reference strain rate, ε_O. For a power hardening matrix material the function $g(\varepsilon^p)$ is taken to be

$$g(\varepsilon^p) = \sigma_o \left[\frac{E\varepsilon^p}{\sigma_o} + 1 \right]^n \quad g(0) = \sigma_o \tag{6}$$

with strain hardening exponent n, Young's modulus E and reference stress σ_O. The material parameters for the ductile layer appearing eqs.5 and 6 were chosen as E=500σ_o, υ=0.3, n=0.1, m=0.01 and the reference strain rate $\dot{\varepsilon}_O$=2x10^{-3}. The parameters appearing in eq.6 for void nucleation were taken as f_N=0.04, s_N=0.1 and ε_N=0.3. For accelerated void growth, the parameters appearing in eq.2 were chosen as, f_f=0.25, f_c=0.10 and f_u*=1/1.25. Also, q_1=1.25 and q_2=q_1^2 were selected for eq.1. In the analysis, the behavior of the crack containing brittle layers is assumed to be elastic. The thickness ratio of the ductile layers to the brittle layers(W/a) was 0.2 (Fig.1). The Young's modulus of the brittle layers was 1.5 times the modulus of the ductile layers, and both layers had the same Poisson's ratio.

Because of symmetry, only one quarter of the geometry seen in Fig.1a is analyzed in plane-strain condition with the FEM mesh shown in Fig.1b. The crack tip is located at the interface, and it is modeled with a hole having a radius of 10^{-4} of the crack length as shown in Fig.1c. The smallest element size around the crack tip region was 1.8x10^{-4} of the crack length; to accommodate this large difference in element size and preserve the aspect ratio of the elements in the finite element discretization the elements were scaled exponentially to the crack tip. There were 1004 nodes and 932 four-node isoparametric elements in the model. The axial displacement rate was 5$\dot{\varepsilon}_O$ and resulting load values were calculated from the reaction forces. The crack extension is implemented by element vanish technique when f values became greater than the f_f value at all integration points.

RESULTS AND DISCUSSION

First, the fracture behavior of the ductile layer material as homogeneous isotropic case was elucidated. To simulate highly constrained crack growth in the laminated composites, the fracture behavior of the ductile matrix material is studied in a deeply cracked double edge notch geometry (DEN) by using the same mesh shown in Fig.1b. The resulting load and load-line displacement curve for this case is shown in Fig.2. As can be seen from the Fig.2 the analysis was carried out well beyond the maximum load. Since the size of the elements increases away from the original crack tip(Fig.1b and c), the observed step like behavior in the load-load line displacement curve is associated with the failure of the large elements away from the near crack tip region which required some strain accumulation to meet the failure condition at all integration points. The evolution of the maximum void volume fraction at the third ring elements (Fig.1c) with increasing axial displacement is summarized in Fig.3. As can be seen from Figs.2 and 3, the failure of the elements in the third row and the crack extension took place in the linear region of the load and load-line displacement curve.

In the next simulation the extension of the collinear cracks into the ductile layers in the laminated composites were analyzed. The resulting load and load-line displacement curve for this case is also given in Fig.2. Although for both cases the uncracked ligament were the same, in the case of the composite, larger load levels were required to achieve the same amount of axial displacement. For this case, the first crack extensions also occurred in the linear region, however, the fracture behavior was very unstable after the peak load and very fast reductions occurred in the load carrying capacity of the uncracked ligament. The evolution of the maximum void volume fractions at the third ring elements for this case is also shown in Fig.3. Due to the development of a larger scale triaxial stress state, as can be seen, for the collinear cracks the damage accumulation in the ductile layer occurred at a much faster rate.

In the next two simulations the role of the interface debonding behavior, as schematically shown in Fig.1, in the failure behavior of the ductile layers was investigated. In these analyses, the interface separation distance from the original crack tip were 0.1 and 0.2 times the original crack length which are designated as interface crack-I and II respectively. In these simulations, it is also assumed that no further debonding at the interface will take place during the fracture process. The resulting load load-line displacement curves for these cases are also given in Fig.2. As can be seen from the figure, by introduction of the delamination to the interface, the resulting peak load values were much lower than the composites having only collinear cracks and this reduction became larger with increasing delamination length. For small delamination length (interface crack -I), there was almost no stable crack extension after the peak load; on the other hand, in the case of larger delamination length (interface crack-II) the fracture was very stable and it required even larger axial displacements than that seen during the fracture process of the ductile layer as homogenous isotropic case. The variation of the void volume fractions for these cases, at the same distance ahead of the crack tip as in previous cases, are also summarized in Fig.3. As can be seen from the figure for the composite having small delamination length, the axial displacement required for the crack extension from the third ring elements is slightly larger than for the composite having only collinear cracks. The axial displacement required for failure to occur at the same location ahead of the crack tip increases with increasing delamination length and approaches that seen for the crack in the ductile layer as homogenous isotropic case. In addition, for the larger delamination case, the rate of increase in the volume fraction was much slower than for the other crack cases investigated for the laminated composite.

The development of plastic zones at the final load levels are given in Fig.4. In the composite having collinear crack, a large extension of the plastic zone along the interface region can be distinguished (Fig.4b). For collinear crack, except in the interface region, the extent of deformation in the ligament is not as widespread as that seen for the homogenous case (Fig.4a). The evolution of the deformation zone for the delaminated cracks in the composite significantly differs from the ones seen previously and the formation of the very pronounced shear bands and the development of large deformation zone at the center of the uncracked ductile ligament, almost the one usually seen in the uniform tensile specimen, can be easily distinguished in Fig.4 d and e. Although it is not shown here as a separate figure, the distribution of the void volume fraction follows the distribution of the plastic strains since both the nucleation of the voids and their growth rate were assumed to be controlled by the plastic strains via eqs.3 and 4. It is apparent that in the delaminated cracks, large scale damage accumulation also occurs at the center of the uncracked ductile ligament away from the current crack tip. For small delamination lengths the damage zone ahead of the crack tip and the damage zone occurring at the center of the ductile ligament join very quickly, with the formation of the shear bands leading to significant drop both in the peak load and in the load-line displacement as seen in Fig.2. With increasing delamination length similar failure takes place; however, since the distance between these two damage zone is now increased and also the amount of material undergoing large scale deformation is much larger (Fig.4e), the crack extension becomes more stable and requires much larger load-line displacements (Fig.2).

The resulting R-curves for the laminated composite having collinear cracks and cracks with delamination are compared with the R-curve behavior of the crack in the ductile layer as a homogenous isotropic case in Fig.5. For the collinear cracks in the laminated composite in which the crack tip is located at the interface, in agreement with Fig.2, the energy required for the initial crack extension into the ductile layer is much smaller than the energy required for the initial crack extension in ductile layer as homogeneous isotropic case. In the laminated composite, as the collinear cracks extend further into the ductile layer, the influence of the stress field of the interface region diminishes and the crack driving force becomes very similar to the one seen for the crack in the ductile layer as a homogenous isotropic case. In the laminated composite having interface separation, fracture behavior is very much dependent upon the delamination length. For short delamination length the energy required for crack initiation was similar to the collinear crack case; however, the energy required for crack extension in the ductile layer was much lower due to the localization of deformation (Fig.4). With increasing delamination length, the energy required for the crack initiation becomes similar to the crack in the ductile layer material, but further crack extension requires much larger driving force due to spread of the deformation into much larger regions and loss of stress triaxiality resulting from the development of a large zone in the center of the uncracked ligament having significant amount of void formation (Fig.4).

Although the results presented above are based on the continuum scale, they clearly indicate the role of the crack morphology, the extent of the delamination and crack tip location on the fracture behavior of ductile layers and their possible contribution to overall composite toughening behavior.

CONCLUSIONS

In this study the failure of the ductile layers in a laminated composite system was studied by using a constitutive relationship that accounts for strength degradation resulting from the nucleation and growth of voids. The results indicate that:

1. In the laminated composites, the void nucleation and growth ahead of the cracks occur at a much faster rate due to evolution of much higher stress values at the interface region.

2. For delaminating crack cases, the fracture behavior is strongly influenced by the delamination length. The resistance of the ductile layers to the crack extension can be reduced significantly by short delamination lengths. For large delamination lengths the resistance to crack extension becomes greater than that seen for the ductile material.

ACKNOWLDEGEMENT

This work was performed for the United States Department of Energy by Iowa State University under contract W-7405-Eng-82. This research was supported by the Director of Energy Research, Office of Basic Energy Sciences.

REFERENCES

1. J.T. Beals and V.C. Nardone, J. Mat. Sci. **29**, 2526, (1994).
2. L. Shaw and R. Abbaschian, Acta Metall. Mater. **42**, 213, (1994).
3. H.C. Cao, B.J. Dalglesih, H. Dave, C. Elliot, A.G. Evans, R. Mehrabian and. G.R.Odette,Acta Metall. Mater. **38**, 2969, (1990).
4. V.V. Krstic, P.S. Nicholson and R.G. Hoagland, J. Am. Ceram. Soc. **64**, 499, (1981).
5. J. Kajuch, J. Short and J.J. Lewandowski, Acta Metall. Mater. **43**, 1955, (1995).
6. L. Xiao and R. Abbaschain, Metall. Trans. **24A**, 403, (1992).
7. T.C. Lu, A.G. Evans, R.J. Hect and R. Mehrabian, Acta. Metall. Mater, **39**, 1853, (1991).
8. W.O. Soboyejo, F.Ye, L-C. Chen, N. Bahtishi, D.S. Schwartz and R.J. Lederich, Acta. Mater. **44**, 2027, (1996)
9. J. Kajuch, J.D. Rigney and J.J. Lewandowski, Mater. Sci. Engng. **A115**, 59, (1992).
10. C.K. Syn, S. Stoner, D.R. Lesuer and O.D. Sherby, "Influence of volume fraction and interlayer bond strength on fracture toughness of multilayer Al6090-25%SiC and Al5812 laminates" in "High performance metal and ceramic composites" ed. K. Upadhya, (TMS Warrandale, PA, 1994), p.125.
11. W.H. Hunt Jr., T.M. Osman and J.J. Lewandowski, J. Metals, **45**, 30, (1993).
12. M.F. Ashby, F.J. Blunt and M. Bannister, Acta Metall. Mater. **37**, 1847, (1989).
13. A.G. Evans and R. McMeeking, Acta Metall. Mater. **34**, 241, (1986).
14. S.B. Biner, Mater. Sci. Engng. **A187**, 125, (1994).
15. Ravichandran, Acta Metall. Mater. **40**, 1009, (1992).
16. Budinasky, J.C. Amazigo and A.G. Evans, J. Mech. Phys. Solids, **36**, 167, (1988).
17. F. Erdogan, P.F. Joseph, J. Am. Ceram. Soc, **72**, 262, (1989).
18. A. L. Gurson, PhD thesis, Brown University, (1975).
19. A. L. Gurson, J. Eng. Mater. Technol. **99**, 2, (1977).
20. V. Tvergaard, Int. J. Frac. **17**, 389, (1981).
21. V. Tvergaard, Int. J. Frac. **18**, 237, (1982).
22. V. Tvergaard and A. Needleman, Acta Metall. Mater. **32**, 157, (1988).
23. C.C. Chu and A. Needleman, J. Eng. Mater. Technol. **102**, 249, (1980).

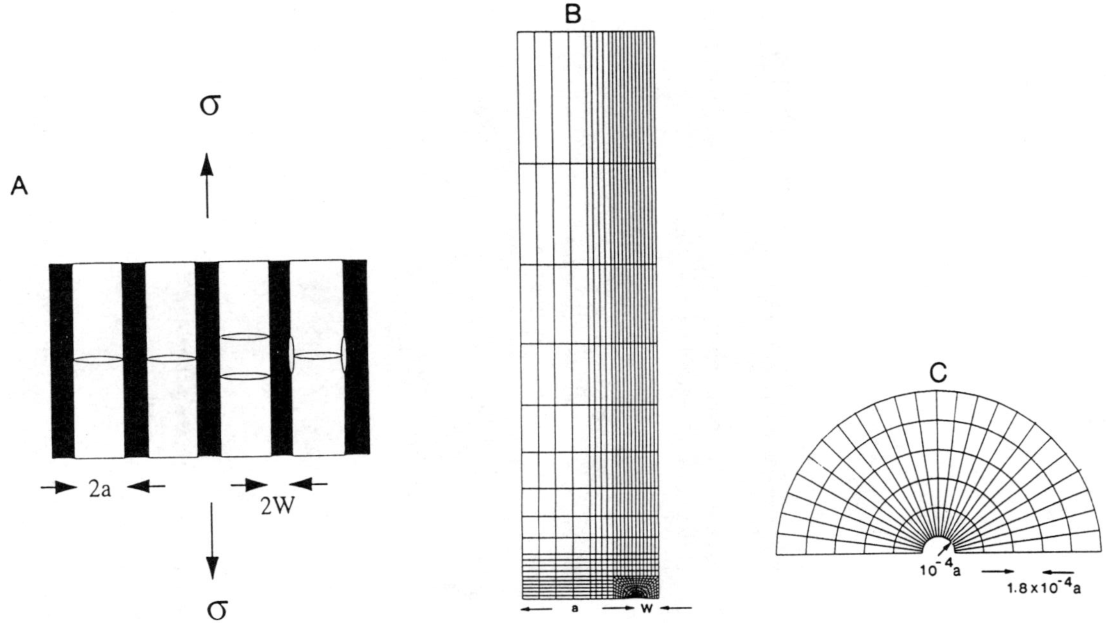

Fig.1 a) Schematic represantation of cracks at the interface of the ductile layer in the laminated composite, b) The FEM mesh used in the analysis and c) the details of the crack tip region

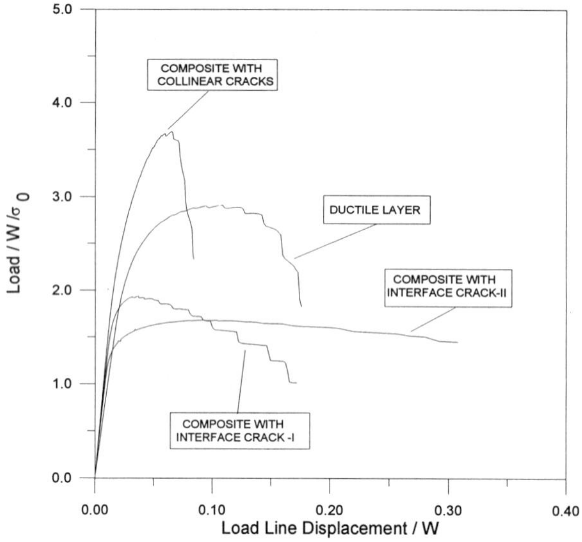

Fig.2 Load and load-line displacement responses of the composites having collinear and delaminating cracks.

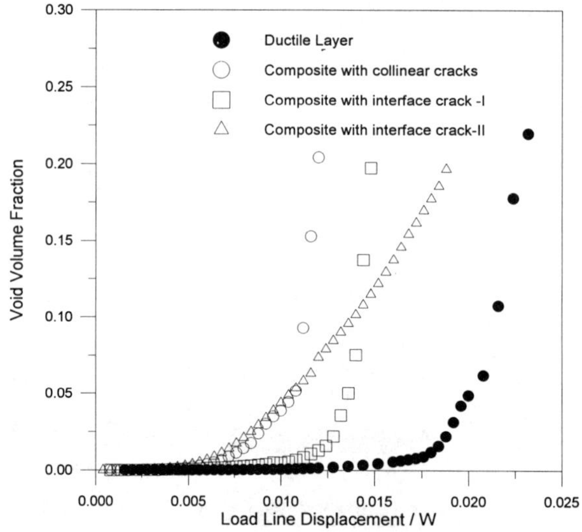

Fig.3 Evolution of the void volume fractions ahead of the collinear and delaminating cracks.

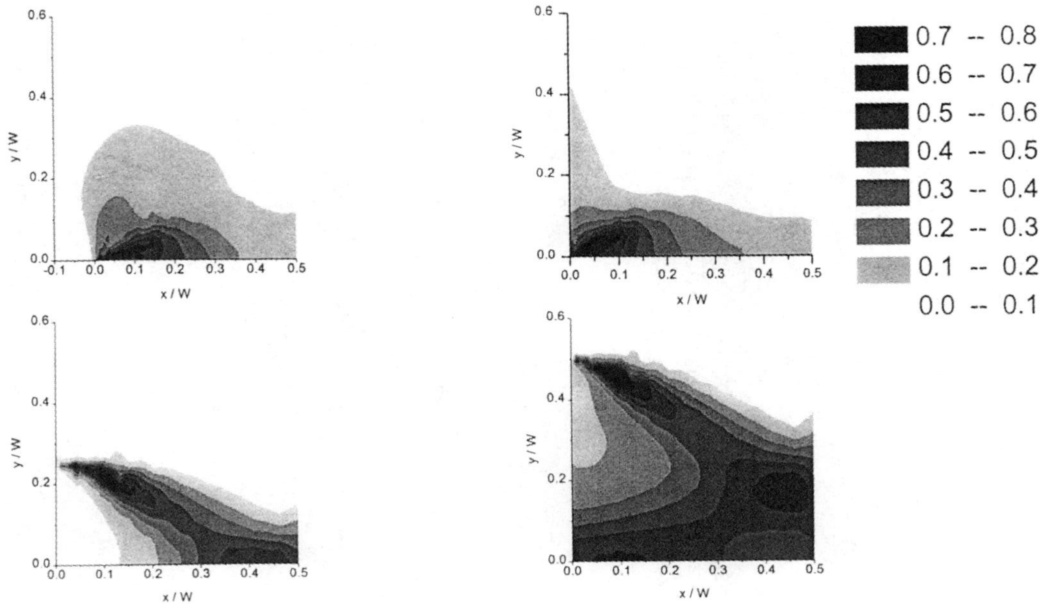

Fig.4 Comparison of the plastic zone sizes at the final load levels. a) crack in ductile layer as a homogeneous case, b) collinear crack, c) collinear crack with delamination, delamination length 0.1 times the original crack length, d) collinear crack with delamination, delamination length 0.2 times the orginal crack length.

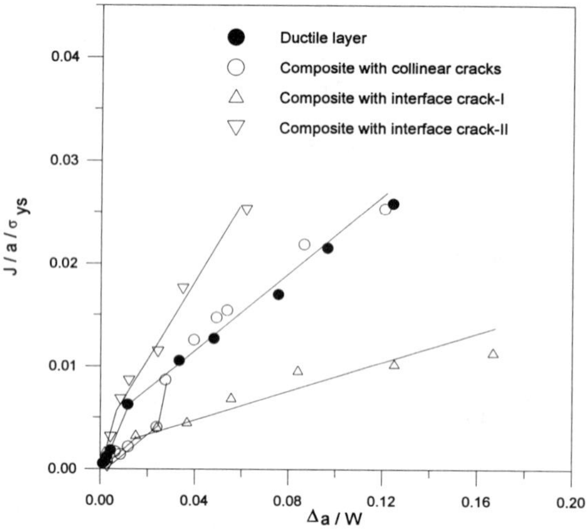

Fig. 5 Crack resistance curves for collinear and delaminated cracks in the laminated composite and comparison with the R-curve behavior of the ductile layer material as homogeneous isotropic case.

GROWTH BEHAVIOR OF BRIDGED CRACKS
AT CREEP REGIME

S. B. BINER

Ames Laboratory ,
Iowa State University,
Ames, Ames IA 50011 U.S.A

ABSTRACT

In this study, the growth behavior of bridged cracks in the creep regime is investigated numerically. In the analysis, crack growths resulting from the growth of prenucleated creep cavities with diffusional and dislocation assisted mechanisms are fully accounted for. The bridging elements are assumed to be elastic. The results indicate that the bridging tractions significantly relax even with the creep deformation of the composite alone. A much weaker stress singularity than the ones described by K or C^* occurs ahead of the bridged cracks in the creep regime. In this weak singularity region the cavities grow at comparably high rates to each other.

Modeling the Mechanical Response of
Structural Materials
Edited by E.M. Taleff and R.K. Mahidhara
The Minerals, Metals & Materials Society, 1997

INTRODUCTION

Composites are under development to meet the requirements of advanced structural applications which will operate at elevated temperatures. To achieve these objectives different avenues are being explored, including efforts to increase the fracture toughness of normally brittle ceramics and intermetallics and to increase stiffness and creep resistance of metals by the addition of reinforcements in the form of whiskers or fibers [1-6]. The mechanism of toughening at room temperature due to crack bridging caused by the reinforcements is now well understood, and models describing this toughening behavior have been developed with various degree of details[7-14].

Recent experimental studies [15-18] suggest that crack bridging by reinforcements also occurs at creep regime. However, in contrast to the room temperature case, the crack bridging and the crack growth at elevated temperatures are time- or rate-dependent processes. First it should be realized that, whether or not crack growth occurs, the stress-strain fields ahead of a crack are time dependent in the creep regime. The initial response of a homogeneous crack (i.e. without any bridging), upon application of load in the creep regime, is essentially elastic. During this initial period, creep strains are negligible (i.e. less than elastic strains) everywhere except in the immediate vicinity of the crack tip; the elastic stress intensity factor, *K*, provides an adequate description of the crack tip stress-strain field; i.e

$$\sigma_{ij} = \frac{K}{\sqrt{2\pi r}}\tilde{\sigma}_{ij}(\theta) \tag{1}$$

where r is the radial distance from the crack tip and $\tilde{\sigma}_{ij}$ contains the angular dependency of the crack tip stress field. The stress and strain rate response of the materials undergoing creep deformation is generally modeled using a power-law constitutive equation:

$$\dot{\varepsilon} = \dot{\varepsilon}_0\left(\frac{\sigma_e}{\sigma_0}\right)^n \tag{2}$$

where $\dot{\varepsilon}$ and σ_e are the von Mises equivalent strain rate and stress respectively, σ_0 is the reference stress, $\dot{\varepsilon}_0$ is a temperature dependent reference strain rate and n is the creep exponent. The analytical studies[19], supported by the numerical analyses[20-22], show that once a steady state has been achieved, the stress-strain fields near the crack tip have an *HRR* singularity[23,24] with the amplitude given by a path independent integral C^*, the rate dependent *J*-integral, i.e.:

$$\sigma_{ij} = \left[\frac{C^*}{\dot{\varepsilon}_0 I_n r}\right]^{\frac{1}{n+1}}\tilde{\sigma}_{ij}(\theta,n) \tag{3}$$

In eq.3 the angular dependency of the stress field $\tilde{\sigma}_{ij}$ and integration constant I_n also depend on the creep exponent n. Between the two limiting situations of K and C^* control lies a transition period in which the creep strains are higher than the elastic strains over an extensive region ahead of the crack tip. During this period, the integral which defines C^* is path dependent. The duration of this transition period (i.e. from the initial loading up to development of a path independent C^* controlled *HRR* field) has been estimated [19] as approximately:

$$t = \alpha\frac{K^2(1-\upsilon^2)}{(n+1)EC^*} \tag{4}$$

where E is the Young's Modulus, ν is the Poisson's ratio and α=0.96 for plane-strain and 1.05 for plane-stress.

In polycrystalline materials (including their composite derivatives) the crack growth at high temperatures mainly occurs by the nucleation and growth of the cavities at the grain boundaries and their coalescence with the crack tip. In addition to the above time dependent stress-fields in homogenous cracks, in the case of bridged cracks the other rate dependent processes may take place simultaneously at creep regime. Whether or not crack growth occurs, the bridging tractions in the bridging zone can change with time by at least three mechanisms: first, because of the time dependent deformations at the interface layer between the reinforcements and the matrix; secondly, with the creep deformation of the reinforcements or their degradation; and finally, with the creep deformation of the matrix alone. Of course, these time dependent changes in the bridging tractions control the rate of evolution of the bridging length. Once crack growth begins, from the above discussion it is clear that, three rate dependent processes: the rate of relaxation of the stress field ahead of the crack tip, the rate of creep damage evolution leading to crack growth, and the rate of changes in the bridging tractions, all will be influenced by each other.

In this study, in order to elucidate the effects of the aforementioned three rate dependent processes, the growth characteristics of bridged cracks at creep regime is studied numerically. It is assumed that crack growth occurs as a result of the growth of the pre-nucleated creep cavities in the composite without detailing the nature of the reinforcements. The bridging reinforcements are assumed to behave in an elastic manner; however, the bridging tractions vary with time due to the overall composite creep deformation.

CONSTITUTIVE RELATION FOR CREEP CRACK GROWTH AND DETAILS OF THE FEM ANALYSES

In this study, to simulate the growth of the bridged cracks resulting from the growth of the cavities, a layer of dimensionless spring elements in the crack plane ahead of the crack tip was utilized similar to that used to study the grain boundary facet cracking [25,26]. For cavities located ahead of a bridged crack as schematically shown in Fig.1, if the cavity radius is c and the average spacing is $2b$, the volumetric growth rate is expressed as[27]:

$$\dot{V} = \dot{V}_1 + \dot{V}_2 \tag{5}$$

where $\dot{V}_1$ is the contribution to the volumetric growth rate due to grain boundary diffusion and $\dot{V}_2$ is the contribution due to the power-law creep deformation of the adjacent grains. Based on the numerical studies of grain boundary cavity growth;

$$\dot{V}_1 = 4\pi D \frac{\sigma_n - (1-f)\sigma_s}{\ln(1-f) - (3-f)(1-f)/2} \tag{6}$$

where

$$f = \max\left[\frac{c^2}{b^2}, \frac{c^2}{(c+1.5L)^2}\right] \tag{7}$$

in which D is the grain boundary diffusion parameter, σ_n is the average stress normal to the current orientation of the grain boundary in the vicinity of the cavity, σ_s is the sintering stress. The parameter L serves as a stress and temperature dependent length scale governing the coupling between diffusive and creep contributions to the cavity growth. $\dot{V}_2$, the contribution of to the volumetric growth rate due to dislocation creep in eq.5, is expressed by:

$$\dot{V}_2 = \begin{cases} 2\pi\dot{\varepsilon}c^3 h(\Psi)[\alpha_n + \beta_n]^n & if\left|\dfrac{\sigma_m}{\sigma_e}\right| \le 1 \\ Sign(\sigma_m) 2\pi\dot{\varepsilon}c^3 h(\Psi)\left[\alpha_n\left|\dfrac{\sigma_m}{\sigma_e}\right| + \beta_n\right]^n & if\left|\dfrac{\sigma_m}{\sigma_e}\right| > 1 \end{cases} \tag{8}$$

where

$$\alpha_n = 3 / 2n, \qquad \beta_n = (n-1)(n+0.4319) / n^2$$
$$h(\psi) = [(1+\cos\psi)^{-1} - 0.5\cos\psi] / \sin\psi \qquad (9)$$

in which σ_m is the average mean stress, $\dot{\varepsilon}$ is the steady state creep rate in eq.2 and ψ is the cavity tip angle as shown in Fig.1. With these definitions, the parameter L in eq.7 is expressed as:

$$L = (D\sigma_e / \dot{\varepsilon})^{1/3} \qquad (10)$$

L serves as a stress and temperature dependent length scale, as discussed in ref.[26]. For $c / L \leq 0.1$ cavity growth is dominated mainly by diffusion, whereas for higher values of c / L the contribution of dislocation creep becomes significant. The other details of the constitutive model and the computational formulation can be found in refs. [27-28].

The evolution of creep deformation and crack growth behavior near the initial crack tip was analyzed using small scale yielding boundary layer approach. The analyses were carried out with the FEM mesh shown in Fig.2. Only half of the region was analyzed because of the symmetry. There were 1366 quadrilateral elements, each consisting of four crossed constant strain triangles, and a total of 2888 nodes in the mesh. Traction free conditions were imposed on the initial crack surface, and tractions corresponding to mode I plane strain elastic stress fields were imposed along the circular remote boundary at time t=0 and held constant thereafter. During the analyses, the creep behavior of the composite was assumed to obey eq.2. The value of n was taken as n=3. $\dot{\varepsilon}_0$ provides a natural time scale during the solution (i.e. $1/\dot{\varepsilon}_0$), therefore, all the time dependent quantities were normalized by $\dot{\varepsilon}_0$ and also all the stress values were normalized with σ_0. Furthermore, Young's modulus of the composite relating the elastic strain increments was taken as $E_c = 10^3\sigma_0$. It was assumed that all the cavities are prenucleated; initial values for the parameters which appear in the constitutive model were chosen as: $b / x_0 = 0.1$ where x_0 is the width of the elements in the crack plane in Fig.2, $c / b = 0.1$ and c / L values were 0.0125 and 0.025. The sintering stress σ_s appearing in eq.6 was omitted during the analyses. The initial stress intensity factor for the applied traction was chosen as $(K_I / \sigma_0)^2 / a_0 = 0.467$.

RESULTS AND DISCUSSION

In the first set of analyses, the growth behavior of the cavities ahead of the crack tip was assumed to be dominated by diffusion by using $c / L = 0.0125$ in the cavity growth model. First, the creep crack growth behavior in the absence of any crack bridging was studied; the variation of the crack growth rate with increasing crack length is shown in Fig.3. In the next simulation, it was assumed that there is no initial bridging. As the crack grows, due to the cavity growth, the bridging develops at the crack wake and this bridging length increases continuously. The modulus of the bridging elements, E_b, was assumed to be larger than the composite modulus; here it was taken as $E_b = 1.5E_c$. For this case, the resulting crack growth rates are also given in Fig.3. As can be seen from the figure, even though the mechanism for cavity growth was the same, the growth rate is considerable slower for the crack having bridging elements. There was an initial reduction in the growth rate with increasing crack length, opposite to that seen for the crack without any bridging, owing to the initial development of the bridging zone.

In the next set of simulations, the role of the cavity growth rate on the growth behavior of the bridged cracks was investigated. In these analyses c / L was taken as 0.025 which increases the influence of the dislocation creep to cavity growth. The other material parameters, the magnitude of the applied load and the initial crack tip location were kept the same as in the earlier simulations. Again, growth behavior of the crack without any bridging was investigated first. The resulting variation in the crack growth rate with increasing crack length is summarized in Fig.4. A factor of two increase in the c / L value increased the failure time by about two orders of magnitude compared with the one seen for diffusional cavity growth (Fig.3). This strong dependency of cavity growth on the c / L ratio was also seen in earlier studies [25-27]. In the next simulation the growth behavior of the fully bridged cracks was considered. The change in the crack growth rate with increasing crack length is also shown in Figs.4. For this case, although the crack growth rates were now much slower, the variation of the growth rate with increasing crack length was qualitatively very similar to that seen for the fully bridged crack growing with diffusional cavity growth in Fig.4.

Although the behavior of the bridging elements was assumed to be elastic, significant changes in the magnitude of the bridging forces were observed during the course of the crack growth. For the bridged cracks growing with different rates, at two different bridging locations, the variations in the magnitude of the bridging forces with time are summarized in Fig.5. As can be seen, the bridging forces initially increase to some maximum value, then gradually decrease to a steady state value. Although the peak values of the bridging forces were different, both the initial rate of increase and the subsequent rate of decrease were not influenced by the crack growth rate (Figs 3 and 4). In Fig.6 the distribution of the bridging forces were compared after the same amount of crack extensions. It also appears that the rate of crack growth also has a large influence on the distribution of the bridging forces. A slower growing crack, with the dislocation assisted cavity growth, has a much more uniform distribution of the bridging forces at a significant distance behind the crack tip, as can be seen in Fig.6.

In the following figures after the same amount of crack extension, the distribution of the normal stress ahead of the cracks were compared with both the K and C^* stress fields (eqs.1 and 3). In these figures the stress values were taken directly from the centroids of the elements ahead of the crack tip without any interpolation and smoothing. Fig.7 shows the distribution of the stresses ahead of the cracks growing without any bridging. As can be seen from the figure for the crack growing with diffusional growth, due to fast cavity growth there was no significant stress relaxation ahead of the crack tip; the observed stress field correlates with the K stress field for almost the entire ligament ahead of the crack tip. On the other hand, for the crack growing with the dislocation driven cavity growth, due to much slower growth of the cavities the crack tip stresses relaxed significantly; and the observed stress distribution correlates with the C^* distribution in the form of *HRR* stress field a significant distance ahead of the crack tip. Fig.8 compares the observed normal stress distribution ahead of the bridged cracks with the K and C^* controlled stress fields. While the stresses at significant portion ahead of the cracks correlates well with C^* in the form of *HRR* stress field, the occurrence of a much weaker singularity at the near tip region can be seen from the figure.

The distributions of the creep damage ahead of the growing cracks after the same amount of crack extension are shown in Figs.9 and 10. As can be seen from the figures, for the cracks growing without any bridging the creep damage decays very rapidly from the crack tip for both cavity growth mechanisms. On the other hand, for the bridged cracks, growth of the cavities occurs at a comparable rate to each other in an extensive region ahead of the crack tip and then decays exponentially, similar to that seen for the cracks growing without any bridging. The sizes of this high growth region also correlates with the weaker stress singularity regions that are seen in Fig.8.

CONCLUSIONS

In this study, the growth behavior of the bridged cracks in creep regime was investigated numerically. The results indicate that:

1. Bridging tractions significantly relax even with the creep deformation of the composite alone. While the rate of this relaxation is not influenced by the rate of the crack growth, both the maximum value and the distribution of the tractions in the bridging zone are affected by the rate of crack growth.
2. A much weaker stress singularity than the ones described with K or C^* occurs ahead of the bridged cracks in the creep regime. In this weak singularity region the cavities grow at comparably high rates to each other.

ACKNOWLEDGEMENTS

This work was performed for the United States Department of Energy by Iowa State University under contract W-7405-Eng-82. This research was supported by the Director of Energy Research, Office of Basic Sciences.

REFERENCES

1. P.F. Becher, *J. Am. Ceram. Soc.* **6**, 255, (1991).
2. L. Ewart and S.Suresh, *J. Mater.Sci.* **27**, 5181, 1992
3. T.C.Lu, A.G. Evans, R.J. Hect and R. Mehrabian, *Acta Metall. Mater.* **39**, 1853, (1991)
4. M.K. Ferber and MG, Jenskins, *J. Am. Ceram. Soc.* **75**, 2453, 1992
5. S.M. Wiederhorn, B.J. Hockey, D.C. Cranmer and R. Yeckley, *J. Mater. Sci.* **28**, 445, 1993
6. M.N. Gungor, E.J. Lavernia and S.G. Fishman (Eds) *"Advanced Metal Matrix Composites For Elevated Temperatures"* ASM, International, Materials Park OH (1991).
7. B. Budianksy, J.W. Hutchinson and A.G. Evans, *J. Mech. Phys. Solids*, **34**, 167, (1986).
8. B.N. Cox, D.B. Marshall and M.D. Thouless, *Acta Metall. Mater.* **37**, 1933, (1989).
9. A.G. Evans and D.B. Marshall, *Acta Metall. Mater.* **37**, 2567, (1989).
10. A.G. Evans and R.M. McMeeking, *Acta Metall. Mater.* **34**, 2435, (1986).
11. D.B. Marshall and B.N. Cox, *Acta Metall. Mater.* **33**, 2013, (1985).
12. S.V. Nair, *J. Am. Ceram. Soc.* **73**, 2839, (1990).
13. S. Namet-Nasser and M. Hori, *Mech. Mat.* **6**, 245, (1987).
14. F. Zok, O. Sbaizero, C.L. Hom and A.G. Evans, *J. Am.Ceram. Soc.* **74**, 187, (1991).
15. U. Rammaurty, T. Hansson and S. Suresh, *J. Am. Ceram. Soc.* **77**,2985, 1994.
16. S.Y. Liu and i.W. Chen, *Acta Mater.* **44**, 2079, 1996.
17. C.H. Henager, Jr. and R.H. Jones, *J. Am. Ceram.* Soc. **77**, 2381, 1994.
18. C.H. Henager Jr. and R.H. Jones, *Mater. Sci. Engn.* **A166**, 211, 1993.
19. H. Riedel and J.R. Rice, in "*Fracture Mechanics*" edt. P.C. Paris, ASTM STP-700, Philadelphia, (1979).
20. J.L. Bassani and F.A. Mclinctock, *Int. J. Solids Struc.* **17**, 479, (1981).
21. S.B. Biner, D.S. Wilkinson and D. Watt, *Eng. Fract. Mech.* **21**, 315, (1985).
22. F.Z.Li, A. Needleman and C.F. Shih, *Int. J. Fract.* **38**, 241, (1988).
23. J.R. Rice and G.F. Rosengreen, *J. Mech. Phys. Solids*, **16**, 1, (1968).
24. J.W. Hutchinson, *J. Mech. Phys. Solids*, **16**, 13, (1968).
25. E. van der Giessen and V. Tvergaard, *Int. J. Fract.* **48**, 153, (1991).
26. S.B. Biner, *Acta Metall. Mater.* **44**, 1813, (1996).
27. V. Tvargaard, *Acta Metall.* **32**, 1977, (1984).

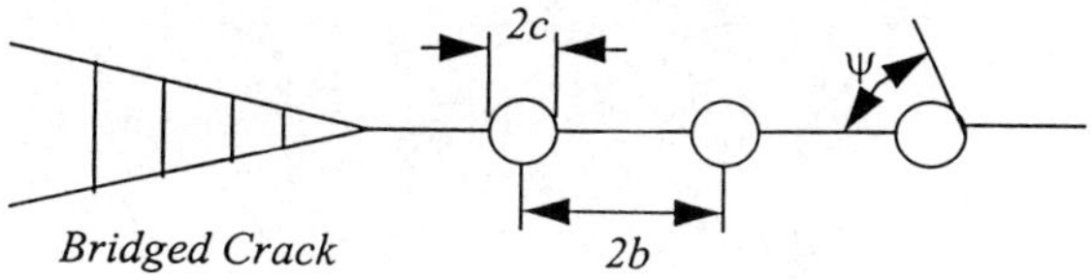

Fig.1 Schematic representation of the creep cavitation ahead of the bridged cracks and the definitions of some of the parameters used in the constitutive model.

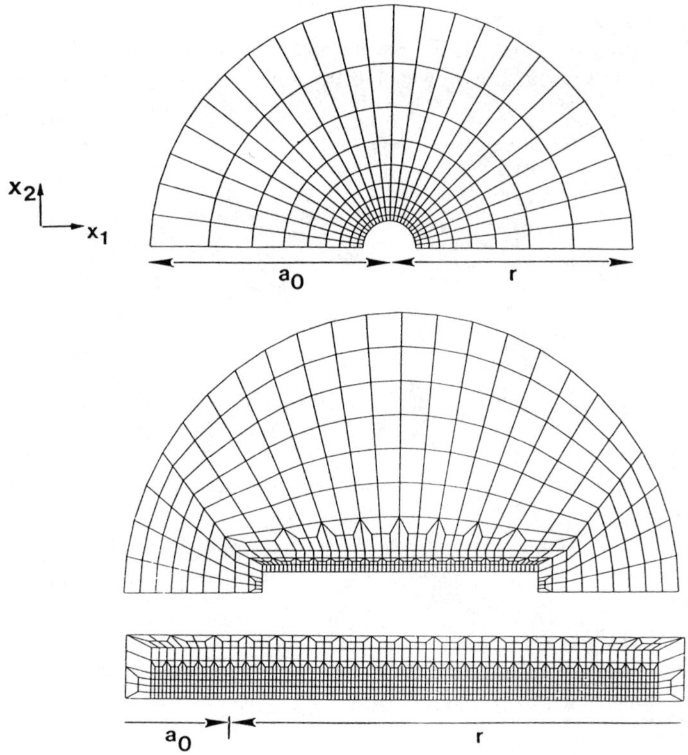

Fig. 2 The FEM mesh used in the analysis.

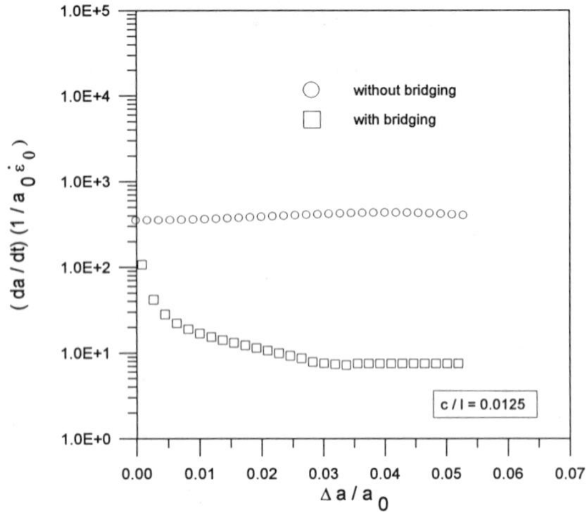

Fig. 3 The variations of the crack growth rates with increasing crack lengths for the cracks growing with diffusional cavity growth.

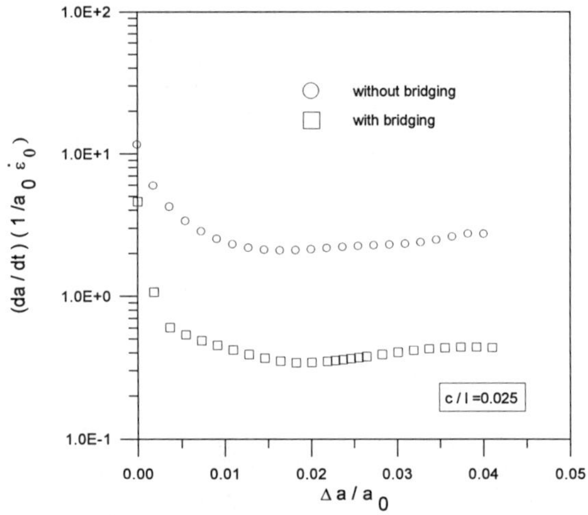

Fig. 4 The variations of the crack growth rates with increasing crack lengths for the cracks growing with dislocation assisted cavity growth.

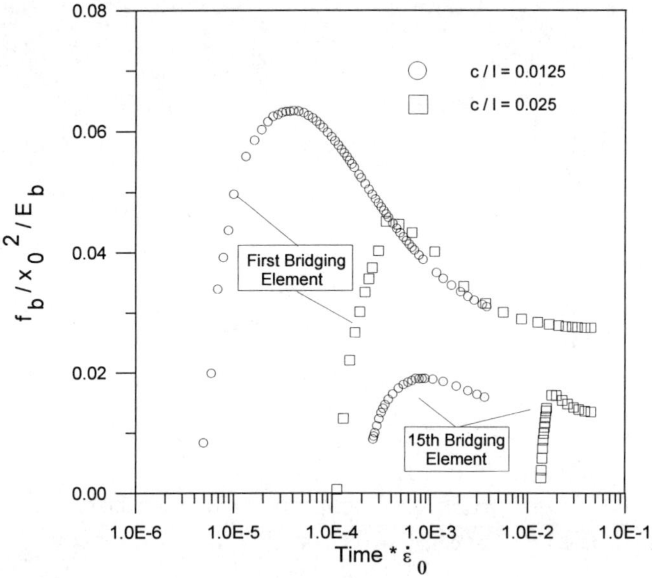

Fig. 5 The variation in the magnitude of the bridging tractions with time for the fully bridged cracks growing with diffusional and dislocation assisted cavity growth mechanisms.

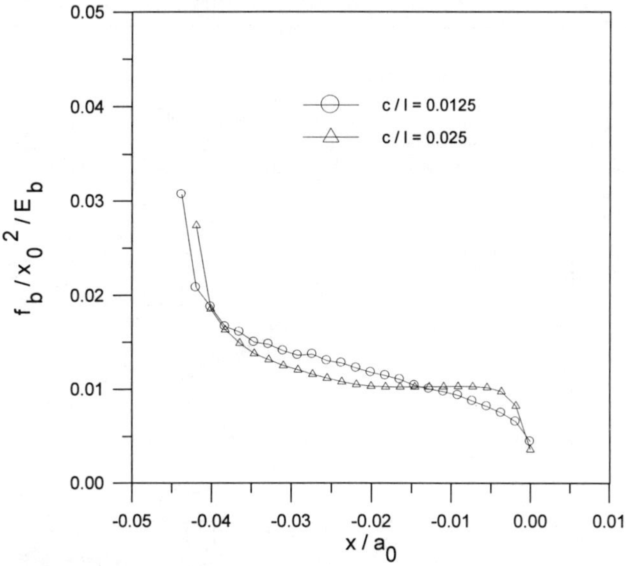

Fig. 6 The distribution of the bridging tractions in the bridging zones for the cracks growing with diffusional and dislocation assisted cavity growth mechanisms.

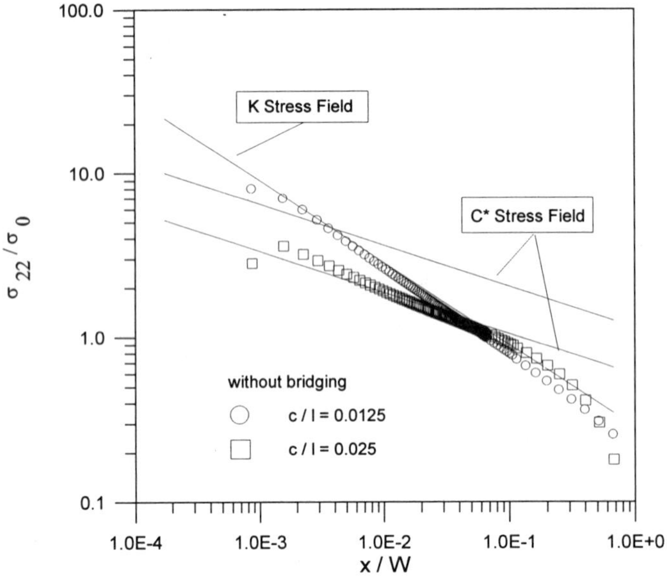

Fig. 7 The normal stress distribution ahead of the homogeneous crack(without bridging) growing with diffusional and dislocation assisted cavity growth mechanisms.

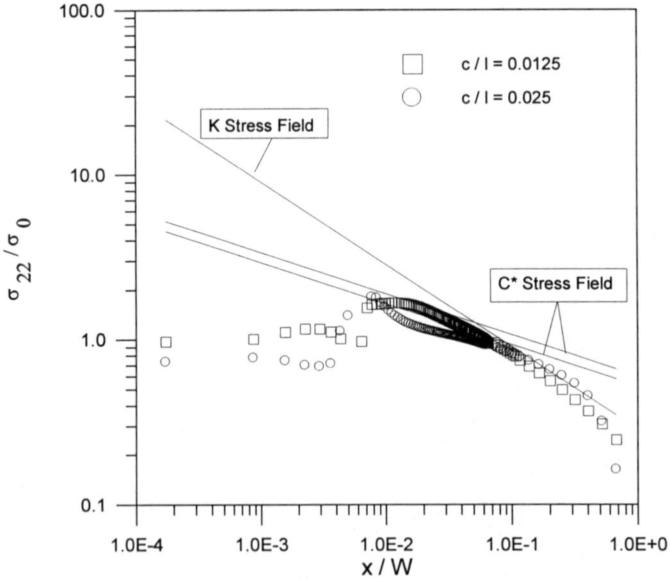

Fig. 8 The normal stress distribution ahead of the bridged cracks growing with diffusional and dislocation assisted cavity growth mechanisms.

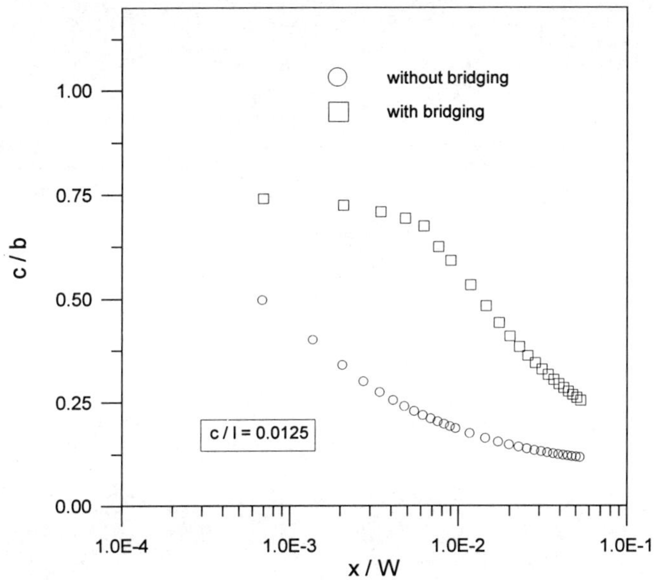

Fig. 9 Comparison of the creep damage evolution for the cracks growing with diffusional cavity growth mechanism.

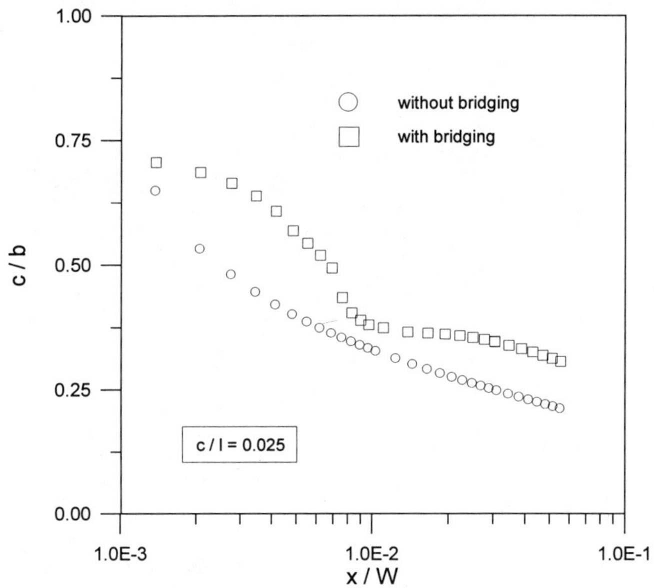

Fig. 10 Comparison of the creep damage evolution for the cracks growing with the dislocation assisted cavity growth mechanism.

SUBJECT INDEX

AUTHOR INDEX